THE IMPACT OF CLIMATE CHANGE ON DRYLANDS
With a Focus on West Africa

ENVIRONMENT & POLICY

VOLUME 39

The titles published in this series are listed at the end of this volume.

The Impact of Climate Change on Drylands

With a Focus on West Africa

Edited by

A.J. Dietz

Department of Geography and Planning, University of Amsterdam, The Netherlands & CERES, Utrecht University, The Netherlands

R. Ruben

Department of Social Sciences, Development Economics Group, Wageningen University and Research Centre, The Netherlands

and

A. Verhagen

Plant Research International, Wageningen University and Research Centre, The Netherlands

KLUWER ACADEMIC PUBLISHERS

DORDRECHT / BOSTON / LONDON

A C.I.P. Catalogue record for this book is available from the Library of Congress.

ISBN 1-4020-1952-1 (HB)
ISBN 1-4020-2158-5 (e-book)

Published by Kluwer Academic Publishers,
P.O. Box 17, 3300 AA Dordrecht, The Netherlands.

Sold and distributed in North, Central and South America
by Kluwer Academic Publishers,
101 Philip Drive, Norwell, MA 02061, U.S.A.

In all other countries, sold and distributed
by Kluwer Academic Publishers,
P.O. Box 322, 3300 AH Dordrecht, The Netherlands.

Printed on acid-free paper

Printed in the Netherlands.

Contents

Contributing Authors

Beek, Wouter van: Department of Anthropology Utrecht University and African Studies Centre Leiden

Born, Gertjan van den: RIVM (Rijks Instituut voor Volksgezondheid en Milieuhygiëne) Bilthoven

Boxel, John van: Institute for Biodiversity and Ecosystem Dynamics University of Amsterdam

Breusers, Mark: AGIDS/Department of Geography and Planning University of Amsterdam

Broekhuis, Annelet: Department of Geography Utrecht University

Brons, Johan: Development Economics Group/Wageningen UR

Bruijn, Mirjam de: African Studies Centre Leiden

Diallo, Asaïta: Université de Ouagadougou

Dietz, Ton: AGIDS/Department of Geography and Planning University of Amsterdam and CERES/Research School for Resource Studies for Development, Utrecht University

Dijk, Han van: African Studies Centre Leiden

Dittoh, Saa: University for Development Studies Tamale N. Ghana

Geest, Kees van der: AGIDS/Department of Geography and Planning University of Amsterdam

Hilhorst, Thea: Institute of Environment and Development London and Koninlijk Instituut voor de Tropen Amsterdam

Jellema, Pleuntje: AGIDS/Department of Geography and Planning University of Amsterdam

Jong, Ali de: Department of Geography Utrecht University

Hazel, Peter: Institute for Food Policy Research Washington

Kersbergen, Lianne: Development Economics Group/Wageningen UR

Keulen, Herman van: Plant Research International/Wageningen UR

Konaté, Mama: Direction Nationale de la Météorologie. Bamako, Mali

Kruseman, Gideon: Development Economics Group/Wageningen UR

Kuijvenhoven, Arie: Development Economics Group/Wageningen UR

Leemans, Rik: RIVM (Rijks Instituut voor Volksgezondheid en Milieuhygiëne) Bilthoven and Wageningen UR

Millar, David: University for Development Studies Tamale N. Ghana

Obeng, Francis: University for Development Studies Tamale N. Ghana and AGIDS/ Department of Geography and Planning University of Amsterdam

Ofori-Sarpong, Edward: Department of Geography/University of Ghana at Legon

Ouédraogo, Boukary: Centre d'Etude, de Documentation et de Recherche des Sciences Sociales, Ouagadougou

Put, Marcel: AGIDS/Department of Geography and Planning University of Amsterdam

Reij, Chris: Vrije Universiteit Amsterdam

Ruben, Ruerd: Development Economics Group/Wageningen UR

Saley, Hassane: National Council on the Environment and Sustainable Development. Niger

Schaeffer, Michiel: RIVM (Rijks Instituut voor Volksgezondheid en Milieuhygiëne) Bilthoven

Sissoko, Keffing: Institut d'Economie Rurale, Bamako

Sylla, Ndiaye Cheikh: Ministère de l'Environement et de la Protection de la Nature. Senegal

Ton, Peter: Organic cotton consultancy and research, Amsterdam

Touré, Mohammed: Institut d'Economie Rurale, Bamako

Veldhuizen, Els: AGIDS/Department of Geography and Planning University of Amsterdam

Verhagen, Jan: Plant Research International/Wageningen UR

Zaal, Fred: AGIDS/Department of Geography and Planning University of Amsterdam

'VOICES FROM THE SAHEL'

- mobility in response to ecological insecurity -

***Camera-sound-editing**: Josee van Steenbrugge*

***Translation**: Mirjam de Bruijn*

***Technical advice**: Videcam*

***Music**: Fulbe/ Hayre*

This film is about practical dilemmas of people living in a harsh environment.

For people in the Sahel and Sudan zone, mobility has always been an important way to cope with the insecurities and hardship of the environment they live in. Hardship is often caused by climatic circumstances: drought and irregular rainfall.

The stories they tell about their lives and the difficulties they encounter often feature migration to other places. Some move from the north to the south of Mali. Some, like the Dem family, eventually stay in the north but make expeditions to other regions. For others, like the Sankare family, life becomes too strenuous and they decide to move to the south where they ultimately stay as life there offers more economic opportunities than life in the north. However, especially for women, the migration experience leads to feelings of being ill at ease, in spite of the fact that the migration brought more economic security. Even those returning to their home village seem to be at a loss.

Recent findings show that migration is becoming an ever-important strategy to survive in the Sahel countries.

This raises questions about the future of these people caught between their home village and their new destination.

The film was made with the financial support of the project Impact of Climate Change on Drylands and the African Studies Centre Leiden.

The film is available in CD-Rom or DVD – version at the African Studies Centre in Leiden, The Netherlands. For further information see: http://asc.leidenuniv.nl/climatechange/. The web site also contains a preview of the film.

Foreword

Climate change has, in recent years, evolved from an environmental issue to a complex development issue. Climate change is not a peripheral issue for development. This is especially true for the arid and semi-arid regions of the world. Today already, the natural variability of rainfall and temperature are among the main factors behind variability in agricultural production, which in turn is one of the main factors behind food insecurity. The availability and quality of water are closely related to the amount and frequency of rainfall. The dryland areas of the world are among the most vulnerable to climate change. At the same time the resilience of human and natural systems in the dryland areas has been remarkable over the last three decades and the West African Sahel in particular.

Since the devastating drought of the early 1970s the emerging environmental and socio economic crisis has been subject of study and debate. The signaling of impacts and defining actions to overcome the problems has most likely prevented a far worse situation developing. Apart from the considerable efforts to improve land and water management and agricultural production the strong resilient population of the Sahel has been critical in the survival of people in the Sahel.

Climate change is an additional stress to this region which is already under stress from other pressures. A timely signaling of impacts of climate change, including changes in climate variability, and defining adaptation strategies in this complex environment will prove important for the development of the Sahelian region.

Clearly, adaptation is not new, changes and variations in climate and other environmental factors have occurred naturally. Both human and natural systems had to adapt to these new conditions. The Impacts of Climate Change on Dryland project (ICCD) tried to draw lessons from the past with the objectives to understand the current situation and define successful adaptation strategies to future changes in climate.

This book presents the findings of the (ICCD) project. The multidisciplinary research group aimed to identify different climate related risk-coping strategies at farm household and individual level for sub-Saharan West Africa. Taking the complex nature of adaptation to climate change as a starting point societal, cultural, geographical, biophysical and economic vulnerability were included in the analysis.

There are three parts in this book. The first part deals with climate change in the Sahel region in a broad regional context. Issues related to demographic development, climate change, its impacts and adaptation strategies are addressed. The second part describes local case studies were for different environmental and socio-economic situations various livelihood strategies are described. The third and final part deals with the policy consequences and climate change preparedness.

To understand farmers' behaviour in West African drylands in preparing ('insuring') for dryer conditions and for agro-climatological droughts, in coping with droughts and adverse production conditions, and in adapting to changed conditions afterwards, we looked at their performance before, during and after drought years in the past identifying several adaptation strategies and policy recommendations.

The conclusions don't look very grim, contrary to the much-painted 'picture of doom' for Africa. West Africa's shock experience in the 1970s and 1980s did have the result that it became much better prepared for possible new drought shocks, and that its agricultural production performance in the 1990s (when rainfall became considerably better) improved.

The future for the Sahel is not necessarily gloomy. However, system breakdown can occur during droughts. One may fear that in those situations religion will be used as a major catalyst for political support to exclusionist claims (Islam versus Christianity and religious sub-groups versus sub-groups) and may result in massive violence and rapid deterioration of all local livelihood options, resulting in a large death toll and mass migration southward and overseas.

Understanding the effects of climate change and defining successful adaptation strategies are still in their infancy. Development of sector and location specific technologies and methodologies are needed. For sub-Saharan West Africa social security mechanisms, understanding migration strategies and regulating land and water entitlements in the light of changing environmental and social conditions are crucial elements in the development of such technologies and methodologies. Careful attention is needed for potential conflicts when resources become scarce. Local government and non-government organisations need support to monitor economic changes and to implement local policies. Agricultural research plays an important role in developing technologies that perform well under drought conditions. International agreements on climate change implications may be exploited for example by redefining subsidy practices.

This book would not have been possible without the contribution of a large number of people and institutes. The Dutch National Research Programme on Climate Change is acknowledged for providing the funding of the project. Centre Techniques de Coopération Agricole et Rurale (CTA) and the Wageningen based Interdisciplinary Research and Education Fund (INREF) co financed the final workshop of the project. Thanks are due to the project advisory board: Prof. dr C. Schweigman; Prof. dr J. Hinderink; Dr T van der Zon; Dr H. Jorritsma; Prof. dr A de Ruijter; Prof. dr C.J. Stigter; Prof. dr E.M.A. Smaling; Dr A. Huysman; and Prof. dr R. Feddes.

The following research groups and institutes contributed in various ways. In Burkina Faso: Centre d'Etude, de Documentation et de Recherche des Sciences Sociales (CEDRES); Institut de Développement Rural (IDR); Programme d'Excecution du Développement Integré (PEDI); Centre Nationale de Recherche Scientifique et Technologique, Institut Nationale de Recherche Agronomique (CNRST-INERA); Antenne Sahèlienne (Research programme of the Wageningen Agricultural University and the University of Ouagadougou). In Ghana: Tamale University of Development Studies; University of Ghana at Legon; CECIK, Bolgatanga. In Mali: Institut Supérieur de Formation et de Recherche Appliquée (ISFRA); Faculté des Langues, Arts et Sciences Humaines (FLASH); Centre National des Recherches Scientifiques et Techniques (CNRST) Institut d'Economie Rurale (IER); Compagnie Malienne pour le Développement Textiles (CMDT); Equipe Systèmes de Production et Gestion de Ressources Naturelles. (ESPGRN). In Niger: CADRES, Niamey, Niger: Université de Niamey, Anthropology and économic history. In The Netherlands: Research School for Resource Studies for Development (CERES); Amsterdam Research Institute for Global Issues and Development Studies, University of Amsterdam (AGIDS); African Studies Centre, Leiden (ASC); Faculty of Geographical Sciences, Institute of Development Studies Utrecht, Utrecht University; Faculty of Social Sciences, Utrecht University; Research Institute for Agrobiology and Soil Fertility (AB-DLO) (Plant Research International, Wageningen UR); Development Economics Group, Department of Economics and Management, Wageningen University (DE-Wageningen UR); National Institute of Public Health and the Environment (RIVM); Winand Staring Centre for Integrated Land, Soil and Water Research (SC-DLO) (Alterra, Wageningen UR).

Thanks to Turner Translations for the language correction. Figure 23.1 and maps 13.1, 15.1, 18.1, 23.1-3 were prepared by UvA-Kaartenmakers.

Besides the contributing authors a large number of scientists contributed to the study: Mohamane Ali, Lieke Berghauser Pont, Ilja Booms, Rebecca Borgdorff, Liesbeth van den Bos, Aline Brandts, Joost Brouwer, Alpha Boureima Gado, Paul de Bruin, Mamadou Djibo, Hamady Djouara, Abdou Salan Fall, Marcel Gerrtis, Selma Griep,

Cheick Gueye, Leo de Haan, Annemieke van Haastrecht, Willemijn van Helden, Gerard Heuvelink, Saskia Jordens, Minette Kits-Nieuwenkamp, Cindy Klompenhouwer, Daniel van Kraailingen, Froukje Kruijssen, Malam Kiari Lawan, Pieter Maas, Serigne Mandour Tall, Maximo Menenti, Jan-Piet van der Mijl, Matta Moustapha, Susan Muis, Ouendenda Nikiema, Karin Nijenhuis, Ahmadou Ongoiba, Peter Osei-Adjei, Kwadwo Owusu, Annemarie Poldermans, Jacoline Pomp, Laura de Pundert, Petra Rijkes, Frank Rouers, Frank de Ruijter, Arjen Schijf, Maaike Snel, Cindy van Soest, Josée van Steenbrugge, Welmoed Soepboer, Pieter van Teeffelen, Peter Uithol, Janneke Vader, Sjoerd de Vos, Richard Yeboah, and Renate Zondag. Without these and others this book would not have been possible.

Ton Dietz, Ruerd Ruben and Jan Verhagen
Amsterdam/Wageningen
September 2003

PART A

CLIMATE CHANGE IN THE SAHEL REGION

Chapter 1

THE ICCD RESEARCH

Ton Dietz and Jan Verhagen

Abstract: The background, the objectives and the research methodology of the "Impact of Climate Change on Dryland" or ICCD project are discussed in this chapter. A framework was developed and livelihood strategies were defined, both were used as general guidelines to facilitate communication between several disciplines and provided a platform for cross-disciplinary research on a high abstraction level at various spatial and temporal scales. Central in the approach is the decision making process.

1. INTRODUCTION

"The impact of climate change on water availability, agriculture and food security in semi-arid regions" was the title of a project that started in 1997. It was the result of a merger of two proposals, submitted separately by the research schools CERES and the Wageningen-based DLV consortium. The first concentrated on societal, cultural and geographical vulnerability whereas the latter focussed on the biophysical and economic vulnerability of semi-arid regions to climate change. At the time, climate change was not an issue in the developing world and long-term planning processes focussed mainly on the growing population and the degrading natural resource base. The issue of how climate change might interact with and affect natural and societal systems was not included in the decision-making process. The project was renamed Impact of Climate Change on Dryland[1] (ICCD). The objectives were ambitious:

1. to analyse the effects of climate change and rainfall variability on drought risk and yield potential;

[1] Drylands are defined as all semi-arid and sub-humid areas in accordance with the UNESCO (1977) classification

A.J. Dietz et al. (eds.),
The Impact of Climate Change on Drylands: With a Focus on West-Africa, 3–17.

2. to develop a comprehensive framework to analyse regional scale impacts and;
3. to identify different risk-coping strategies at farm household and individual level.

In order to address these objectives, three phases were defined:

Phase one, the identification, aimed at

a) A characterisation of the dryland regions on a global scale and at placing West Africa in this global perspective in terms of the various components that are susceptible to climate change, i.e. water availability, environmental sustainability, food security, population dynamics and socio-economic change;
b) A comparison of existing climate scenarios for the dryland regions and dryland West Africa in particular, covering the period 1990-2050.

Phase two, the core of the project consisted of case studies for sub-Saharan West Africa in which we tried

a) to establish a link between rainfall variability and yield variability using a drought index and crop growth simulation models;
b) to gain insight into the driving forces of changes in land use and agricultural production;
c) to establish geographical and socio-cultural differences in existing coping strategies, to assess the adaptive capacity of groups and individuals.

In phase three, the methodology developed was to be applied to assist regional planners to deal with policy consequences of the research findings.

2. FRAMEWORK

Climate shaped both the bio-physical and the socio-economic environments in sub-Saharan West Africa. High temperatures, low and erratic rainfall patterns and nutrient-poor soils with limited vegetation characterise the Sahelian drylands. The socio-economic structures, notably the farming systems, have adapted to this environment. Variation is an inherent aspect of climate and this variation is important for the functioning of the dryland regions. Changes in variations need to be addressed to be able to estimate possible impacts. Analysis of past and present system behaviour is the basis for increasing our understanding and helps us to define sustainable systems in the low-productive dynamic environment.

Water and food security is determined by a combination of bio-physical and socio-economic factors. An impact assessment of climate change on water and food security should integrate these factors. When addressing the future food security situation, aspects of global change and climate change should be part of the analysis.

Availability and accessibility in combination with the quality of food determine whether an individual or group is food secure. Availability, or the amount of water or food that is in stock, is closely related to food production and food import levels. Accessibility, or the possibility of actually acquiring the water or food products, is closely tight to socio-economic factors. The quality or nutritional value of these products is the result of the food production and processing chain, starting in the farmer's field and ending at the point where the consumer acquires the product. Generally, and more acutely in times of drought, food security depends on both the availability and the accessibility of food. Even if food is available, it may not be accessible if people have no entitlements to food, that is, they are unable to access food based on an understanding that they have the right to obtain it. This may be through own production, trade, employment or other resources available to the household, or through social relationships. An important entitlement is obtained through trade, when a product is traded and bartered or sold to obtain food products. This market entitlement can disappear during droughts. Pastoralists are squeezed out of their pastoral existence when prices of food grains soar and animal prices drop as all pastoralists try to sell animals to buy food at the same time. Increasing numbers of low-quality animals on the markets further accelerate this process. Sen, in his 'Essay on entitlement and deprivation' presented examples of this when he discussed the droughts in the Sahel (Sen, 1981). Though Sen pointed at the dramatic consequences of hundreds of thousands of deaths during those crises, this type of crisis does not often occur. Less dramatic but more frequent severe dry seasons may have a more fundamental impact on dryland society and in particular on poorer households. When starting with very unfavourable initial conditions, as is the case in sub-Saharan West Africa, problems will be persistent when trying to alleviate water and food insecurity.

Using the concept of vulnerability, this study tries to learn from the past and so find ways of coping with future situations. Climate change is regarded as an additional stress on the system. The vulnerability of production systems and societies to climate change is tied to the stability and resilience of these systems. In agricultural science, resilience is a component of sustainability (Fresco & Kroonenberg, 1992). It is the ability of a system to absorb changes and restore production levels after a disturbing event. Farmers in unstable production environments have developed a range of livelihood strategies to cope with the inherent variability of the system and to increase food security.

Resilience in social sciences may be defined as the capacity of a person or group to anticipate, cope with, resist and recover from the impact of a disturbing event. People's responses to risk are not only determined by expectations but also by their values, preferences, culture, religion, etc. (Vogel, 1998). It is clear that responses of groups and mechanisms leading to social instability are extremely complex. Some of the short-term behavioural patterns of individuals and groups are predictable, but these are generally not the most interesting. Long-term behavioural patterns are more important for an understanding of the consequences of decisions and pathways individuals and groups may follow. However, these remain unpredictable even when one has perfect knowledge of all the important facts.

In order to analyse the vulnerability of water and food security to climate change we need to address the existing capacities and existing resources, both natural and societal, that can be drawn on to deal with environmental changes. Here too the factors of availability, accessibility and quality are essential elements. Technical, agronomic solutions can only address part of the problem. The same, however, holds for economic and socio-cultural solutions. Strategies for obtaining security in unstable environments need to be integrative.

A framework was developed in which the complex man environment relationship can be studied. The framework facilitates communication between several disciplines and provides a platform for cross-disciplinary research on a high abstraction level at various spatial and temporal scales.

A central entity in the framework is the Decision-Making Unit (DMU). The DMU shapes both the ecological and socio-economical environments. Decision-making is a combination of individual and collective choices, that is interactions between the environments are therefore crucial for an understanding of the interaction between the decision-making unit and its environments.

The layout of the framework is presented in Figure 1.1. The environments in which the DMU operates and which are shaped by it are the biophysical and the socio-economical environments. The three environments and their interactions are discussed below starting with the DMU.

The bio-physical environment is subdivided into items such as: climate, water, terrain, soil. This level is already quite specific and it is here that properties and functions can be quantified or described. Quantitative and qualitative indicators and thresholds can be used to evaluate the status of a system. Soil fertility can be determined and provides a clear indication of prospects for crop production. The slopes of the terrain in combination with soil texture and soil organic matter provide information on erodability. Other commonly used indicators are the distance to markets, rainfall levels, and number of livestock.

The quality of institutes, government etc. can also be ascertained using this approach. Size, the level of education of the staff and how its target group perceives the organisation are all indicators of the quality of an institute or organisation.

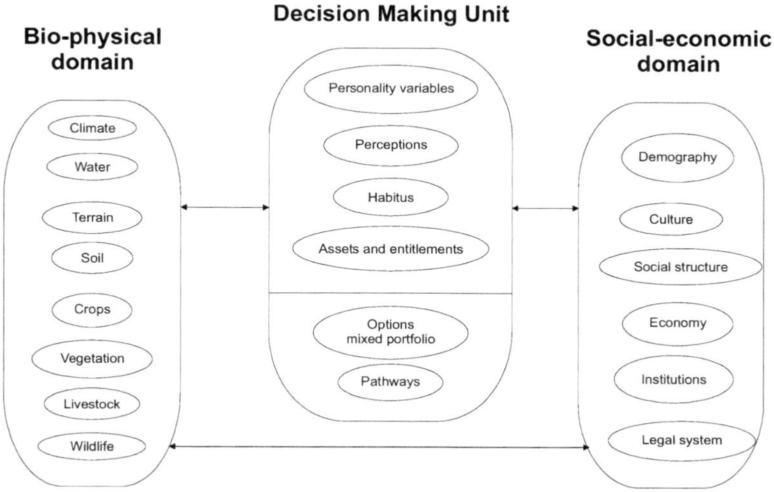

Figure 1.1. Layout of the research framework.

These different entities and items influence each other on different time and spatial scales. In the framework, their actions and interactions are reflected by the functions ascribed to them. Once this aspect of the model is understood, we may select a level in time and in space, an object or item to focus attention on and a set of relations between similar items or between different items. We may consider the importance of the age structure of a certain level (households) for the relationship between the economy and culture in a nation for the next few decades, for example.

3. MALTHUSIAN DOOM OR BOSERUPIAN TRANSITIONS?

The predicted population explosion in conjunction with a possible deterioration in climatic conditions for agricultural production is a quite alarming scenario for dryland West Africa in particular. The expected rapid land use changes may be an added threat as a result of land degradation which threatens to give lower crop yields and lower possibilities to cover firewood energy needs, especially near urban centres of energy demand. Growing water shortages may add to the gloomy picture. On the other hand, very negative scenarios predicted during the big Sahel drought of the early 1970s did not materialise. Many observers point to the considerable efforts to improve land and water management and agricultural production and they have presented a picture of a resilient population in the region. It is also possible to refer to areas with equally problematic agricultural production circumstances (e.g. most of India) which have much higher population densities and use natural resources much more intensively, without necessarily moving towards Malthusian doom (cf Walker & Ryan, 1990 and Put & Van Dijk, 2000). Stress can also force a transition to more intensive resource use, and even to higher sustainability of that resource use. There is room for 'Boserupian optimism' (Mortimore, 1998 and see Tiffen *et al.*, 1994, for an introduction with an East African emphasis). For parts of dryland West Africa recent observers indeed report numerous examples of 'transitions to a more sustainable land use' and better land and water management technology, resulting in higher yields and more profitable agriculture. They also show a major diversification of the economy and a rapid impact of urbanisation and of a more urban-based economy. However, part of that recent optimism may be based on the slightly improving rainfall situation during the 1990s, rebuilding strength during 'relatively fat years'. If the upward short-term trend is not only part of a normal cyclical movement, but also part of a long-term downward dynamic trends, there is reason for caution.

'Normal' intensification theory, which only looks at population, land and other resources and technology, and which neglects the impact of climate change or even climate variability, differentiates between a number of options which can all assist in solving problems related to increasing population densities in a situation with scarce resources.

Decision-making units (individuals, households, and institutions) can use a portfolio of relevant options to obtain food or to increase the possibilities for obtaining food (or energy or water). During periods of relative rainfall abundance, the portfolio looks different when compared to periods of relative drought. Assuming that future scenarios for dryland West Africa should be based on lower rainfall and higher temperatures, for the ICCD research it is particularly relevant to study past portfolios of options, which were chosen during relatively dry years.

For all decision-making units in a particular research area together, the total portfolio of experienced options can be seen as the 'portfolio of theoretical options'. Possibilities to use parts of this portfolio of theoretical options depend on characteristics of the individual decision making unit, on the type of decision-making unit (based on particular typologies; e.g. 'household types based on assets', or 'ethnic groups'). For a particular type of decision-making unit a particular selection of options can be regarded as 'the portfolio of relevant options'. So, 'relevant options' are a subset of all theoretical options that are experienced by the combination of all DMUs in a particular area. What a portfolio of relevant options is will depend on the type of decision making unit (many assets or hardly any assets; many entitlements or few entitlements; crop habitus or livestock habitus; pathway experiences that are traumatic or not, etc.).

Each DMU at each particular time (or period) has selected a particular 'portfolio of experienced or chosen options'. For each DMU we can compile a portfolio history of chosen options through time. We call this 'a pathway of chosen options'. It is probable that decision-making units, which look alike (or which are in the same class of a typology), all have pathways which look alike as well.

Portfolios of theoretical, relevant and chosen options, as well as pathways can be studied at the level of individual decision-making units (individual men and women, households, functionaries in organizations). However, if all the information on relevant individual decision-making units in a certain region is put together, the regional grid cell can also be seen as a (spatial) unit of analysis and all options which are 'internal' can be treated as a 'black box', focusing on all relationships with the outside world: all options which (partly) depend on exchange relationships with decision-making units outside the region.

We have made a list of nine major elements of 'livelihood strategies', which are all part of the portfolio of theoretical options. We will briefly look at each of these nine major elements of the portfolio of options for the acquisition of food (although the same approach can also be followed if we would focus on energy or drinking water resources):

What can people do if livelihood stress increases?

OPTION 1: **Food acquisition by using stored food or selling other stored assets and buy/barter food in return.**

Storage of food, which has been produced in years with relatively abundant harvests, may be an age-old practice, but it may also be a new or reinvented strategy, either at the level of individuals/households or at the level of communities or even as a result of national policies. Better storage technology and lowering food losses during the harvests, during food preparation and during consumption can also result in a major improvement because of the rather large losses that occur. Many 'coping strategies' during problematic times can also be based on investing in various types of assets and selling those in dire times. Assets can be in the form of accumulated wealth (gold, or other jewellery; animals; bank savings; land; house; household utensils) or in the form of productive capital (e.g. agricultural inputs, agricultural and non-agricultural machinery). The sale of small ruminants (goats, sheep) is a widespread 'first aid' strategy in case of trouble.

OPTION 2: **Direct food intensification:** increasing the total food production is a possibility.

- by extending the food crop acreage into 'wastelands', forest areas, game- or bio-reserves or pasture and range-lands; of course this depends on the positions of relative strengths of competing land users: cultivators versus charcoal burners, game wardens, forest departments, pastoralists, etc., as well as 'pioneer' cultivators competing among themselves, often with an ethnic element in claims and counter-claims;

- by using a variety of geographical niches more efficiently, and by distributing crops and varieties more efficiently over seasons;
- by limiting the fallow period and - if done in a sustainable manner - by maintaining or improving nutrient banks by adding manure, fertilizer, compost, etc.;
- by improving soil and water management; this partly depends on the long-term security of access to these improved lands and this often depends on indigenous, religious and/or state-defined land, water and other natural resource rights;
- by adopting crops and varieties with higher or more secure yields per ha. and by improving access to 'improved' seeds of these crops and varieties;
- by developing irrigation, either using river water, canal or piped water or groundwater reserves; this may also increase the number of growing seasons in a year;
- by intensifying animal husbandry practices (e.g. better veterinary care, higher-yielding species, stall feeding, better husbandry); or by sending part of the animals to places which have better supplies of feed and water;
- by investing more labour and care in food crop cultivation and in animal husbandry practices;
- by investing more labour and care in gathering 'wild food', and by accepting a diet that in 'normal' circumstances is regarded as inferior ('poor men's food', or 'hunger food') or taboo.

Land use models, which include differences in soil resources, technology/management, and farming systems approaches, can be used to predict various food production scenarios and its requirements in terms of inputs (Verhagen, 1998). National economic modelling can also shed light on the driving forces of changes in land use from a market and state intervention perspective (Brons *et al.*, 1999).

OPTION 3: **Investing in indirect food intensification**.

Another possibility is to invest in indirect food intensification, producing crops, raising livestock or collecting tree products for an external market and buying or bartering food in return. The possibilities depend on net terms of trade for the actual producers. This approach has been applied recently to cotton (for the sub-humid zone), groundnuts (for the semi-arid zone), charcoal (for both) and meat (for both, but with higher 'caloric terms of trade' in the semi-arid zone. Zaal (1998) shows surprisingly high and even improving caloric terms of trade for a region in the north of Burkina Faso. This possibility depends both on the price developments on the world market and at regional markets for export products, on the price developments of food on these external markets and on the marketing infrastructure, and transaction costs (including taxes, bribes, monopoly profits, etc.) in the areas of demand. This option only 'works', when the net impact of producing for the market gives the producers/consumers a higher result in food terms than if they were to have produced the food themselves. A country like Senegal has for quite some time clearly chosen for a groundnuts-for-cereals exchange, importing one third of its food requirements during the 1990s. International and national market dynamics and government policy all play a role in determining likely outcomes, with a shift taking place away from policy-driven adjustments to market-driven adjustments after the early 1990s (de Haan *et al.*, 1995).

OPTION 4: **Food acquisition by marketing non-agricultural products and buying/bartering food**, by changing a basically agricultural economy into one in which the production of mining products, handicrafts or industrial products becomes more important and where non-agricultural producers receive higher net rewards for their labour time than if they were to have produced their own food. This option very much depends on niches in the world market (e.g. 'ethnic art'), or on making use of comparative advantages of the region, e.g. low wages, low environmental sink costs, or non-existence of effective anti-pollution policy. A breakthrough of 'tiger'-type industrialisation is far away, though, given the extremely low education level in the area (with primary net enrolment percentages of below 50% in Senegal to below 20% in Mali, and secondary enrolment percentages below 10% for men and below 5% for women in most areas; World Bank, 1995).

OPTION 5: **Food acquisition by selling services and buying/bartering food**, by attracting tourists and other visitors from elsewhere, either to enjoy nature and landscape adventurism, culture (either in its 'ethnic' form or as a hybrid popular culture -e.g. west-African popular music-), or sexual services and the 'image of care'.

OPTION 6: **Food acquisition by selling labour and buying/ bartering food**. If the scale of analysis is the individual decision-making unit, all income acquired by selling labour for wages (permanent, seasonal, casual; for government agencies, Non Governmental Organisations or the private sector; nearby or far away) is relevant. If the scale of analysis is the region, the relevant option is the acquisition of remittance money sent by people who come from the region but who work elsewhere, either on 'hunger trips', on seasonal trips or for long periods, and who send food, money or material goods either regularly or irregularly. Remittance networks may link the area to many parts of the globe, although mostly the links are rather bilateral. In dryland West Africa the links with the (urban) coastal zone, although fluctuating in destination, have become extensive, but many Sahelians have now also taken up residence in less densely populated humid and sub-humid areas towards the south, but still far from the coast. A growing number of Sahelians are migrating to Europe and are sending back considerable sums of money. Figures for 1990 show a total remittance sum (net transfers at the national level) of Burkina Faso and Mali together of 557 m$, while total export income was 769 m$ and total official development aid 800 m$ (World Bank, 1992). In recent years, the net transfers have become less important, though, because of a deepening crisis in Ivory Coast, which was the major receiver of migrants from the north. In 1998, total net transfers of remittances for these two countries were 313 m$, while total export income was 1036 m$ and total official development aid 754 m$ (World Bank, 2000).

OPTION 7: **Food acquisition by social security arrangements,**

- either directly, in the form of social care or in the form of food aid by local governments or non-governmental agencies or by foreign donors of development assistance or by international non-governmental organisations,
- or indirectly in the form of income support, for which food can be bought: pensions, social security payments, bank or trader's/moneylender's credit.

Food aid (mostly cereals) has been a structural phenomenon in Sahelian countries since the famine of 1973-'74. Some cereal imports are obtained via food aid (in 1993 40% in Mali, 25% in Burkina Faso, 20% in Ghana and 12% in Senegal), but in most years it only covers a small part of all food needs (in 1993 less than 5%; World Bank, 1995). It is possible, though, that food aid could become a more structural phenomenon as development assistance has already become a structural phenomenon in most study areas, with official development aid between 8 and 25% of Gross National Income and a major element of government budgets (World Bank, 1997).

OPTION 8: **Food acquisition by stealing.**

This can either be direct (stealing food from other people's fields or storage places, or by stealing/raiding animals) or it can be indirect through the stealing of material goods and the exchange of these for food, or by demanding bribes, contributions, etc. On a regional scale, waging a war or raiding neighbours and transporting the 'booty' back home is a possibility.

OPTION 9: **Lowering food demand.**

This can also be a 'solution' to a pressing or threatening food problem:

- at the level of scale of individual decision-making units one strategy can be to accept lower food intake; eating less meals a day or eating food of a lower quality;

- at the level of households the strategy may be to send members (children mostly) away, e.g. place them in the care of relatively well-to-do families or to 'marry daughters away'; another strategy involves breaking up (joint) families and accepting responsibility for a lower number of household members;
- at the level of scale of regions it can be acquired by:
- *Out-migration.* Out-migration not necessarily results in remittances and out-migration should not only be judged by its capacity to generate remittance income. Out-migration can also lower the burden on those staying behind and lower the overall food needs. In West Africa, there is a considerable spatial mobility and not only by people who are traditionally mobile, like the Fulani (or Peulh). (cf. de Bruijn & Van Dijk, 1995 and Breusers, 1998). Many research areas experience a population in flux, with large numbers of people moving out (towards the south mostly), large numbers of people moving in (from the north generally) and quite a number of people moving through. This considerably changes entitlements to make use of land and other natural resources, with 'identity' being an important attribute to claim rights and 'identification' an important means to gain negotiated access.
- *Diminishing natural population growth by lower birth rates.* Family planning is regarded in some circles as the most obvious thing to do to bring down the high birth rates which are a result of very high total fertility rates (close to 6 in Mali and Burkina Faso) and better health conditions; birth rates can also diminish as a result of severe body stress due to malnutrition, resulting in higher abortions and infertility;
- *Diminishing natural population growth due to higher death rates.* Crude death rates are still high by world standards, but much lower than in the 1960s and they are expected to go down further, despite the impact of diseases like AIDS. There might , however, be a chaotic explosion of violence, as elsewhere in Africa, and that of course could have a major impact on food demand. A violent disruption of livelihoods will also result in a disruption of agricultural production, land management and distribution channels and in a destruction of food and capital stocks (including existing investments in natural resources and infrastructure) it also has a negative impact on local food supply, with long-term effects that may completely undermine what can be cynically called 'gains' of lower food demand.

4. MERGING INTENSIFICATION THEORY WITH CLIMATE CHANGE THEORY

Intensification theory, which takes into account the deteriorating climatic conditions, should deal with the likely changes in the 'portfolio of options' and the more severe restrictions in 'room to manoeuvre' (see Feenstra *et al.*, 1998). For the West African drylands it will be evident that the possibilities for storing food for 'lean years' may reduce when there will be less 'years of abundance', if nothing else changes (OPTION 1). On the other hand, individual and collective attempts to save more food and store it for bad times may become far better organised than during recent decades. Direct food intensification (OPTION 2) may be endangered, as a result of more extreme and severe drought risks and generally lower yields of 'normal' food crops and animals. There will also be a shift of land use combinations to more marginal types of land use (in yield terms), e.g. from maize back to sorghum, with lower average yields, or from crops to livestock (with lower average caloric production per ha) (see Ribot *et al.*, 1996). As far as the indirect food intensification option is concerned, (OPTION 3) it will be clear that in the northern zones of groundnut or cotton cultivation this may no longer be possible; with the groundnut and cotton zones both moving south. Its economic impact depends on relative price developments between groundnuts, cotton, and other export crops on the one hand and imported food on the other. The same is true for livestock. If the caloric terms of trade for livestock against grains remain as good as they currently are it might well result in a situation in which far less intensive land use in direct food acquisition terms (e.g. cattle instead of groundnuts) results in higher indirect food acquisition. All other options (4 to 9) will become more important, probably resulting in a tendency of 'de-agrarianisation' of livelihood portfolios. As this will also result in a change of settlement patterns, with urban settlements becoming far more important (Harts-Broekhuis & De Jong, 1993; Club du Sahel, 1995), it will probably cause a major increase in the demand for, and price of, charcoal. In the short run, charcoal income could increase its share in livelihood portfolios; in the long run, if not properly managed, it could contribute to a damaged environment, with an additional negative impact on micro-climates and land qualities and may even contribute further to diminishing rainfall. All this could partly be counter-balanced by improved agro-biological conditions as a result of higher CO_2 concentrations in the atmosphere.

5. LOOKING FOR PATTERNS OF RESPONSES

Individual people and households (or other types of micro-decision-making units of individuals) who are confronted with these elements of resource stress in dryland areas that become dryer have all developed 'coping strategies', both to overcome difficult resource conditions during 'average years' and to survive drought years (Dietz & Van Haastrecht, 1997). The word 'coping' in fact stands for three different elements of behaviour to improve the security of life and livelihood in a situation of high climatic risks: insurance (prepare for bad times), coping (actually deal with the crisis when it happens) and adaptation/recovery (trying to rebuild after a crisis) (see van der Geest, 2000). The vulnerability of a household, or region to drought stress or to deteriorating climatic circumstances depends on its ability to manage the insurance, coping and adaptation variables: adaptive capacity. In the drylands, households have adapted to seasonal fluctuations which result in relatively good months (for pastoralists during the rainy season, for cultivators after the harvests) followed by bad months (the 'hunger season', the 'soudure', for pastoralists towards the end of the dry season, for cultivators during the rainy, cultivation season). Each household has a variety of 'normal' responses (in terms of insurance, coping, and recovery/adaptation) to the adverse months in each year. In drought years the relatively good months become shorter or do not happen at all, the bad months last longer and the 'usual' problems become more severe (there is a host of literature about 'coping strategies in dryland environments', see Reitsma *et al.*, 1992 for a summary of literature until the early 1990s; important later additions are Blaikie *et al.*, 1994; Bohle *et al.*, 1994; de Bruijn & Van Dijk, 1995, Davies, 1996; Davies & Hussain, 1997, Ellis, 1998; Reenberg, 1998, Ribot, 1995; Ribot *et al.*, 1996; Scoones, 1996; Watts & Bohle, 1993).

At the level of micro-units, we often find a mixture of options being utilised, a broad and diverse portfolio, that is often flexible and 'adaptive' (Millar, 1996). The combination of elements may be continuously changing and portfolios of chosen options in 'bad years' can look completely different from those in good years. Understanding people's behaviour necessitates a long-term perspective, looking at production and consumption behaviour, but also at people's social security networks. Local elements then mingle with a variety of extra-local elements, and agricultural ones with non-agricultural ones. Individuals make use of a variety of fallback mechanisms, which include the 'household' and the 'family', but which also contain other linkages. Women make use of different sets of actors than men, and older people use different sets to those used by younger people. Investments in social and symbolic capital ('networks' and 'identity') can bear fruit soon, or only after many years (de Bruijn & Van Dijk, 1995). In the ICCD research (see ICCD, 1998) we focus on people's long-term 'pathways', particular portfolio histories, with particular time-space combinations. Where long-term household data exists (see Brons *et al.*, 2000ab) we try to reconstruct 'patterned responses'. We think we can learn most about behaviour in future periods with more livelihood stress by zooming in on 'past performance' during years with extreme to moderate drought risks. The likely importance of intra-family networks during difficult times and over time necessitates an approach that goes beyond household surveys. It is useful to reconstruct family genealogies and to understand diverse responses to drought risks, and diverse pathways, by members of the same lineage. We also try to find out how internal social security arrangements within a lineage link up with external social security behaviour via the market, the state and non-governmental organisations. It is also useful to find out how drought risk buffers operate via assets and how entitlements to assets are handed over from one generation to the next in the various cultural groups who live in the various west African case study areas that were studied (e.g. Hesseling & Ba, 1994; Breusers, 1998). Results can be compared with existing policy-oriented frameworks, like Ribot *et al.*, 1996; Kok & Heij, 1998, and Reenberg, 1998.

It is expected that the following variables will influence the pathways of chosen portfolios of options at the level of individuals, households and networks of (clan-) families. At the level of specific areas (e.g. the regional cells of the ICCD research) a historical-geographical analysis of these variables for all current inhabitants is useful, as long as it is acknowledged that many people in these types of areas are very mobile and the geography of their 'insurance, coping and adaptation mechanisms' does in fact often entail a much larger geographical space than the regional cell alone. These variables play a role at both social and geographical levels of analysis:

- the existing natural capital, and in particular the availability of high-quality niches of agricultural production, and of risk-aversive fall-back niches in times of drought;
- access to this natural capital: entitlements to natural resources;
- the endowment of physical and financial-economic capital and actual entitlements to these endowments in times of need; this includes existing mechanisms to insure against hardship and risks, and to borrow money and/or goods; especially important are entitlements to (formal and informal) insurance payments and access to (formal and informal) banking institutions during and after periods of crisis;
- the quality of human capital, like health, fertility, knowledge and labour power that can be mobilised at places and times where and when it is needed;
- cultural capital: the use of identity; the existence of behavioural taboos and their flexibility in times of crisis; psychological 'stamina' to endure hardship; codes for the payment of labour and codes for 'sharing';
- social and political capital: existing networks of support and of support possibilities in times of need; its social and geographical extent;
- the possibilities for mobilising this social and political capital: entitlements to social resources;

- the availability of markets and the transaction costs to reach these markets; the quality of the organisation of these markets, in particular during times of (drought) crisis;
- existing mechanisms of taxation, tribute, rent, interest, gifts and 'gifts' and possibilities to evade these mechanisms, in particular during periods of stress, like droughts.

Chapter 2

THE WORLD'S DRYLANDS: A CLASSIFICATION

Ton Dietz and Els Veldhuizen

Abstract: In this chapter a global overview of drylands is presented. The study region and case study areas are selected using a few simple criteria. Starting with the aridity class and degree of degradation areas were classified based on the population density and on the presence of at least one large city in the area.

1. THE WORLD'S TROPICAL AND SUBTROPICAL DRYLANDS

The world's drylands can be characterised as semi-arid and sub-humid areas, with average annual P/ETP between 0.20 and 0.75 (UNESCO, 1977). Average rainfall conditions restrict rain-dependent agriculture to mainly sorghum, millet and marginal maize cultivation as food crops and groundnuts and cotton as crops for agro-industry and for export. Animal husbandry based on mainly cattle, goats and sheep can also add to the food supply and cash income. In general, rain-fed agriculture generates relatively low crop yields per hectare. Yearly rainfall variation can be considerable. Rainfall unreliability results in relatively high risks of crop failure due to drought (and occasionally excessive rainfall). Yet, the world's drylands support a considerable part of the world's population. According to the World Atlas of Desertification (UNEP, 1997, p. 106), in Africa 40% of the population lives in areas prone to drought and susceptible to desertification. In Asia the figure is almost as high (39%), while in South America it is 30%.

In West Africa, UNESCO's map of the world distribution of arid regions (UNESCO, 1977) shows a band of semi-arid conditions from Dakar in Senegal, via Ouagadougou in Burkina Faso to Niamey in Niger

A.J. Dietz et al. (eds.),
The Impact of Climate Change on Drylands: With a Focus on West-Africa, 19–26.

and further to Kano in Nigeria, the Sahel proper. South of this zone there is a band of sub-humid conditions.

Other tropical and subtropical drylands in the world can be found in the Americas (Mexico and the southern United States, northern Venezuela, north-eastern Brazil, western Ecuador, Peru, parts of Chile, southern Bolivia, western Paraguay and northern Argentina), in northern Africa, and in major parts of eastern and southern Africa, in West Asia (Israel, Jordan, Syria, Lebanon, south-western Arabia and Yemen, major areas of Iran and Afghanistan) and in the Indian subcontinent (major parts of Pakistan, western and southern India), in small areas of China and in large areas of Australia.

Table 2.1. Number of tropical/subtropical semi-arid and sub-humid cells per major world region (one cell = 1°×1°).

Region	Semi-arid	Sub-humid	Total	Total as % of region
West Africa	86	53	139	20
Eastern & southern Africa	272	328	600	57
North Africa	40	3	43	9
West Asia	45	13	58	12
Indian subcontinent	85	101	186	55
China	44	46	90	12
South & Central America	197	212	409	25
USA	52	32	84	11
Australia	211	82	293	46
Total	1032	870	1902	17

Source: for aridity: UNESCO, 1977, for area: World Bank, 1998; the assessments and interpretations were made by a *leeronderzoek* (student research) group of the Dept. of Geography and Planning at the University of Amsterdam, supervised by Ton Dietz and Els Veldhuizen, and consisting of Marcel Gerrits, Minette Kits Nieuwenkamp, Karin Nijenhuis, Annemarie Poldermans and Laura de Pundert (see Dietz & Veldhuizen, 1998). Later, a population update was produced, to which also Janneke Vader contributed (see Veldhuizen & Dietz, 1999). Together they form the basis of Map 2.1 to 2.3.

UNESCO's aridity assessment was based on rainfall (P) and evapo-transpiration (ETP) conditions mostly for 1930-1960. Semi-arid conditions are defined as between P/ETP of 0.20 and 0.50, as for sub-humid conditions the boundaries are set at a P/ETP between 0.50 and 0.75. Combined with assessments of land degradation and population density it formed the basis for the ICCD typology of all drylands in the tropics and subtropics (Dietz & Veldhuizen, 1998). Land degradation assessments were based on the so-called GLASOD map (Oldeman *et al.*, 1991). Population density was first assessed by using data for the situation in the 1960s, mainly based on the World Atlas of Agriculture (IGDA, 1971) and differentiating between 'high density areas' (with more than 50 inhabitants per square kilometre) and 'low density areas' (<50 inh. km^{-2}). Later, it was possible to update this population density data for the situation of the mid-1990s (Global Demography Project, CIESIN, ESRA, NCGIA, see UNEP/GRID, 1998). Urbanisation assessments were added for the situation around 1960 and around 1990, based on the United Nations Demographic Yearbooks, and using the existence of urban administrative areas of more than 100,000 inhabitants as a threshold for differentiating 'urban' from 'rural' areas. All the data was mapped by using a grid system of 1°×1°, roughly covering 110×110 km.

Table *2.2.* Detailed profile of the world's tropical and subtropical drylands: semi-arid.

Region	**N**	**High degradation**				**Low degradation**			
		High pop. density		*Low pop. density*		*High pop. density*		*Low pop. density*	
		urban	*rural*	*urban*	*rural*	*urban*	*rural*	*urban*	*rural*
West Africa	86	7	12	1	37	0	7	0	22
East & Southern Africa	272	3	8	4	90	0	4	1	162
North Africa	40	4	4	0	16	7	9	0	0
West Asia	45	3	5	1	14	8	5	1	8
Indian subcontinent	85	14	19	2	5	9	8	1	27
China	44	0	0	0	0	0	0	0	44
South/Centr. America	197	7	7	6	52	4	6	10	105
USA	52	3	1	0	18	3	0	2	25
Australia	211	0	0	0	10	1	0	0	200
Total	1032	41	56	14	242	32	39	15	593
West Africa/total	8%	17%	21%	7%	15%	0%	18%	0%	4%

In total the world has 1902 1°×1° 'cells' in the tropics or subtropics, which were completely or mainly semi-arid and/or sub-humid, covering an area of 23 million km², or 17% of the world's land mass. Table 2.1 gives an overview for the world's major regions, showing that the West African drylands, the focus of this study, had the fifth largest expanse, after eastern & southern Africa, south & central America, Australia and the Indian sub-continent. The West-African drylands cover 20% of the total West African land area. Most of the remainder is arid, like in North Africa, West Asia, and Australia, unlike all the other world regions where most of the non-drylands are humid.

Maps 2.1-2.3 show the results for Africa, Central/South America and Asia/Australia: semi-arid (SA) versus sub-humid (SH) areas; high land degradation (HDeg) versus low land degradation (LDeg); high population density (HPopdens) versus low population density (LPopdens) in 1994 and rural versus urban in 1990. Table 2.2 and Table 2.3 show the detailed results per world region for semi-arid and sub-humid conditions.

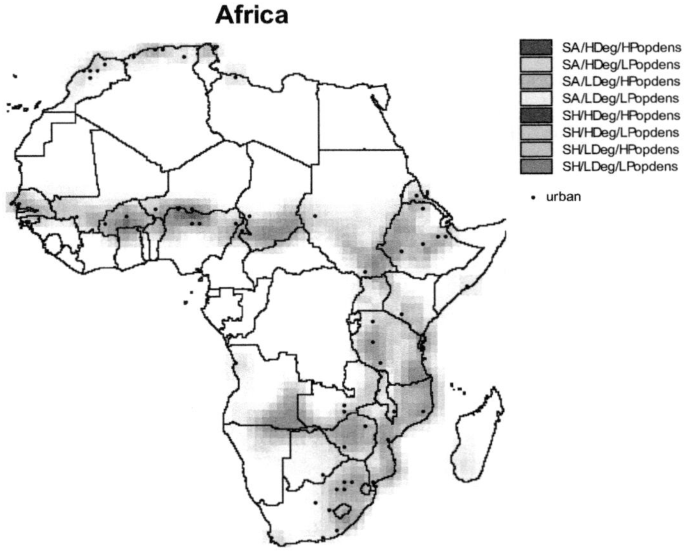

Map 2.1. Drylands of Africa *(See colour section, p. 451).*

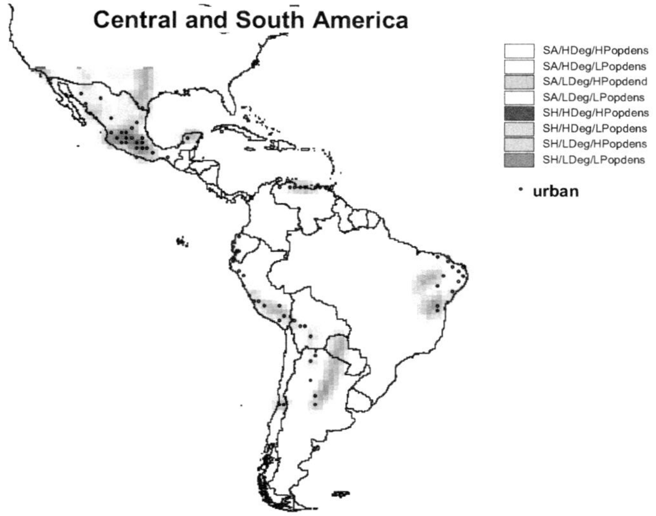

Map 2.2. Drylands of Central and South America *(See colour section, p. 452).*

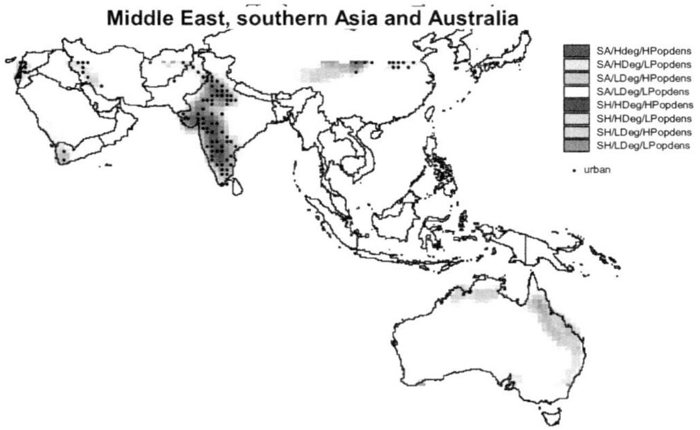

Map 2.3. Drylands of Asia and Australia *(See colour section, p. 452).*

Table 2.3 Detailed profile of the world's tropical and subtropical drylands: sub-humid.

Region	N	High degradation				Low degradation			
		High pop. density		*Low pop. density*		*High pop. density*		*Low pop. density*	
		urban	*rural*	*urban*	*rural*	*urban*	*rural*	*urban*	*rural*
West Africa	53	1	4	0	4	2	8	1	33
East & Southern Africa	328	9	16	4	67	11	11	5	205
North Africa	3	1	0	0	0	1	1	0	0
West Asia	13	0	0	0	0	4	5	0	4
Indian subcontinent	101	25	26	0	2	24	21	0	3
China	46	3	5	0	2	10	4	0	22
South/Centr. America	212	10	1	11	65	5	1	21	98
USA	32	1	0	0	4	5	1	2	19
Australia	82	0	0	0	5	1	2	1	73
Total	870	50	52	15	149	63	54	30	457
West Africa/total	6%	2%	8%	0%	3%	3%	15%	3%	7%

West Africa's dryland profile, compared to the world's tropical and subtropical dryland profile as a whole, is portrayed in Table 2.4. West Africa's drylands comprise only 7% of the world's total. However, the most problematic areas, semi-arid areas with relatively high degradation and a relatively high population density, are considerably over represented in West Africa: they consist of 20% of the world's total in that type of drylands.

For West Africa's drylands, the distribution of the types per country is shown in Table 2.5. Four categories of drylands are most prominent in West Africa:

- The semi-arid areas with high degradation but relatively low population density have 38 cells, many of them in Niger. One of these cells is urban: Maiduguri in Nigeria's north-eastern corner;
- The least problematic type in terms of probable resource stress (sub-humid, with low degradation and low population density), is represented by 34 cells, many of those in Mali. One of those cells is urban: Bobo-Dioulasso, Burkina Faso's second city ;
- The category of least problematic semi-arid areas (those with low degradation and low population densities) consists of 22 cells, many of those in Mali and non of them urban;
- The category of most problematic semi-arid areas comes fourth, with 19 cases: three cells, two of them urban, in Senegal (urban: Dakar and Thiès/Kaolack), three in Burkina Faso, at the Mossi plateau, including Burkina Faso's capital city Ouagadougou, four in Niger (including the urban cells of Niamey and Maradi in Niger), and nine in northern Nigeria, including the urban areas Kano and Sokoto.

Table 2.4 West Africa's dryland profile, compared to the World's dryland profile (tropics and subtropics): number of 1°x1° cells and percentage

Type of dryland	West Africa, number of cells	World, number of cells	West Africa as % of World
Total Drylands	139	1902	7
Semi-arid	86	1032	8
High degradation	57	353	16
High density	19	97	20
Urban	7	41	17
Rural	12	56	21
Low density	38	256	15
Urban	1	14	14
Rural	37	242	15
Low degradation	29	679	4
High density	7	71	10
Urban	0	32	0
Rural	7	39	18
Low density	22	608	4
Urban	0	15	0
Rural	22	593	4
Sub-humid	53	870	6
High degradation	9	266	3
High density	5	102	5
Urban	1	50	2
Rural	4	52	8
Low density	4	164	2
Urban	0	15	0
Rural	4	149	3
Low degradation	44	604	7
High density	10	117	9
Urban	2	63	3
Rural	8	54	15
Low density	34	487	7
Urban	1	30	3
Rural	33	457	7

Source: Based on Table 2.2 and 2.3.

Four other types of dryland areas are less prominent in West Africa:

- The category of sub-humid areas with low degradation but high densities is represented by 10 cells, including two urban ones: Bamako, capital city of Mali, and Maroua, administrative headquarters of Cameroon's Extreme North Province;
- There are seven cases in the category of semi-arid areas with low degradation but high densities , almost all of those in Nigeria and non of them urban;
- The category of most problematic sub-humid areas, with high degradation and high population densities, comprises five cases, including one urban cell: Katsina in Nigeria;
- Finally the category of sub-humid areas with high degradation and low densities has four cells, non of those urban.

Table 2.5 Types of drylands in West Africa, country details, number of 1°x1° cells.

Country	Total	SAHD HP	SAHD LP	SALD HP	SALD LP	SHHD HP	SHHD LP	SHLD HP	SHLD LP
Mauritania	4		4						
Senegal	13	3(2urb)	3		5				2
Gambia	3							2	1
Guinea Bissao	1								1
Guinea	1								1
Mali	27		4		12		1	1(1urb)	9
Burkina Faso	20	3(1urb)	7		2	1	1	1	5(1urb)
Cote d'Ivoire	1								1
Ghana	6					1			5
Togo	1						1		
Benin	5								5
Niger	18	4(2urb)	13		1				
Nigeria	32	9(2urb)	5(1urb)	6	2	2(1urb)		5	3
Cameroon	5		2	1		1	1	1(1urb)	1
Total	139	19(7urb)	38(1urb)	7	22	5(1urb)	4	10(2urb)	34(1urb)

SA = Semi Arid SH = Sub-humid (according to UNESCO map);
HD = High degradation LD = Low degradation (according to GLASOD map);
HP = High population density LP = Low population density (around 1994);
urb = with at least one city of more than 100,000 inhabitants in 1990.

For the purposes of in-depth analysis, five rural and two urban areas were selected as examples of their dryland types (see Table 2.6). For some other areas (e.g. north-western Ghana, Niamey in Niger and Dakar in Senegal) additional information was gathered.

Table 2.6 Selected case study areas in West Africa.

Type	number in West Africa	rural area selected	urban area selected
SAHD HP	19	Kaya (Burkina Faso)	Ouagadougou (Burkina Faso)
SAHD LP	38	Douentza/Bandiagara (Mali), Gorom-Gorom (Burkina Faso)	
SHHD HP	5	Bolgatanga (Ghana)	
SHLD HP	10		Bamako (Mali)
SHLD LP	34	Koutiala/Sikasso (Mali)	

Chapter 3

CLIMATE CHANGE IN DRYLAND WEST AFRICA?

The empirical evidence of rainfall variability and trends

Marcel Put, Jan Verhagen, Els Veldhuizen and Pleuntje Jellema

Abstract: A short analysis of rainfall trends for West Africa is presented in this chapter. Changes in aridity class and the related drought risk are calculated.

1. RAINFALL TRENDS IN WEST AFRICA

For all West African drylands, rainfall data for the period 1960-1990 shows a rather dramatic decline in average rainfall conditions Map 3.1 gives the changes in aridity between 1930-60 and 1960-1990. Some of the regions in the northern zone with semi-arid conditions in 1930-60 had clearly become arid (on average) in the 1960-1990 period, with unsuitable conditions for millet or sorghum production in most years. A considerable part of the sub-humid zone in the period 1930-1960 had become semi-arid in 1960-1990 with considerable drought risks, certainly for crops which are less adaptable to drought stress (maize, cotton). For 172 rainfall stations in dryland West Africa rainfall and drought risk dynamics have been studied for the last 40 years (based on data of the Agro-meteorology Group FAO-SDRN, NOAA, and AGRHYMET). The results for two of these stations are presented here: Koutiala as an example of a sub-humid area according to the UNESCO typology and Douentza as an example of a semi-arid area: see Figure 3.1. Background material on the conceptualisation of 'drought' and 'drought risks' has been presented in Put & Dietz, 1998. The drought index (for millet and sorghum) runs from no risk (0), via very light risk (1), light risk (2), moderate risk (3), and severe risk (4) to extreme risk (5) (see Chapter 6 of this book). Indeed, in both Koutiala and Douentza the precipitation trend is downward during the 1960-1985 period. However, the situation after 1990 shows an improvement in both areas. For all 11-

A.J. Dietz et al. (eds.),
The Impact of Climate Change on Drylands: With a Focus on West-Africa, 27–32.

year periods, all 172 rainfall stations averages have been calculated for annual rainfall and for drought risks (Put, 1999). Maps 3.2 to 3.5 display the average rainfall situation for 1960-70, 1975-85, 1983-1993 and a comparison between 1960-70 and 1975-85 of the 500 and 800 rainfall iso-lines.

2. DROUGHT RISKS IN WEST AFRICA

Maps 3.6 to 3.8 show the average drought risk situation for 1960-70, 1975-85 and for 1983-1993. This is another way of presenting the rainfall deterioration and drought risk increases between the 1960s and the years thereafter. More or less the same trend could be discerned for all rainfall stations in the research area: considerable deterioration between 1960 and 1985 and after 1985, or in some places a major recovery somewhat later but to a level below the 1960-1970 average.

Map 3.1 Changes in aridity in West Africa (ICCD-regions) between 1930-'60 and 1960-'90 (aridity classes are based on UNESCO, 1977).

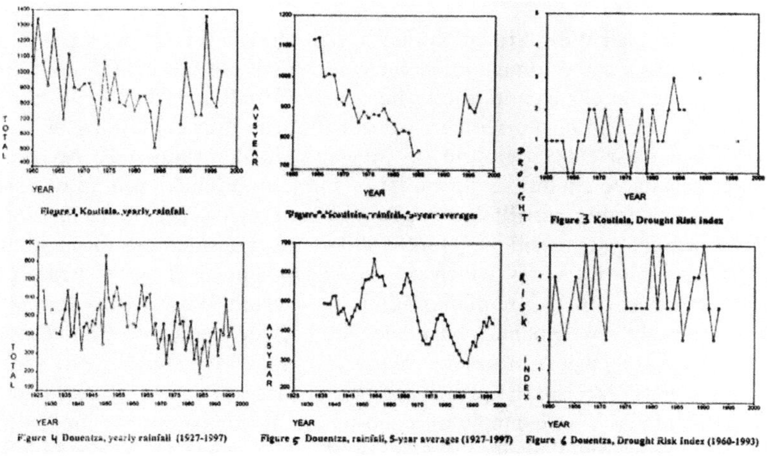

Figure 3.1. Rainfall and drought risk trends for Koutiala (1,2,3) and Douentza (4,5,6).

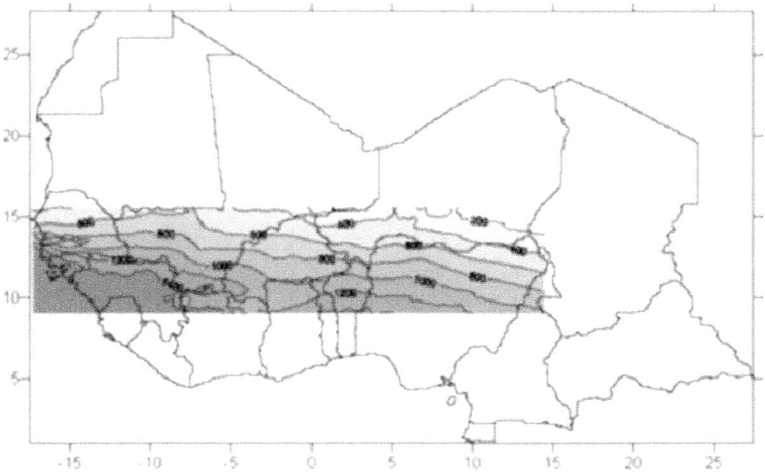

Map 3.1. Dryland West Africa: Rainfall situation 1960-'70 *(See colour section, p. 453).*

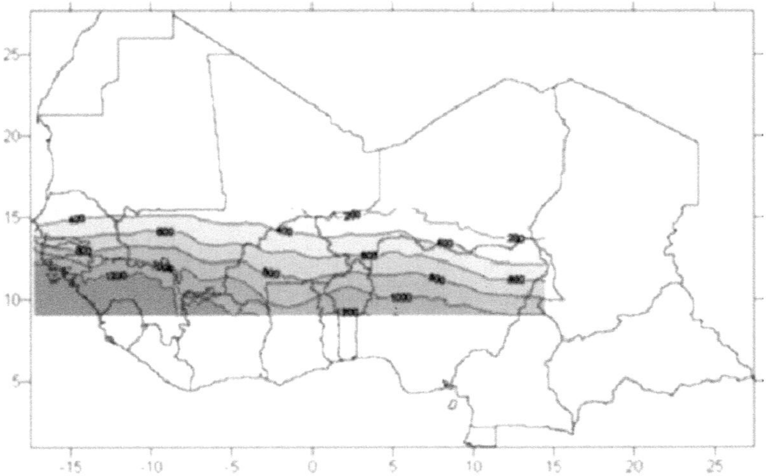

Map 3.2 Dryland West Africa: Rainfall situation 1975-'85 *(See colour section, p. 453).*

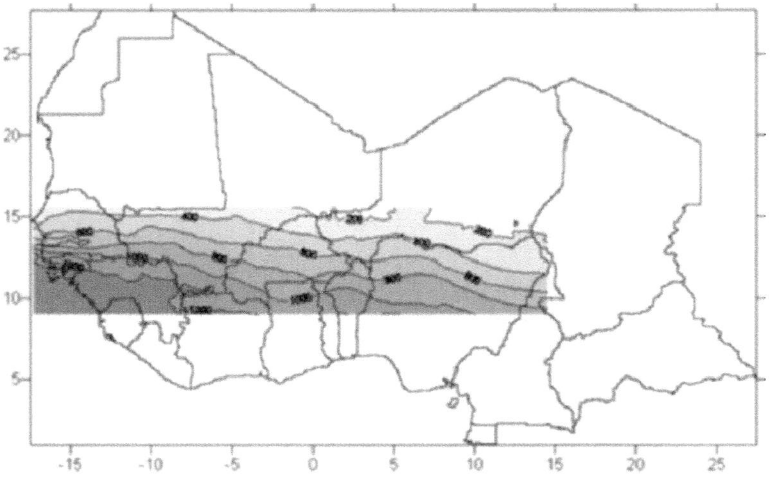

*Map 3.3 Dry*land West Africa: Rainfall situation 1983-'93 *(See colour section, p. 454).*

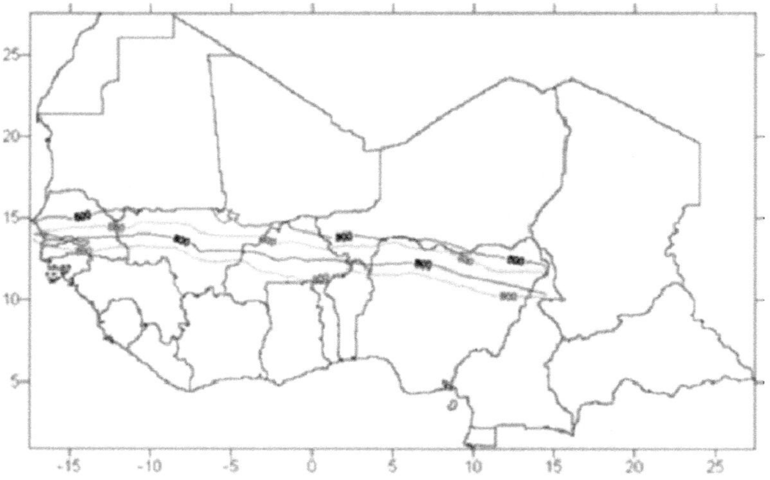

Map 3.4. West Africa: changes in 500 and 800 rainfall iso-lines, 1960-70 and 1975-'85 *(See colour section, p. 454).*

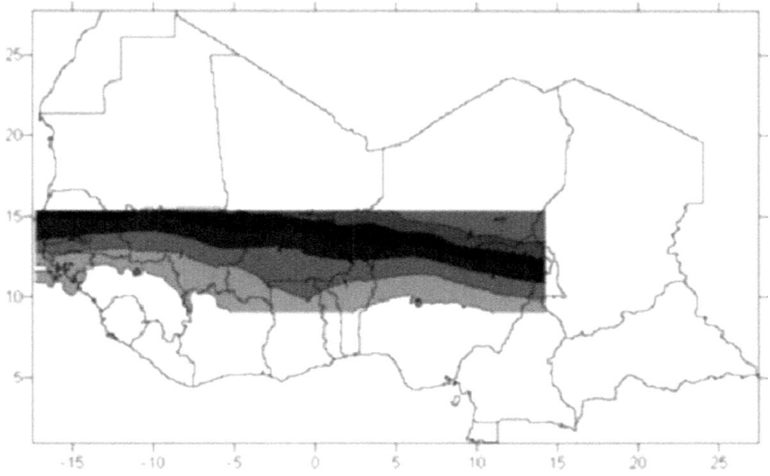

Map 3.5. Dryland West Africa: drought risks for millet/sorghum production 1960-'70 *(See colour section, p. 455).*

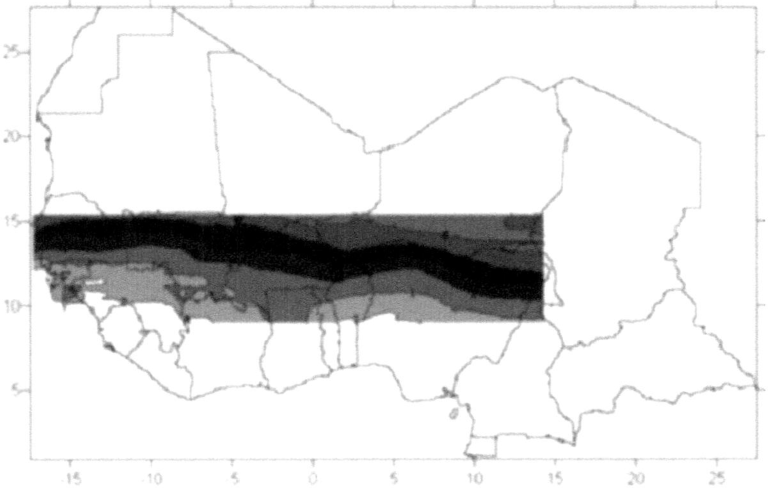

Map 3.6. Dryland West Africa: drought risks for millet/sorghum production 1975-'85 *(See colour section, p. 455).*

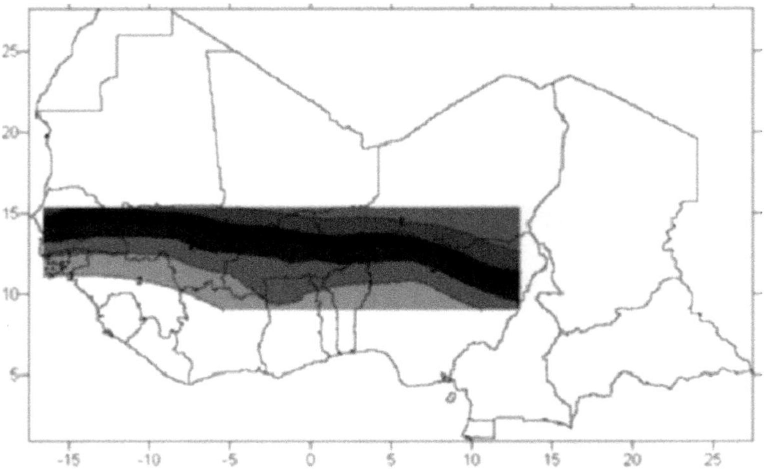

Map 3.7. Dryland West Africa: drought risks for millet/sorghum production 1983-'93 *(See colour section, p. 456).*

Chapter 4

UNCERTAINCIES IN MODELLING CLIMATE CHANGE

John van Boxel

Abstract: Regionally there are large differences between the climate models, especially in the projections for future precipitation amounts. On a regional scale, one model predicts an increase in precipitation whereas another model predicts a decrease. We should therefore conclude that the outcome of the climate models is not yet accurate enough for a regional interpretation.

1. INTRODUCTION

Concentrations of carbon dioxide and other greenhouse gases are rising and there is a growing concern about the effect this may have on the global climate and on communities around the world. The lack of an adequate ancient analogue for future climates means that we ultimately must use and trust climate models, evaluated against modern observations and geologic records (Kump 2002). However there still are a number of uncertainties that affect the results of the climate models. Some of these uncertainties are the result of the limited knowledge of the processes involved or the coarse grid spacing of the models, others result from the boundary conditions imposed upon the models. These uncertainties are discussed below, with an emphasis on precipitation, since precipitation is a very important agro-climatic variable (e.g. it has a large impact on food production) and is a difficult variable to model accurately.

Due to the presence of clouds and greenhouse gasses in the atmosphere, the average surface temperature on the surface of the earth is about 30°C higher than the equilibrium temperature of the earth (IPCC 1990, Ruddiman 2000, Houghton 2002). The main greenhouse gasses are water vapour, carbon dioxide (CO_2), methane, nitrous oxide,

A.J. Dietz et al. (eds.),
The Impact of Climate Change on Drylands: With a Focus on West-Africa, 33–42.

chlorofluorocarbons (CFCs) and ozone. During the 20th century, CO_2 concentrations increased by 35%, methane concentrations more than doubled and CFCs were introduced. The combined effect of the greenhouse gasses (with the exception of water vapour) is often referred to as radiative forcing. The increasing radiative forcing potentially enhances the greenhouse effect. Currently the radiative forcing is mainly determined by CO_2 (60%), but there is also a significant contribution from methane (20%).

Over the last decades a debate has been going on among meteorologists and climatologists as to whether the increased concentrations of greenhouse gasses would severely enhance the greenhouse effect. Nowadays, however, there is a general consensus that the increasing concentrations of greenhouse gasses will affect the global climate, that the climate changes in the last decades of the 20th century probably are the first symptoms of this global climate change and that more changes are to be expected in the 21st century (e.g. IPCC 2002a). Clarke (2001) summarizes the consensus as follows: "Lindzen is one of the few qualified meteorologists who regularly challenges the greenhouse model of global warming".

During the Pleistocene epoch (the last two million years) the earth climate was characterized by alternate ice ages and interglacials. Atmospheric carbon dioxide concentrations over this period ranged from roughly 210 ± 30 ppm (parts per million) during the ice ages and peaked at about 280 ppm during the interglacials (IPCC 2002a). During the Pleistocene period the correlation between temperature and carbon dioxide concentration on the one hand and between temperature and methane concentration on the other hand was high. Nevertheless these greenhouse gasses are not the sole causes for the major climate changes during the Pleistocene. It is believed that the ice ages were mainly triggered by variations in the amount of solar radiation reaching the earth, due to orbital changes and variations in solar activity. A number of positive feedbacks caused the changes to be rapid and large. These feedbacks include changes in the thermohaline circulation in the ocean, the snow-albedo feedback, but also the response of the main greenhouse gasses (water vapour, CO_2 and methane) to temperature changes and vice versa.

Much longer ago, during the Early Cretaceous period, carbon dioxide concentrations in the atmosphere were several times higher than the contemporary concentrations. At that time, the mean temperature on earth was about 5°C higher than it is now (Barret 2003), despite a lower input of solar radiation. When the Indian subcontinent joined with Asia and the formation of the Himalayas started, huge amounts of silicates were released, which reacted chemically with carbon dioxide, causing a fall in the carbon dioxide concentrations. The onset of the formation of the Antarctic ice sheet, 34 million years ago, occurred largely because of this fall in atmospheric carbon dioxide concentration (Harret 2003,

DeConto and Pollard 2003). So these large changes in greenhouse gas concentration did have an enormous impact on the global climate.

Over the last 1000 years (with the exception of the 20th century), carbon dioxide concentrations have fluctuated ±10 ppm around 280 ppm (IPCC 1996). We have to go back to about 23 million years in the earth's history to find carbon dioxide concentrations that compare to the contemporary concentration of 375 ppm. Currently the carbon dioxide concentration in the atmosphere is rising by about 2 ppm per year.

Over the last glacial-interglacial cycle (some 150,000 years) methane concentrations varied from 0.3 to 0.7 ppm (Chappellaz *et al.* 1990). From the 15th to the 18th century, concentrations were around 0.8 ppm. The current concentration is 1.8 ppm and is rising by approximately 0.01 ppm annually (Prather *et al.* 2002).

2. UNCERTAINCIES IN EMISSION SCENARIOS

The amount of greenhouse gases in the atmosphere controls the radiative forcing and is therefore an important boundary condition for assessing future climates. There is, however, a lot of uncertainty regarding the extent to which the radiative forcing will increase, since the production of greenhouse gases depends on many uncertain factors, such as the willingness to develop and apply new technologies that prevent the emission of greenhouse gasses, the growth of the world population and the economic development on a world wide scale.

To cope with these kinds of uncertainties, IPCC developed a number of emission scenarios, which are described in the Special Report on Emission Scenarios (SRES) (IPCC 2000). These SRES-scenarios involve a number of different possibilities, like high or low population growth, favourable or unfavourable economic development and whether we are going to develop and apply new technologies preventing greenhouse gas emissions or proceed the way we are doing now (business as usual scenario).

Figure 4.1a shows the emissions of carbon dioxide (the main greenhouse gas) according to six of these SRES-scenarios and the older IS92a scenario (IPCC 2002a). The current emission is around 7 Gt C/yr (7x109 tons of carbon per year). In the most unfavourable scenarios (A1Fl and A2) this increases to over 25 Gt C/yr by the end of the 21st century. The two most favourable scenarios assume that many new technologies will be developed. In these scenarios, the emission increases until about 2040 and strongly decreases afterwards to emission levels which are considerably lower than the current ones.

Figure 4.1b shows the carbon dioxide concentrations in the atmosphere that will result from these emissions IPCC 2002a). The figure shows that even the most favourable emission scenarios lead to a

century. In the most pessimistic scenarios the carbon dioxide concentration in the atmosphere increases to over three times the per-industrial levels.

From this example we can learn that the uncertainty in the radiative forcing resulting from the emission of greenhouse gases does not have to do with whether it will increase, but rather by how much. Any realistic scenario will produce a doubled radiative forcing (compared to the pre-industrial level) by the end of the century. However a tripled radiative forcing is also possible.

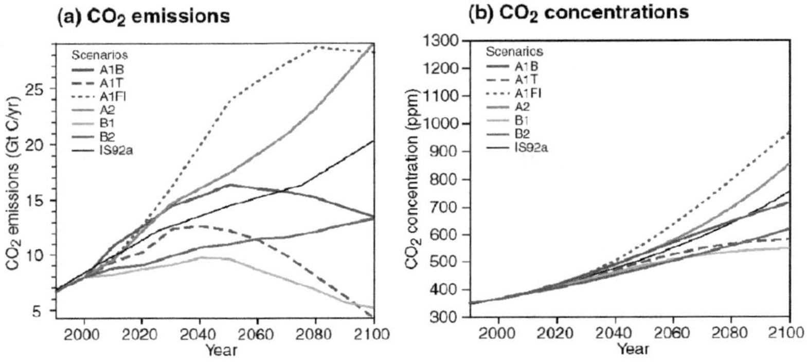

Figuur 4.1.Carbon dioxide emissions according to six SRES scenarios and the IS92a scenario (a) and the resulting concentrations of carbon dioxide (b). (Source: IPCC 2002a) *(See colour section, p. 456).*

3. DEVELOPMENTS IN CLIMATE MODELS

In the 1990 IPCC report, Bretherton *et al.* (1990) state "Though the atmospheric models are highly developed, the treatment of the ocean is quite primitive". At that time the coupled ocean atmosphere models were restricted to coarse resolution models (large grid spacing) because of limitations in computing power and because the models contained simplified presentations of physics of the ocean circulation. The 1990 IPCC report mentions the following key areas of scientific uncertainty (IPCC 1990):

- Clouds: formation, dissipation radiative properties and the way in which they influence greenhouse forcing
- Oceans: exchange of energy between ocean and atmosphere and between surface layers and deep ocean
- Greenhouse gases: quantification of release, uptake and chemical reactions.
- Polar ice sheets: which affect predictions of sea level rise.

Since 1990, computing power has increased tremendously. We know much more about the physics of the oceans and coupled ocean-atmosphere models are now commonly used. Most models now take into account the role of sulphate aerosols and the most sophisticated models now also include atmospheric chemistry. A lot is known about the physics of cloud formation, but it is still difficult to use this information in the cloud models, since the formation of convective clouds in particular takes place at scales which are much smaller than the grid spacing of the climate models. The projections of sea level rises are still affected by our limited knowledge of the behaviour of the ice caps on Greenland and Antarctica. In the Third Assessment Report (IPCC 2002a) the technical summary (Albritton *et al.* 2002) produces a nice overview of the major developments in climate models over the last decades and McAvaney *et al.* (2002) discuss the evaluation of the climate models.

Figure 4.2 illustrates that modern climate models are now perfectly capable of modelling the mean global temperature over the 20th century.

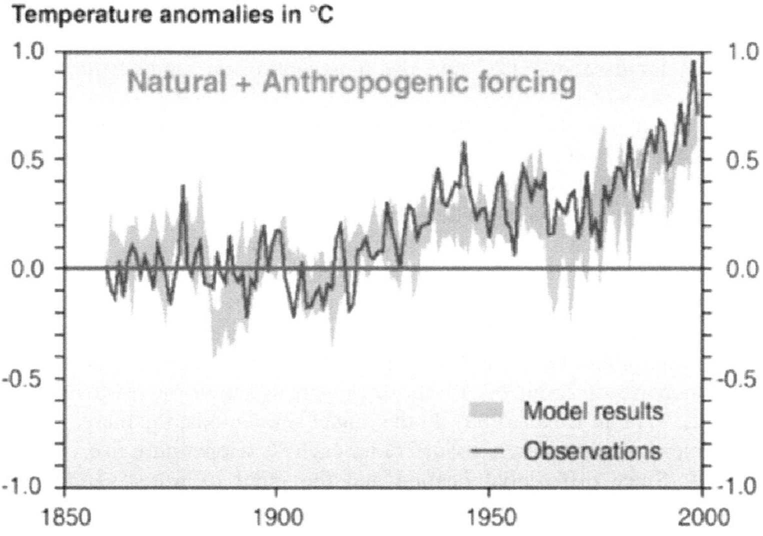

Figure 4.2. Observed and modelled variation in globally averaged temperature anomalies. The model used both natural forcing (e.g., variations in solar activity, volcanism) and anthropogenic forcing (increased concentrations of greenhouse gases). (source: IPCC 2002b) *(See Colour Section, p. 457)*

4. CLIMATE PROJECTIONS

Figure 4.3 presents the projections for the global mean temperature for the 21st century calculated using a number of different climate models and emission scenarios.

In the worst scenario (A1Fl) the most pessimistic model forecasts a temperature increase of 6.8°C by the year 2100 and the most optimistic model an increase of 3.3°C. These are very large changes if we take into account that the difference in global mean temperature between the last ice age and now is about 6°C. During the 20th century the carbon dioxide concentration increased by about 100 ppm and the global mean temperature increased by about 0.7°C, of which about 0.5°C should be attributed to the increased radiative forcing. In this light, it is not unlikely that an increase in atmospheric carbon dioxide concentrations of more than 600 ppm compared to the pre-industrial levels produces temperature increases of more than 3°C. We could make a very rough estimate that each 100 ppm increase in atmospheric carbon dioxide levels results in an increase of around 0.5 to 1.0°C in global mean temperature.

For the best emission scenario (B1) the mean of all model forecasts is an increase of 2.0°C and the most optimistic model still predicts an increase of 1.4°C.

All models calculate that the temperature increase will vary with latitude. In the tropics, where water vapour is the main greenhouse gas, the temperature increase will be much less than the global mean, whereas in the polar regions, where the main greenhouse gas is carbon dioxide, the increase can be a multiple of the global mean.

The surface temperature has a large influence on evaporation from the oceans, which are the source of about 90% of the water vapour in the atmosphere. Therefore, higher temperatures coincide with higher evaporation. Over the last century, based on a global mean temperature increase of about 0.7°C, the mean precipitation on earth increased by 2.5% (Van Boxel 2000). In the model simulations, the increase of global mean precipitation is about 2% for each °C temperature rise.

Since differential heating and the input of water vapour into the atmosphere have large implications for the general circulation it is to be expected that the general circulation will change. Therefore the increase in precipitation will also not be evenly spread over the earth surface. Over the 20th century arid and semi arid regions in the southern hemisphere mostly received more precipitation towards the end of the century, whereas the semiarid and arid regions in the northern hemisphere, including the Sahel, generally became drier (Van Boxel 2000). Most of the temperate latitudes on the northern hemisphere received more precipitation towards the end of the 20th century whereas there was generally little change in the wet tropics, with the exception of east Africa, which got wetter.

The next section demonstrates the regional changes in precipitation as predicted by four climate models when radiative forcing levels doubles compared to the pre-industrial levels.

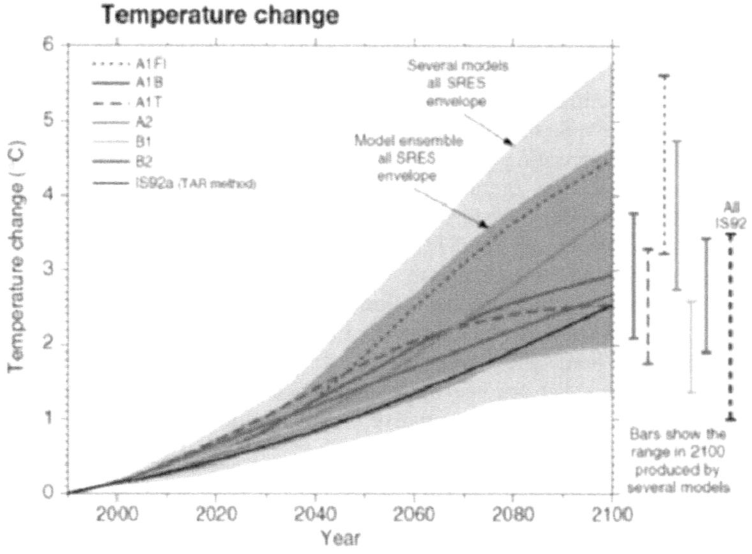

Figure 4.3. Predictions of global mean temperature using different climate models and emission scenarios (Source: IPCC 2002a). *(See Colour Section, p. 457)*

5. REGIONAL DIFFERENCES BETWEEN CLIMATE MODELS

When assessing the impact of climate change researchers are usually not so much interested in the global mean change, but rather in the changes in the local or regional climate (e.g., Palutikof *et al.* 2002, Van den Born *et al.* 2000). One approach involves the direct use of the outcome of the climate models (e.g., Van den Born *et al.* 2000) or the incorporation of a higher resolution model into the climate model (e.g., Palutikof *et al.* 2002).

Since there are enormous regional differences between the various climate models, however, it is still almost impossible to use the outcome of climate models for these regional interpretations. Because precipitation is a very important climate variable, but also is a difficult one to predict, we will use precipitation as an example to illustrate this.

Figure 4.4 shows the relative increases in precipitation as predicted using four different climate models when radiative forcing is doubled compared to the pre-industrial levels and taking into account the expected changes in sulphate aerosol concentration.

The US will become much drier according to the CCSR-98 model, but the HADCM2 model says it will become wetter. The other two models predict more precipitation in the western US and less in the eastern US. The Sahel will get drier according to the CSIRO and CGCM1 model, whereas the other two models predict a more favourable (wetter) climate. Many more differences can be found when comparing the models. So on a regional scale there are large differences between the models concerning the precipitation climate.

Nevertheless there are also several similarities between the models. All the models predict an increase in the global mean precipitation ranging from 2.5 to 5.1 %. The amount of increase in global mean precipitation is related to the predicted increase in global mean temperature. Most models agree that the Mediterranean will become drier in a warmer climate and that most of the temperate latitudes on the northern hemisphere will receive more precipitation. In all models there are regions that become wetter and regions that become drier. In general, both changes will have negative impacts on the short term.

6. CONCLUSIONS

Over the 20th century, the concentrations of greenhouse gases in the atmosphere have increased. Global mean temperature increased by about 0.7°C and global mean precipitation increased about 2.5%.

Over the 21st century, carbon dioxide concentrations in the atmosphere will continue to rise due to the use of fossil fuels and possibly ongoing deforestation. Further climate changes are to be expected. Major uncertainties as to the extent to which the climate will change are linked to uncertainties regarding future greenhouse gases emissions. There are differences in the global means of temperature and precipitation increase between the various climate models, but the trends are all in the same direction.

Regionally there are large differences between the climate models, especially in the projections for future precipitation amounts. On a regional scale, one model predicts an increase in precipitation whereas another model predicts a decrease. We should therefore conclude that the outcome of the climate models is not yet accurate enough for a regional interpretation.

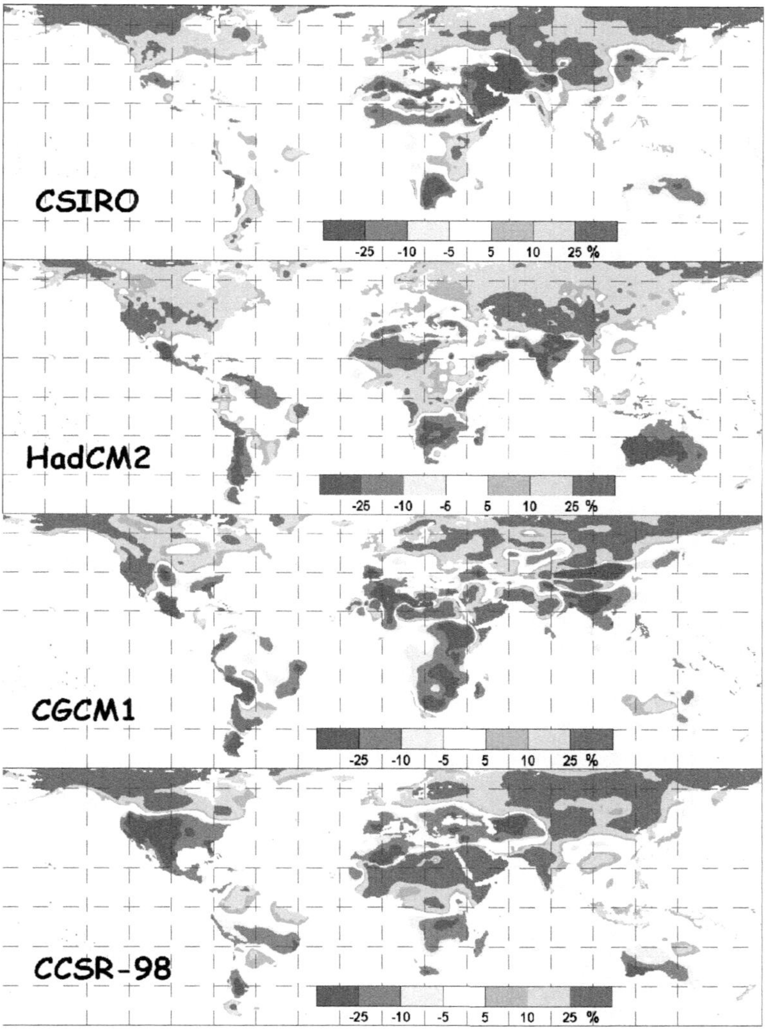

Figure 4.4 The relative increases in precipitation as predicted using four different climate models *(See colour section, p. 458).*

Chapter 5

CLIMATE CHANGE SCENARIOS FOR DRYLAND WEST AFRICA, 1990-2050

Gertjan van den Born, Rik Leemans and Michiel Schaeffer

Abstract: IMAGE 2 was used to define three climate change scenarios for dryland west Africa. Changes in precipitation and the resulting shift in aridity class are shown using maps. The three scenarios cover a wide range of possible changes in climate and related impacts.

1. INTRODUCTION

Reliable climate change scenarios are needed to assess impacts. Climate scenarios have become more closely linked to the changes in concentrations of CO_2 and other GHGs. There are several ways of creating such scenarios (Carter *et al.*, 1994). The simplest approach is the arbitrary or systematic prescription of a specific climatic change attainable by varying the temperature and precipitation (or other) model input. This synthetic approach is a simple sensitivity analysis that highlights the vulnerability of the system modelled. The major disadvantage of this approach is that the scenarios do not represent a realistic future climate. Although popular in simple and rapid assessments, the approach is no longer used in more thorough assessments. Recent developments in this approach involve linking such changes to weather generators. Such algorithms describe the current variability of local and regional climate, and statistically produce climatic series that resemble realistic climatic patterns. They can be used for future climate as well, but the limitations of synthetic approaches remain to a large extent.

A more advanced approach is to use an analogue climate, either from a past warmer or cooler period, or from a region with a climate assumed similar to the future climate. For example, the warmer period in the

A.J. Dietz et al. (eds.),
The Impact of Climate Change on Drylands: With a Focus on West-Africa, 43–47.

Holocene (the so-called climatic optimum) has often been proposed as a good analogue for early GHG-induced climate change. Unfortunately, the position/inclination of the earth relative to the sun was different and this generated climatic conditions which differed considerably from a future greenhouse climate (Mitchell, 1990). The other approach is to use single warm or dry years from the historical record. In principle, this is a valid approach for testing the vulnerability of ecosystems and/or agriculture, but single years may not show long-term impacts. The disadvantage of the regional analogue is that non-climatic factors, such as soil types, land uses and socio-economic parameters differ. These factors could influence local crop and ecosystem patterns more strongly than climatic factors. Conclusions from analogue studies can therefore be misleading.

Both the linked synthetic-weather generator and analysis of historical data records are examined in the regional studies.

The most frequently used approach is to derive scenarios from three-dimensional Atmospheric Global Circulation Models, the GCMs. These full three-dimensional climate models simulate global climatic patterns on a grid between 2 and 9 degrees longitude and/or latitude and several vertical layers thick, by simultaneously solving the energy-balance equations of each grid cell. The temperature exchanges with the oceans are prescribed and there are no systemic interactions between the oceans and atmosphere. Generally, such equilibrium GCM simulations perform well for current climate conditions. Seasonal temperature patterns and pressure fields are similar to observed climate. However, the characteristic patterns of precipitation, critical for agriculture and ecosystems, only resemble reality at the highest grid resolutions of 2 to 3 degrees. These resolutions are very computationally demanding. GCMs are disturbed by changed GHG concentrations.

2. CLIMATE SCENARIOS USING IMAGE 2.1 AND GCMS

In projections of climate change and assessment of impacts, a balance has to be found between detail and consistency on the one hand and efficiency and comprehensiveness on the other. The first two qualities can be supplied by a combination of GCM (General Circulation Model) data and sophisticated downscaling methods. Downscaling is needed to transform the GCM data to a high enough resolution to be of use in an impact assessment. Several considerations brought forward in this paper prohibit the use of GCMs when a comparison of several scenarios is required. GCMs also lack a more comprehensive treatment of the interaction of atmosphere and ocean with the biosphere and the human intervention herein. Such an approach can be supported by an Integrated Assessment Model (IAM) such as IMAGE 2.1 (Integrated Model to

Assess the Greenhouse Effect). IAMs, however, lack a capability to handle geographical detail for regional studies. IMAGE uses GCM generated data in an attempt to combine virtues.

As an integrated assessment model, IMAGE, provides a comprehensive treatment of the global climate change issue (Alcamo *et al.*, 1998). A system of coupled models projects pathways of changes in atmospheric conditions, climate, land use and land cover. Many feedback processes between atmosphere, ocean, biosphere and human systems are represented. Like all other Integrated assessment models (IAMs) for climate change, IMAGE 2.1 covers the entire globe but, in addition, also performs a lot of calculations on a global terrestrial grid of 0.5o longitude and latitude, for 10o latitudinal climate zones and for 13 world regions. These spatial and regional resolutions increase model testability against measurements, allow an improved representation of feedback and provide more detailed information for climate impact analysis. Moreover, the sub-models of IMAGE 2 are, in general, more process-oriented and contain fewer global parameterisations than previous models.

To draw conclusions within a region, like for example sub-Saharan West Africa, one would prefer to make use of a method of downscaling the GCM data rather than linear interpolation between GCM grid cell centres. Results from the meso-scale climate model mentioned above can probably be readily applied if a climate change run has been performed using this ensemble of global and regional climate models. The method of applying this regional data will then be fully analogous to the use of GCM data in IMAGE in general as described. Applying climate change data from a nested regional model would be no different from the standard method of applying different GCMs in IMAGE.

A baseline and three scenarios are defined using two different GCMs: GFDL and Max Planck (MPI-GCM). The results for sub-Saharan West Africa are presented in Table 5.1.

The Baseline A (medium population and medium economic growth) in combination with MPI-GCM, leads to a CO_2 concentration of about 500 ppmv by 2050, which is about 150 ppmv above current concentration. In sub-Saharan West Africa this will lead to a temperature increase of about 1.5 to 2.5 °C. and a decrease in precipitation ranging from 100 to 400 mm yr^{-1}. Current vegetation zones in the area shift more towards the south. Subsequent to these changes, the aridity shifts towards the south.

Table 5.1 Changes in climatic zones in sub-Saharan West Africa over the period 1990-2050, (a) Baseline A, MPL GCM and (b) GFDL GCM.

climate zone	In 1990	in %	In 2050	in %	Change 1990-2050
MPL GCM					
Humid	66.2	12%	56.5	10%	-15%
sub humid	114.8	21%	101.5	18%	-12%
semi arid	171	31%	166	30%	-3%
Arid/hyper arid	203.3	37%	231.4	42%	14%
Total	555.4	100%	555.4	100%	0%
GFDL GCM					
Humid	67.4	12%	61.9	11%	-8%
sub humid	113.9	21%	119.1	21%	5%
semi arid	177.5	32%	182.3	33%	3%
Arid/hyper arid	196.5	35%	192	35%	-2%
Total	*555.4*	*100%*	*555.4*	*100%*	*0%*

About one-third of the area is expected to change in aridity, mainly to a higher aridity. Using the GFDL GCM the change in aridity is quit different. The increase in temperature by 2050 is substantially lower and the precipitation pattern does not show a decreasing trend in precipitation. As a consequence, there is almost no shift in vegetation and aridity classification. The results for Baseline C (medium population and high economic growth) in combination with MPI-GCM, differ substantially if compared to the results for Baseline A and the same GCM. The emissions in Baseline C are high. The increase in temperature is about half a degree higher than baseline A and the precipitation decreases more strongly. These differences are a direct consequence of the implemented scaling approach. The difference between Baseline A and C in achieved aridity, however, is small. The results show that in a substantial part of West Africa aridity levels are increasing. This increase, in effect a change in seasonality (start and length of growing season), affects the agricultural production systems and could indirectly trigger major socio-economic changes as well.

The three scenarios cover a "wide" range of possible changes in climate and related impacts. The range seems "wide", but is probably small from the point of view of vulnerability. The shift in precipitation and shift in climate for the baseline and most distinctive scenario are presented in Map 5.1 and Map 5.2.

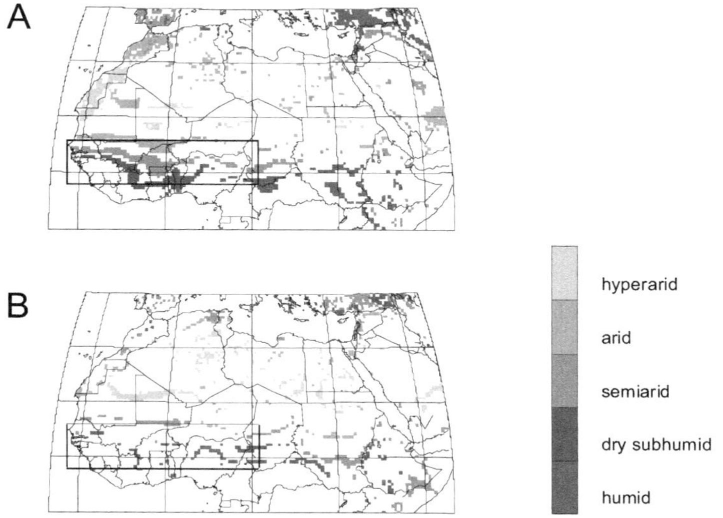

Map 5.1. Change in aridity classification, 1990-2050, for A: baseline A scenario, MPI-GCM; B:baseline A scenario, GFDL-GCM (ICCD study area is indicated by the box, coloured cells subject to change and show classification by 2050).
(See Colour Section, p. 459)

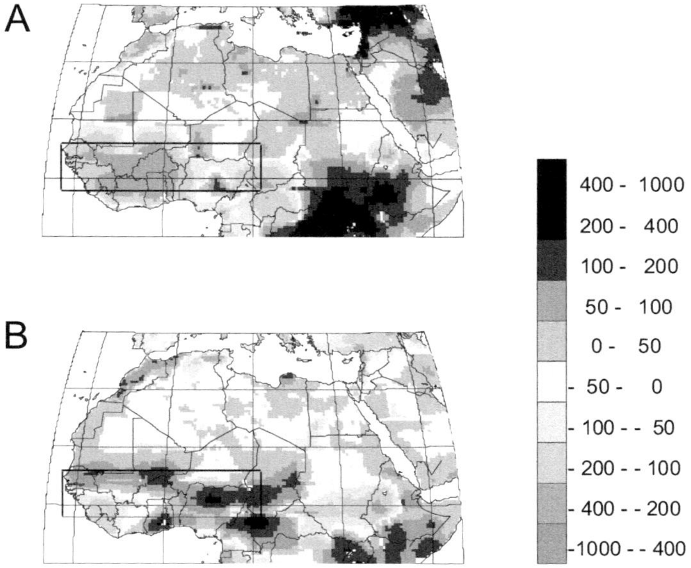

Map 5.2. Change in precipitation (mm), 1990-2050, for A: baseline A scenario, MPI-GCM and B: baseline A scenario, GFDL-GCM (ICCD study area is indicated by the box). *(See Colour Section, p. 459)*

Chapter 6

CLIMATE CHANGE AND DROUGHT RISKS FOR AGRICULTURE

Jan Verhagen, Marcel Put, Fred Zaal and Herman van Keulen

Abstract: Changes in rainfall patterns and risk of crop failure are discussed in this chapter. Agriculture is by far the most important economic activity in the region. The success of this economic activity relies heavily on water availability during the growing season. For rainfed production systems timing and the amount of precipitation determine to a large extent the success of the growing season. The projected changes in the variability and total amount of rainfall are expected to worsen the situation.

1. CLIMATE CHANGE AND CLIMATE VARIABILITY

Agriculture is by far the most important economic activity in sub-Saharan West Africa (SSWA). It is still the main source of income and livelihood in this region. A large section of the population is active in subsistence agriculture. In sub-Saharan West Africa the arid and semi arid regions are among the harshest and vulnerable production environments in the world. Combined with a fast growing and mobile population, the pressure on the natural and societal resource base will increase.

The rainy season is crucial for agricultural production; it is during this period that conditions determine whether there are going to be food shortages or not. As indicated by the GCM-scenarios, climate change may confront dryland West Africa with even lower and more variable precipitation, higher temperatures and higher evaporation. This may result in even higher risks for crop production, with the ultimate consequence being a decrease in food availability. However, consequences of climate change on agricultural production are highly

A.J. Dietz et al. (eds.),
The Impact of Climate Change on Drylands: With a Focus on West-Africa, 49–59.

speculative. Partly due to the uncertainties related to the magnitude of climate change, but also due to highly dynamic socio-economic and institutional context that has direct and indirect effects on agricultural production. Whatever the causes, less food, water and fuel means that the local population is forced to adopt and develop alternatives and that all kinds of institutions and organisations will have to adapt to this changing situation as well.

Shifting cultivation, land clearing for crop production, is the main system. Some crop rotation is customary but the intrinsic low soil fertility does not allow continuous cropping without external inputs such as fertiliser. Increasing soil fertility will enhance the production capacity of the land and increase water use efficiency. Labour availability is another constraint in crop production as the continuous fight against weeds and pests is carried out mainly by hand. This section deals with questions related to climate change and climate variability in the region and their effect on crop production with the focus on rainfall, as this is the main driver of the agricultural production systems. The important link between the arable and livestock component in sub-Saharan West Africa is not included in this assessment but will be addressed in the detailed regional studies.

The concept of climate is dynamic and includes short-term variations. Climate variability refers to inter-annual variability of individual climatic parameters around longer-term mean values. Climate variability is inherent in dryland areas (Le Houérou, 1996). Climate variability can occur no matter whether the longer-term climate is stable or changing. For example, inter-annual rainfall values may fluctuate around an overall trend towards wetter or drier conditions.

2. DROUGHT

The encyclopaedia Britannica gives the following definition of drought: it is the lack or insufficiency of rain for an extended period that causes a considerable hydrologic imbalance and, consequently, water shortages, crop damage, stream flow reduction and depletion of groundwater and soil moisture. Drought is a normal, recurrent phenomenon, which occurs in virtually all climatic zones, although its characteristics vary significantly from one zone to another (Hulme, 1995).

The interaction between the natural event and the demand people place on water supply determines the impact of a drought event. Recent droughts in both developing and developed countries have underscored the vulnerability of all societies to this natural hazard, although some groups are more vulnerable than others. It is possible to differentiate the drought impact according to a number of disciplinary perspectives:

meteorological, hydrological, agricultural and socio-economic drought. (see National Drought Mitigation Center: www.enso.unl.edu)

Meteorological drought

Meteorological drought is usually defined on the basis of the degree of dryness and the duration of the dry period. Thus, Hulme (1995) defines meteorological drought as a reduction in rainfall supply compared with a specified average condition over some specified period. Definitions of meteorological drought must be considered as region-specific since the atmospheric conditions that result in deficiencies of precipitation are highly variable from region to region. This study uses a definition based on actual precipitation departures from average amounts on a monthly, seasonal, or annual time scale.

Hydrological drought

Hydrological drought is associated with the effects of periods of precipitation shortfalls on surface or subsurface water storage systems (i.e. streams, lakes, reservoirs, and ground water) (Wilhite, 1993). Although the origin of all droughts is a deficiency of precipitation, hydrologists are more concerned with how this deficiency plays out through the hydrologic system (a watershed or river basin). Hydrological droughts are usually out of phase with meteorological droughts since it takes longer for precipitation deficiencies to show up in components of the hydrological system

Agricultural drought

The driest climates with sparse vegetation that is adapted to aridity are characterised by permanent drought. In these areas agriculture is impossible without continuous irrigation. Semi-arid regions are characterised by seasonal drought during the so-called dry season. In areas such as sub-Saharan West Africa the growing season for rain-fed agriculture is restricted to the rainy season. More commonly, the drought is related to abnormal rainfall failure, which may occur in all climate zones but is regarded as a characteristic of the humid and sub-humid climate zones. Such failures are erratic and hard to predict and the effect is often limited to small areas. The so called invisible drought is characteristic for areas which appear to have enough water but in which crops suffer water shortage related stress due to high temperatures resulting in high evaporation and transpiration rates.

Socio-economic drought

Socio-economic definitions of drought associate the supply and demand of some economic good with elements of meteorological, hydrological and agricultural drought. It differs from the aforementioned types of drought because its occurrence depends on the time and space processes of supply and demand to identify or classify droughts. Socio-economic drought occurs when the demand for an economic good exceeds supply as a result of a weather-related shortfall in this good.

3. CLIMATE DATA

Long-term average (1960-1990) monthly data of cloudiness, temperature and precipitation taken from the Leemans & Cramer (1991) dataset were used for the baseline calculations. The spatial resolution of this global dataset is 30×30 arc minutes. The degrees of cloudiness in combination with the geographical position were used to calculate solar radiation. The evapo-transpiration was calculated using Makkink's method. The monthly data could be used directly to calculate the Drought Index. The climate scenarios, MPI-GCM and GFDL-GCM were delivered in the same format as the Leemans & Cramer (1991) dataset.

To arrive at the daily climatic information required for the simulation model the monthly data had to be converted. Linear interpolation was used for radiation and temperature data. Daily rainfall was derived from the combination of the number of rainy days and the total amount of rainfall. The 'number of rain days' was taken from the database compiled by Müller (1996) and subsequently translated to a 30×30' grid.

In rain-fed agriculture, the timing and distribution of rain is essential for crop growth and yield formation. Especially in semi-arid regions, this daily variability is crucial when addressing drought risk. However, daily data is not always readily available or easy to obtain. However, monthly totals of rainfall are available in most situations. If the daily dynamic of the system is omitted and average values are used based on monthly totals, stress situations due to low or excess rainfall are averaged out and may result in errors (Nonhebel, 1993).

Using weather generators, daily data can be generated based on monthly totals and the number of rain days. The weather generator used in this study was developed for the West African Sahel Zone (de Ruijter, 1990). The generator is based on the combination of a first order Markov chain and a gamma distribution (Geng *et al.*, 1986). The Markov chain is used to establish successive daily precipitation using transitional probabilities. The probability distribution of the amount of rain is described using a gamma distribution. A random number generator is used to determine the amount of rain for a given wet day. The parameterisation of both the Markov chain and gamma distribution is region-specific. Parameterisation was applied to 7 stations in Mali, using long-term (30 years) daily datasets. After successful parameterisation the rainfall generator was built into a crop growth model

Information on evapo-transpiration is crucial when quantifying the effect of drought stress on crop performance. Evapo-transpiration is a complex process combining transpiration and evaporation. Transpiration is the process of water transport via the root system to the stem and leaves into the atmosphere. Evaporation is the upward transport of water from the soil surface to the atmosphere. Both processes are driven by the potential differences between the water in the soil and the atmosphere. The potential rate of evapo-transpiration is a combination of the demand by the atmosphere and the properties of the soil and crop.

A number of methods are available for calculating the potential evapo-transpiration. Of these, the most important and best known method is the Penman method (Penman, 1948). Less well-known are the Priestley-Taylor (Preistley & Talor, 1972) and Makkink (Makkink, 1957) methods. All the methods are used to calculate evapo-transpiration for non-restricted conditions; a well-watered short crop. The advantage of the latter two methods is that they require fewer parameters, whereas Penman requires detailed readings of radiation, temperature, vapour pressure and wind speed. Makkink and Priestley-Taylor require only radiation and temperature data and assume that a constant relation exists between evaporative demand by radiation and by wind. All three methods are examined and described in Van Kraalingen & Stol (1997). The Makkink method uses incoming short-wave radiation and temperature, whereas the Priestley-Taylor method uses long-wave radiation and temperature. In this study the more practical Makkink method was used.

4. DROUGHT RISK ASSESSMENT

Several techniques are developed to assess the risk of crop failure in relation to rainfall variability. One approach is the drought index, which provides a composite picture based on a time series analysis of the effect of water deficiency on crop performance. It is a static approach often linked to a given region and crop. In this study a drought index for sorghum and millet is used to assess risk based on a discrete classification for the MPI-GCM scenario and the analysis of historical data records. A modelling approach is followed using a weather generator and a simple crop growth model to quantify the effects of climate change on the crop performance. The modelling approach is more generic than the index approach. Finally, a monitoring tool is assessed which relates the Normalised Difference Vegetation Index (NDVI) time series to a long series of satellite images to characterise vegetation dynamics in relation to climate at regional and continental levels. The NDVI was designed in order to understand the status of vegetation using remotely sensed data. It represents the quantity and activity levels of vegetation. The NDVI is related to the proportion of photo-synthetically absorbed radiation and is calculated from atmospherically corrected reflectance data from the visible and near infrared channels.

A basic approach to determining drought is to use a discrete index. Such an index, as an indicator for increasing drought stress, facilitates communication between scientists of different disciplines, managers and policy makers. Such indicators are quantitative and provide insight into the expected success rate for crop production in a given region and may trigger counteractive measures by policy makers. The drought indicator is static and has no strong predictive value.

The drought index used in this study is based on indices made by Bailey (1979) and the FAO (1980). Bailey uses monthly precipitation and mean monthly temperatures to define a moisture index' in which 'wet', 'neutral' and 'dry' months are differentiated.

It is constructed as follows:

$S = 0.18P/(1.045) \times T,$

where S is the dimensionless drought index, P is the precipitation in cm and T is the temperature in °C. Months are 'wet' when $S > 0.81$ and 'dry' when $S < 0.53$. In between 0.53 and 0.81, months are called 'neutral'. The FAO approach, developed to define Agro-Ecological Zones, uses monthly precipitation (P) and monthly potential evapo-transpiration (ETP) and allows soil moisture storage from a 'rain' month for a following 'dry' month. For the FAO, a month was regarded as sufficiently wet for a crop that is adapted to dry conditions when $P > 0.5$ ETP. Also when P>ETP, the surplus rainfall can be accumulated for the next month as soil moisture storage with a maximum storage period of one month. Table 6.1 shows the full range of the six indices.

Using average monthly precipitation, average monthly temperature and average monthly daily potential evapo-transpiration, the index distinguishes between 'very wet' (Bailey's 'wet'), 'moderately wet' (Bailey's 'neutral'), 'slightly wet' (FAO's 'rain'), and 'dry' months. For sub-Saharan West Africa we will focus on the rainy season, starting in May and extending to October. As the method uses average monthly data, high-resolution or daily climate variability is eliminated from the equation.

Using recorded climate data, the drought index can be calculated for a number of years and so provide insight into how the drought risk index behaves over time (see Chapter 3).

Table 6.1 Drought Index after Bailey (1979) and FAO (1980).

Index	Classification	Method	Criteria
0	No drought	Bailey	At least four consecutive 'very wet' months
1	Very light drought	Bailey	At least four consecutive 'moderately wet' months
2	Light drought	Bailey	Three consecutive 'moderately wet' months
3	Moderate drought	FAO	Three consecutive 'slightly wet' months
4	Severe drought	FAO	Two consecutive 'slightly wet' months and a 'dry' month which is sufficiently compensated by the storage of rain from the previous month (storage: P-ETP; available rain = $P_m + P_{m-1} - ETP_{m-1}$).
5	Extreme drought	FAO	Two consecutive 'slightly wet' months without moisture storage in the third month or less than two consecutive 'slightly wet' months

Changes in precipitation, are used to calculate the drought risk. Map 6.1 clearly shows that changes in precipitation as presented for the Baseline A scenario using the MPI-GCM have a dramatic impact on the spatial distribution of the risk prone areas. When compared to Map 3.2, the picture is similar to that of the dry spells of the Seventies.

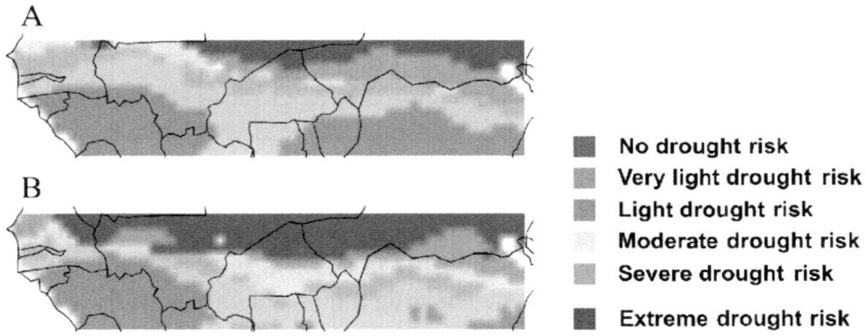

Map 6.1. Changes in drought risk for 1990-2050, A depicts the current situation (based on 30-year average) B displays the projected drought risk for baseline A scenario, MPI-GCM *(See colour section, p. 460).*

5. CROP GROWTH SIMULATION MODEL

Crop growth obeys certain physiological principles. These growth processes can be quantified in response to the environment by mathematical formulae (Spitters & Schapendonk, 1990). The environment crop interaction can be described using simple static models without a description of process rates and dynamic models where state variables change in accordance to fluctuating process rates. The drought index is an example. Static models have the advantage of a small number of parameters and a simple algorithm.

The dynamic approach, however, has the advantage of greater flexibility. In addition, it gives a greater insight into the sensitivity of underlying processes interacting with fluctuating climatic factors. Such a mechanistic dynamic model, enables the prediction of crop growth rates and yields under a variety of environmental and management conditions. Crop growth models are useful for understanding and exploring system responses to environmental and management operations. Crop models may also be used for yield forecasting or can be applied in land use evaluation, e.g. to assess the production potentials of new cropping areas as regards dependence on climatic conditions and the availability of water and fertiliser.

The model used in this study is based on the LINTUL-type models (Bouman *et al.*, 1996), it is a simple general crop model, which simulates dry matter production as the result of light interception and utilisation with a constant light use efficiency. Detailed information on how to construct and use crop growth simulation models can be found in e.g. Penning de Vries & Van Laar (1982), van Keulen & Wolf (1986), and Penning de Vries *et al.* (1989). Reviews of the various approaches followed in crop growth simulation and examples of their application have been given by, among others, Loomis *et al.* (1979), Penning de Vries (1983), Whisler *et al.* (1986), and Wisiol & Hesketh (1987), Spitters *et al.*, 1989.

Temperature is the only factor that determines the development rate of the crop. Crop growth is possible within a temperature range of 5 to 35 °C. In order to complete a full growth cycle, a certain amount of accumulated temperature degrees are required. The required degree-days may vary among crops. In this study a standardised cereal crop (wheat, rice, maize, millet or sorghum) is taken as a proxy for a wide range of crops that could be grown. The crop requires a total of 1200 °Cd, divided in two stages. The radiation interception by leaves increases linearly during the first stage of crop development, from zero at emergence to unity at a development stage of 600 °Cd (van Heemst, 1986). Subsequently, interception remains at unity during 600 °Cd and reduces to zero over the last part of the growing period, because of leaf senescence. Intercepted radiation is converted into biomass by multiplication with a constant radiation use efficiency set at 3.0 g MJ-1 (Monteith, 1981, 1990; Gallaghar & Biscoe, 1978; Goudriaan & Van Laar, 1994).

Radiation and temperature are, obviously, not the only environmental factors determining crop yield. Water, which is absolutely essential for crop growth, is in most cases equally important. Crop growth might be limited by a shortage of water during all or part of the growing period.

The movement of water through to crop via the stomata into the atmosphere is called transpiration. CO_2 is transported in the opposite direction for the formation of carbohydrates. Most of the water used by the crop is, in fact, lost to the atmosphere. When the uptake of water from the soil through the roots is not sufficient to replace the water lost to the atmosphere, the plant reduces the outflow of water by closing its stomata. However, this reaction also hampers inflow of CO_2 and hence plant growth. The two processes are more or less proportional. In other words for each kilogram of plant material produced (under a given set of external conditions) the same amount of water is required.

The potential evapo-transpiration determines the demand for water. For a closed crop canopy, when the soil is completely covered by leaves, this is in fact the transpiration demand. The amount of water needed to realise the transpiration demand can be calculated on the basis of crop-specific transpiration coefficients (Monteith, 1990).

The availability of water depends on rainfall and soil physical properties. Soil water between field capacity and wilting point is assumed to be available (see Table 6.2). When the soil water content falls below the texture-specific threshold, crop transpiration becomes less than the potential and crop growth decreases proportionally.

Table 6.2 Volumetric water content at field capacity and wilting point, and water holding capacity (cm m^{-1}) source: FAO, 1996.

Soil property	Soil type		
	Coarse	**Medium**	**Fine**
Field capacity	13	32	54
Permanent wilting point	4	10	44
Water-holding capacity	9	22	10

The process of water movement into (rainfall) and out of (evapo-transpiration, and drainage) the soil body is calculated with a capacity type, dynamic soil-water-balance model. Total storage was related to soil depth set at a max of 1 meter. Because of the constant relation ratio between growth and transpiration, a constant transpiration coefficient can be used. The transpiration coefficient may vary between 200 and 350 for cereals. A value of 250 is commonly used. This means that for each 250 l of water 1 kg of crop dry matter is produced. This coefficient is used to calculate the demand for water given the increase in biomass. When the water supply is lower than the demand, the growth rate is reduced proportionally. Crop yields result from the multiplication of the accumulated biomass over the growing period by a harvest index, set at 0.43.

Calculations are carried out using a daily time step, on a 0.5°×0.5° spatial grid for the baseline (1960-1990) and the three climate change scenarios for sub-Saharan West Africa. Because a weather generator is used to scale down precipitation figures from monthly to daily values, each soil-weather combination is calculated 20 times. As the start of the growing season is not exactly known (the growing season starts somewhere in May and ends in October), calculations are started at two week intervals from 1 May till the end of October. The resulting database contains the calculated yield for each land pixel of 0.5°×0.5° for 20 rainfall distributions for 12 starting points in the growing season.

The onset of the growing season is a crucial aspect of agriculture. 'False' starts may cause entire seasons to be lost. Using the database containing the simulated database, the starting data that produce the highest production levels are extracted to establish the spatial variability in the onset of the growing season. Map 6.2 clearly reveals, for all scenarios, an increase in variability, indicating an increased risk in assessing the planting date. The diversification could also mean that the risk of crop failure is spread over the region, with it being possible to shift market infrastructures in order to exploit such opportunities. Baseline GCM Is the 1960 – 1990 average. S1 A MPI Scenario 1 uses baseline A with the Max Planck Institute climate model. S2 A GFDL Scenario 2 uses baseline A with the Global Fluid Dynamics Laboratory climate model. S3 C MPI Scenario 1 uses baseline C with the Max Planck Institute climate model. Baseline A combines a medium population growth with a medium economic growth. Baseline C combines a medium population and high economic growth.

Using the 20 calculated yield levels for each soil-crop-weather combination, including some of the rainfall variability, we can assess the risk of given management practices such as planting date. With regard to yield we will concentrate on the probability that a given yield level is exceeded rather than providing 'exact' yield levels. Map 6.3 shows the consequences of not adjusting the planting data for the baseline and S2 A GFDL scenario.

The highest yields are obtained when planting takes place on around day 190 (beginning of July). If this is applied to the changed situation, a large part of the region, notably sandy soils and (semi)-arid regions, experiences higher risk levels resulting in lower production levels. Adaptations, earlier planting or a change in crop variety require other structures to be in place: e.g. seeds need to be available and labour and markets also need to be in place.

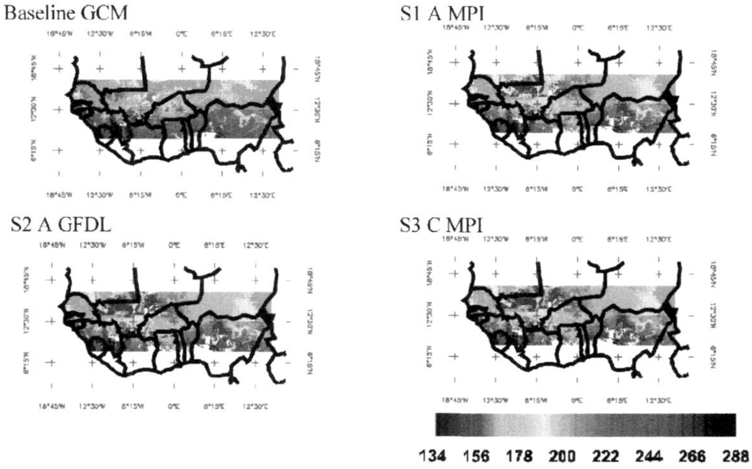

Map 6.2. Start of the growing season, day of year, related to highest water limited production level. Baseline GCM Is the 1960 – 1990 average. S1 A MPI Scenario 1 uses baseline A with the Max Planck Institute climate model. S2 A GFDL Scenario 2 uses baseline A with the Global Fluid Dynamics Laboratory climate model. S3 C MPI Scenario 1 uses baseline C with the Max Planck Institute climate model. Baseline A combines a medium population growth with a medium economic growth; Baseline C combines a medium population and high economic growth *(See colour section, p. 460).*

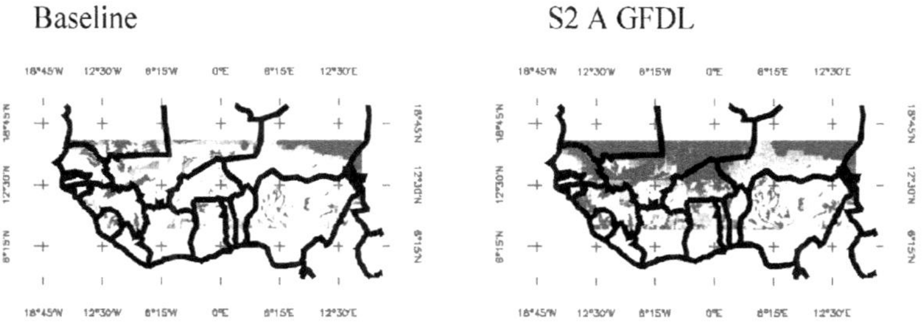

Map 6.3 Probability of yield exceeding 2250 kg ha-1 for the 1960-1990 baseline and S2 A scenario using GFDL GCM, day of planting is 190. (red = 1, white = 0). Baseline Is the 1960 – 1990 average. S2 A GFDL Scenario 2 uses baseline A with the Global Fluid Dynamics Laboratory climate model. Baseline A combines a medium population growth with a medium economic growth *(See colour section, p. 461).*

Chapter 7

SAHELIAN LIVELIHOODS ON THE REBOUND

A critical analysis of rainfall, drought index and yields in Sahelian agriculture

Fred Zaal, Ton Dietz, Johan Brons, Kees van der Geest and Edward Ofori-Sarpong

Abstract: In this chapter an attempt is made to find statistical relations between rainfall, yield levels and the drought index. For the whole of the study region, average yield data was compared with average annual rainfall as derived from Meteorological services databases. Data from all available rainfall stations within such a study region was included to generate this simple average rainfall figure. Two drought indices were added to the analysis. No clear relation could be determined.

1. MODELS AND DATABASES

To test assumptions about a statistical relationship between rainfall and crop yields in a region like the Sahel and the usefulness of a 'drought risk assessment' approach (see Chapter 3 and 6) it is wise to consider the availability and reliability of rainfall and yield data and the level of scale which is regarded as most appropriate for this analysis. Originally, the regional (district) level was considered to be the level at which data would be available. However, data proved to be available at a lower level than previously thought, although the quality of this data was again limited because it had been collected for an entirely different purpose (most often the monitoring of national production by the Ministry of Agriculture of the respective countries).

For the whole of the study region, average yield data (as derived from secondary data sources such as regional, provincial or district Ministry of Agriculture statistical files and annexes) was compared with average annual rainfall as derived from Meteorological services databases. Data from all available rainfall stations within such a study region was included to generate this simple average rainfall figure.

A.J. Dietz et al. (eds.),
The Impact of Climate Change on Drylands: With a Focus on West-Africa, 61–77.

We also developed a Drought Index (DI). When combined, these methods of average and Drought Index together can be compared with that of the national Rainfall Index (RI) developed by Gommes and Petrassi (1994). The RI allows comparisons to be made between years and between countries. It is calculated for each country by taking a national annual precipitation average weighted according to the long-term precipitation averages of all the individual stations. Gommes and Petrassi demonstrate that the national RI correlates well with national crop yield levels in Africa.

The Drought Index as we used it (see Chapter 6) needs data on precipitation, temperature and potential evaporation and only weather stations that measure this data could be included in the analysis. However, since most weather stations do not gauge the last two variables on a year-to-year basis, it was decided to select those stations that have time series on monthly precipitation (P) and monthly averages on temperature (T) and potential evaporation (ETP). To allow for comparisons between data from different weather stations, the only stations selected were those with precipitation data for the period 1960-1997. ETP can be determined in three ways, each having its own advantages and limitations (Dietz et al 1998). To get a reasonable spread of weather stations over our study area, it was decided to include all stations that calculate ETP data, irrespective of the method used.

There are various advantages of the method used here as regards presenting rainfall and drought trends in the form of the RI or DI. The country-size or regional scale is designed to correlate with other country-wide and regional statistics, especially on agricultural production. The RI as used by Gommes and Petrassi is independent of absolute amounts of rainfall, which may be very localized. It allows general comparisons to be made regarding an entire country. Because of the long-term record, a frequency distribution of RI values is available which allows historical comparisons to be made. If the record is not complete for an individual station, the RI can still be calculated without that station. This allows a long-term RI record. There are also several disadvantages. Because the RI is weighted by annual rainfall, those stations in wetter areas of a country (which, by nature, are often the more agriculturally productive areas) have a greater influence on the RI than stations in drier areas. The simple annual average as used by us is not subject to this problem and the regional (provincial/ district) scale permits this if trends are correlated with agricultural data at the same scale. The DI can be regressed with higher-level data only and does not permit conclusions beyond the level of its analysis. Only part of the variability found at the lower levels is directly influenced by these higher-level variables. However, both the RI by Gommes and Petrassi and the DI as used by us may be less useful when examining overall drought conditions and the hydrological, environmental, and societal impacts resulting from it at any other level.

Similar problems are found in relation to agricultural production and productivity data. At the aggregate level these latter variables, however measured, may be representative of regional trends, but at the level of the farm or plot, it may be that only a minor portion of production and yield variability as experienced by the decision-making farmer is actually related to the rainfall situation of any particular season. All other variables, even those related to the actual DI used (soil moisture retention capacity, infiltration capacity and thus the determinants of the soil moisture storage factor in the DI), are actually determined at the much lower level of scale: the farm, plot or even part-of-plot level. At the plot and farm level, labour, technology, inputs and other non-climatic and economic variables play an important role in determining the soil moisture storage factor and this is methodologically difficult to include in the analysis as all these variables change over the years and most importantly within years due to the reaction of decision makers to the rainfall of any particular year. We might therefore find that millet and sorghum crop yields in dry years are better than in wet years simply because more attention is being paid to these crops in dry years (reseeding on different plots, more labour in weeding and soil moisture infiltration, etc) to the detriment of attention for crops that depend on relatively favourable rainfall patterns.

One way of finding out more about this problem of level of analysis and type of variable in relation to predictive value of the regression between rainfall regimes and production and yields is to carry out an analysis at a number of levels in the same region for the same period. We do this retrospectively in this chapter (using all the data collected in the framework of the ICCD study). Data from the three case study areas for which we have detailed information at more than one level (Ghana, Mali and Burkina Faso) are being used to answer the above question.

2. AN ANALYSIS OF RAINFALL AND YIELD DATA FOR NORTHERN GHANA, MALI AND BURKINA

In the three countries concerned, four regions were studied in depth, with the focus on the relationships between rainfall, the drought index and various other variables. In Ghana, the Bolgatanga region in the Northeast, and the Upper West Region in the Northwest were studied. Here, the analysis of rainfall and yields remained at regional level, without additional fieldwork veing carried out to collect data at village or household level as was the case in the other study countries. In Mali, data was collected at village level in the region of Koutiala and Sikasso, but the aggregate data has been used here to carry out a similar analysis as that carried out for the northern Ghana region. Finally, in Burkina Faso, the data was used that had been collected by the Ministry of Agriculture and Animal Resources (MARA/DSAP) in the Kaya region of Sanmatenga and Bam provinces.

2.1 Northern Ghana

Usually, analyses at the higher levels of regions and states are characterised by a substantial correlation between rainfall and production and productivity figures over time (Brons et al 1998). Various lower level sources of variability (individual, household and villages) even out and major trends are isolated and show a high degree of correlation. The fact that these higher-level correlations are so low in the case of Ghana is therefore rather surprising and forces us to think of alternative models for yield predictions.

The data is presented in Table 7.1 and is based on the ICCD report on the northern Ghana region of Bolgatanga (Dietz & Millar 1999) and on an M.A. thesis on Upper West Region (Van der Geest, 2002).

Table 7.1 Yield estimates and correlations with rainfall and DI, Bolgatanga, Ghana, 1987-1997

Kg/ha					**Average**	**Average**		
Year	**Millet**	**Sorghum**	**Groundnuts**	**Maize**	**Rice**	**Average**	**Rainfall**	**DI**
1987	562.4	639.1	1148	748	1052.1	873.98	983	0.7
1988	606.8	962.8	680.6	782	1235.8	873.98	936	0.9
1989	747.4	913	713.4	544	2004	942.72	1158	0.3
1990	518	763.6	606.8	748	668	716.86	808	1.7
1991	532.8	680.6	811.8	1300.5	1586.5	982	1050	1.7
1992	518	697.2	803.6	1003	1703.4	923.08	929	0.9
1993	1050.8	821.7	926.6	952	1636.6	1109.66	836	1.2
1994	747.4	713.8	852.8	850	2004	1001.64	1069	0.7
1995	1169.2	937.9	910.2	977.5	2404.8	1256.96	838	1.7
1996	1073	1203.5	787.2	901	2204.4	1227.5	976	2.6
1997	599.4	796.8	803.6	501.5	1803.6	864.16	864	2.5
Average	740	830	820	850	1670	982	949	1.4
Correlation								
Rainfall	-0.18	-0.03	0.02	-0.02	0.27	0.01		
DI	0.11	0.31	-0.16	0.19	0.16	0.12		
Corrected correlation (1989 deleted)								
rainfall	-0.24	-0.18	0.23	0.38	0.17	0.07		
DI	0.21	0.51	-0.42	0.02	0.33	0.12		

Note: The figures for 1989 were of doubtful quality and have been deleted in the second correlation analysis. Pearson's correlation for rainfall and, Spearman's for DI were used.

The correlation coefficients are not very high and none are significant apart from the correlation between rice yields and rainfall as a growth trend proxy. However, a number of conclusions can be drawn. First of all, the correlation between DI and crop yields is higher on average than between rainfall and crop yields. However, the low coefficients for rainfall and the high but positive coefficients for the DI are surprising. The coefficients were expected to be high and positive for rainfall (more rain means higher yields, ceteris paribus) and high and negative for the DI (higher DI means more risk of crop failure). Moreover, the correlation between DI and sorghum is higher than for any other crop, but again it is positive, which is particularly unexpected since the DI was designed to predict yields for this relatively drought-adaptive crop. In the case of groundnuts, maize and rice, the DI does not appear to be a good predictor at all. On the other hand, the correlation between rainfall figures and crop yields is extremely low and negative for millet and sorghum, and rather high and positive for groundnuts and for rice, which seems to imply that these crops are attracted by high rainfall conditions. This applies to rice as well, even if yield is independent of the soil moisture content over the growing season due to the use of irrigation. However, none of the coefficients is significant at the 0.01 level (nor the 0.05 level).

Another rather interesting finding is that the outliers in the analysis seem to be the result of mistakes in the dataset, as the yields for rice and millet are similar in both 1989 and 1994. As the data for sorghum for 1989 also seems to be too low for the rainfall of that year, it is assumed that the data for that year may be incorrect. Excluding these figures, the correlation coefficients are more extreme as is indicated in the lowest two lines of the above table. A clear negative relationship appears between groundnuts and maize (somewhat less pronounced) and the DI, and a high correlation appears between these crops and rainfall. This is what we would expect, so it seems that the rainfall and DI data can be used to predict crop yields of maize and groundnuts, although for maize the relationship seems not to be so robust as the coefficient shifts between the two correlation analyses. There is a positive and high correlation between rice and the DI in the second analysis. The result for millet and sorghum and their relationship to rainfall and DI is even more problematic now than in the earlier analysis. Not only are the signs the same, they are now more negative (rainfall) and positive (DI) than before. Clearly, if not for the non-significant character of the relationships, there may be some truth in the remark of an old farmer in Bongo, North-eastern Ghana that: 'In a dry year there will be a good harvest. Formerly the young did not understand, but now they do.' The question is: why would this be so?

In order to compare this data with another region in Ghana for which we have similar data, we now turn to the Upper West Region, where similar research was carried out within the framework of the ICCD project (Van der Geest 2002). The data is similar in structure and has been analysed in a similar manner. The results are shown in the following table.

Table 7.2 Yield estimates and correlations with rainfall and DI, Ghana, Upper West Region, 1986-1998.

Year	Millet	Sorghum	Groundnuts	Maize	Yams	Average*	Rainfall	DI
1986	700	700	1300	900	6000	900	865.4	2.67
1987	700	700	1500	700	5100	900	822	1
1988	800	700	1700	1500	5100	1175	850.1	1
1989	600	600	1600	800	5000	900	1065.2	0.75
1990	800	1000	300	1000	7200	775	731.7	3.5
1991	600	700	800	1300	12000	850	1101.2	3
1992	700	900	800	200	13500	650	845.4	1.25
1993	1000	1400	1600	500	13600	1125	1078.1	0.75
1994	1100	1400	1600	1400	10500	1375	1167.1	1
1995	1100	1300	1500	1400	11000	1325	1420.4	1
1996	1100	1300	1200	1100	10800	1175	996.5	3.25
1997	1100	1100	1400	1000	10100	1150	1063.5	1
1998	1000	1200	1400	1100	8500	1175	958.4	2.25
Average	869	1000	1285	992	9108	1037	997.31	1.7
Correlation								
Rainfall	0.49	0.49	0.42	0.39	0.44	0.64		
DI	-0.06	-0.04	-0.854**	0.17	0.1	-0.382**		

Notes
* Average of grain crops and groundnuts, excluding yams.
** significant at the .01 level.
Rainfall and DI data on the basis of four stations, yield estimates from the Ministry of Food and Agriculture.
Pearson's correlation was taken for the analysis of rainfall and Spearman's for DI

Again, there is only one statistically significant relationship, namely that between the DI and groundnut yields. All the other relationships are insignificant at the 0.01 level (though significant at the 0.1 level for millet, sorghum and maize). That is, the relationships are as expected (positive for rainfall-grain yields, negative for DI and grain yields) with the exception of maize and yams. The correlation between average grain yields taken as an aggregate and rainfall is high and significant. Interestingly, however, it appears that there is a steady growth of yields of the most important cereals over time, irrespective of rainfall or DI. If we take the year to be an indication of a trend, there is a high and significant correlation (r = 0.8 at the 0.01 level for both millet and sorghum). A similar relationship was found in the case of rice in the Bolgatanga region (r = 0.7 at the 0.05 level).

These results are less surprising than those obtained in the Bolgatanga region and suggest that the data quality in the Bolgatanga may be faulty, or that other variables are more dominant in this region.

These other variables do not induce a growth trend as there is only a positive and significant correlation between rice and such a trend in the Bolgatanga region, while in the Upper West region all drought crops (millet, sorghum and yams) show a clearly positive and significant correlation.

2.2 The regional level analysis in Mali

The analysis in Mali had a different point of departure, but for reasons of comparability we will first discuss the data at the regional level, before analysing the findings at village level.

Table 7.3 shows the data on yield estimates, rainfall and DI and the correlations at regional level for the Koutiala-Sikasso region in southern Mali.

This region of Koutiala-Sikasso is quite comparable to the northern Ghana region in terms of rainfall and DI levels and average grain yields. The correlation between rainfall and yields of the important cereals is between .67 and .87, which is reasonably high. One of the reasons may be that the data used for the analysis comes from a large area, covering a 200 km North to South zone. This forces the rainfall variable to the fore, while at the lower levels and in smaller areas rainfall is similar for most sites and other variables appear to be important. However, both cotton (coefficient of .21) and rice (with .11) have very low coefficients, indicating that for the region as a whole there is no clear relationship between rainfall and yields for these two crops. We will see later that, within this zone, rice and cotton production tends to be attracted by adequate rainfall conditions. All the correlations between grain crops (and groundnuts) and rainfall are highly significant (at the 0.01 level).

The same does not apply to the relationship between DI and crop yields. Unexpectedly, the correlations between DI and yields for the various products appear to be variable. In the case of groundnuts, it is high and negative as expected, but for the other products the coefficient is either low and negative (maize) or even positive (millet and sorghum). Rice shows a negative and relatively high coefficient, but cotton shows a positive and reasonably high coefficient, which is remarkable. None of the coefficients is significant at the 0.01 or 0.05 level. These findings again point to the inadequacy of the DI as a way of predicting the yields of dryland crops such as millet and sorghum at the level of the region. It implies that some serious thinking will have to be applied to the adaptation of the index if it is to be used in dryland research other than in the area where it was originally developed (in India).

Table 7.3 Yield estimates and correlations with rainfall and DI, Mali: Koutiala and Sikasso, 1984-1998.

Kg/ha			**Average**	**Average**				
Year	**Millet**	**Sorghum**	**Maize**	**Groundnut**	**Cotton**	**Rice**	**rainfall**	**DI**
1984	1047	1233	1328	715	1251	745	914	2.3
1985	1324	1093	1537	939	1240	905	1017	2
1986	1186	1017	1802	1120	1329	1183	1103	1.7
1987	879	953	1385	992	1307	1241	977	1
1988	1028	750	1309	832	1340	905	893	1
1989	889	852	1385	875	1184	1679	993	2.3
1990	1146	1030	1575	768	1352	964	989	3.5
1991	1087	979	1707	1024	1262	1562	1101	1
1992	622	712	1347	768	1095	1343	854	1
1993	543	623	1157	587	1240	1591	833	2
1994	771	674	1347	800	1095	1343	874	
1995	781	801	1556	854	1251	1372	958	
1996	988	1271	1897	1067	1117	1460	1128	1
1997	860		1674	653	1066	1256	934	1.9
1998	948		1812	759	1020	1340		
Average	940	922	1521	850	1210	1259	969	1.7
Correlation								
Rainfall	0.673**	0.703**	0.870**	0.852**	0.21	0.12		
DI	0.23	0.22	-0.18	-0.47	0.09	-0.16		

Note:
Rainfall estimates on the basis of nine stations, DI on the basis of the three stations for which the necessary detailed data are available.
** Significant at the 0.01 level
Pearson's correlation is used for rainfall, and Spearman's for DI

Comparing the analysis in Ghana and Mali, the conclusion may be that the data for Ghana (and especially the Bolgatanga region) is probably incorrect as the correlation coefficients are generally lower than in the case of Mali, while the level of analysis and the general size of this region (North-South) is comparable. The fact that in both cases there does not seem to be a proper relationship between positive rainfall and negative DI correlations indicates that the DI may be inappropriate. The Upper West Region data is between these two extremes.

2.2.1 The village-level analysis in Mali

The village-level analysis in Mali was based on an extensive database including variables related to land use, crop productivity, farm characteristics, village characteristics and economic and bio-physical external conditions (Brons *et al* 1999). The data was restructured to allow an analysis at village level. Principle Component Analysis was applied to reduce the dataset to an number of independent components that systematically reflected village characteristics. Subsequently, for a limited number of important cropping systems, multiple regression analysis was applied. We will follow this route to arrive at an analysis of one of the most important cropping systems in the discussion on rainfall-yield relationships.

The result of the PCA is a reduction for 43 villages of 26 variables into 7 main components or factors (Brons et al 1999: 19-20). The cumulative percentage of the explained variance is 76 percent. The components can be described as follows:

Commercial grain production

This factor is characterised by a substantial village area of diversified grain production for commercial purposes (rice but also maize, millet and sorghum) which generates high gross margins and revenues. Moreover, the joint high yield levels of groundnut points to some diversification taking place into the production of leguminous crops and the positive effects of generally high input levels. It should be noted that this strategy can only be followed when rainfall risks (e.g. low rainfall but also high variability) are limited or controllable.

Commercial cotton production

This factor indicates that high investments for input purchases in commercial cotton production generate a surplus that can be used to satisfy consumption expenditures. Specialisation in cotton production is mainly feasible due to favourable agro-climatic conditions (high rainfall and low rainfall variability) that enable substantial investments for input use.

Livestock farmers

Villages where farmers possess relatively high amounts of cattle and oxen can be found in the Northern region (where livestock production has a comparative advantage) and in the cotton areas, which are characterised by relatively favourable land endowments. Investments in cattle purchases here are usually financed from (earlier) cotton revenues. Inputs for livestock production are limited and returns per hectare are low. Part of the livestock production is located close to villages and specialises in dairy products. The population in these villages relies on commercial purchases of grains to satisfy food security and is very successful in this respect. These villages are not significantly affected by soil degradation. This can be understood from the recent understanding that pastoral and livestock production as a system may be better adapted to local ecological restrictions and a high market involvement than was previously understood.

Commercial rice production

Villages that diversify into rice cultivation and specialise as far as grain production is concerned are strongly dependent on stable rainfall patterns. Rice was introduced originally as a major diversification crop and initial investments are financed out of cotton revenues, which are relatively high per area unit. In the medium term, the farming systems become more dependent on rice as a major crop and this leads to a new type of specialisation at village level. Commercial rice production is mainly selected by farmers with limited land resources; those who are able to exploit a labour-intensive cropping system albeit with low labour productivity. This is once again a reminder of the favourable relationship between market involvement and food security considerations, which depends on a specific set of variables, such as distance to markets, to function.

Extensive grain production

Villages that offer large areas per person may still rely on extensive grain production for food self-sufficiency. The availability of oxen facilitates the required land preparation and weeding activities for an extension of the cultivated area. Reliance on extensive grains production mainly occurs in villages with low, but stable, rainfall patterns. This system guarantees high net revenues (low external input costs) and adequate food security levels. Apparently, the high income levels encourage activities in the non-agricultural sector (as a relationship vice versa could not be established). This is the component with the highest score on the number of non-agricultural activities per 1000 habitants.

Sustainable subsistence farming

This component is not represented by a significant factor score in any of the agricultural indicators or rather, it us the only component which scores high on indicators of what it may not be: it is not degraded, nor close to services, the latter pointing to less favourable socio-economic conditions. With little agricultural development, low land use intensity and a favourable state of the natural resource conditions compared to the other components, it is a component which seems characteristic of an early stage of agricultural development in the region, now only found in remote areas.

Marginal subsistence farming

In the less developed villages where there are relatively few farms equipped with oxen traction and little non-agricultural activity, farmers maintain diversified farming systems without this having any particular implications for yield levels. These farmers have a small number of livestock and often only incomplete equipment. Due to the considerable distance to the cities, off-farm employment options are equally limited. Low levels of education and health also inhibit participation in the labour market. The combined effect of the features in this component is a regular occurrence of food shortage. This component compares unfavourably with the former in that the food security situation is rather bad.

We will present an analysis and estimate the importance of the various inputs and conditions including rainfall in agriculture at village level. A limited number of cropping systems are particularly relevant here. An analysis of all crops and the determining variables of productivity is relevant to all components and in particular the dominant systems of commercial grain production, commercial cotton production, livestock farming, commercial rice production and extensive grain production. The grain producing cropping systems, whether commercial or otherwise, will be analysed separately, as well as the cropping system for maize, which is the only grain for which correlations at the expected level and sign were found in the regional analyses in Ghana and Mali (and Burkina Faso as we will see). We will present a limited number of production functions: gross revenue per ha is estimated for all crops taken together (all the above components), for grain production and for maize. Table 7.4 shows the result.

This analysis shows that, in the complete model, for all cropping systems identified earlier as components, rainfall does not appear as a significant variable that can explain revenues. Instead, inputs in terms of human labour and animal traction are the determinant. Rain does play a role as regards the grain-production based components, but the labour inputs are again important, as are the soil characteristics. This illustrates the labour-intensive character of grain production in this area, while the appearance of the degradation variable and soil characteristics in general may be interpreted as the expression of a process of soil mining taking place during intensive grain production in the absence of adequate input levels. For the maize cropping system in particular, total area cultivated and again the inputs in terms of animal traction and expenditure on fertilisers appear as significant variables. The reduced model shows that rainfall is a significant variable at the level of all cropping systems and for grain-based production in general. This does apply so much to the maize production-based cropping system. Maize, and cereal-based cropping system revenues are generally influenced by input levels (labour, expenses) and soil characteristics.

The relatively lower r-squares show that at the level of the villages studied, the explanatory power of the models is lower than the simple models at the regional levels of Ghana, Mali and Burkina Faso, and even lower than the r-square value of a similar production functions for the study region in Burkina Faso, as we will show. For all cropping systems together, rainfall does appear to be the most important variable in explaining variability in revenues.

Table 7.4 Gross revenue per ha, Cobb-Douglas production functions, complete and reduced model.

Complete model	**All crops**			**Grain**			**Maize**		
	Coef.	*t-value*	*Sign.*	*Coef.*	*t-value*	*Sign.*	*Coef.*	*t-value*	*Sign.*
Intercept	-2.75	1.23		3.85	1.35		1.06	0.35	**
Total Area	0.08	0.83		0	0.04		0.12	0.9	**
Rainfall 1997	0.66	2.18		0.35	1.7	**	0.29	0.7	
Persons per ha #	0.27	1.66	*	0.48	2.37	**	0.51	2.34	
Oxen per ha #	0.41	2.54	**	0.67	1.74	**	0.46	2.07	*
Expenses per ha	0.02	0.14		-0.05	0.32		0	0	**
Degradation index	0.07	1.48		0.12	2.16	**	0.09	1.51	
Clay dummy	0.18	1.52		0.36	2.33	**	0.47	2.87	
Sand dummy	0.19	1.69		0.24	1.62		0.31	2.02	
Distance education	0.02	1.42		0.01	0.86		0.02	1.2	
Drought index 1997	0.01	0.07		0.05	0.53		0.07	0.73	
Adjusted r-square	0.47			0.31			0.29		
Reduced model									
Intercept				5.42	2.41	**			
Rainfall 1997	0.72	3.26	***	0.54	1.85	**			
Persons per ha				0.44	2.28	**	0.5	2.44	**
Oxen per ha	0.37	2.36	**				0.41	1.87	**
Expenses per ha	0.06	1.74	**				0.1	2.13	**
Clay dummy				0.1	1.84	**	0.4	2.51	**
Sand dummy				0.21	1.69	*	0.25	1.69	**
Adjusted r-square	0.47			0.29			0.34		

Notes: Rainfall based on May-October figures (94 percent of total). Clay and sand dummies as opposed to gravely soil. Significance levels are one-tailed.

In all cases, except for maize, rainfall is an important determinant of crop yields and revenues, but sufficient rainfall is most important for farmers specialising in commercially oriented (grain) strategies. Access to external inputs has the greatest influence on aggregated gross revenues and on maize yields, while animal traction is especially important for maize production. Access to land is hardly ever a constraining factor, but access to sufficient labour is the most relevant factor for both commercial and subsistence cereal production.

2.3 The household level analysis in Burkina Faso

The analysis in Burkina Faso was founded on a comparative basis of datasets from two different periods and analysed the relationship between climate and other variables with production, productivity and land use change at household level. Again, for reasons of comparability we will first discuss the data at regional level and continue afterwards with an analysis of the findings at household level.

The following table presents data on yield estimates, rainfall and DI and correlations at this regional level.

Table 7.5 Yield estimates and correlations with rainfall and DI, Burkina Faso, 1984-1998.

Kg/ha			**Average**	**Average**				
Year	**Millet**	**Sorghum**	**Maize**	**Groundnut**	**Cotton**	**Rice**	**rainfall**	**DI**
1984	407	376	109	396	377	2036	490	2.7
1985	533	514	715	505	191	2068	459	3.3
1986	529	618	617	618	568	1437	518	2.7
1987	300	328	153	357	388	1882	451	1.7
1988	776	704	1445	818	534	921	724	2
1989	376	474	617	985	777	1613	603	3.3
1990	428	498	1048	706		197	503	3
1991	639	841	595	453		255	712	3.3
1992	495	765	1055	1398		255	650	2
1993	608	851	1003	880	957	1918	564	2.3
1994		689	794	801	187		943	1
1995	546	709	890	703	431		648	3.3
1996	495	629	1048	1000	500		613	2
1997	288	381	210	308	236	431	554	2
Average	494	598	736	709	468	1183	602	2.5
Correlation								
Rainfall	0.69	0.68	0.39	0.32	-0.05	-0.53		
DI	0.16	0.03	-0.03	-0.15	0.26	0.06		
Correlation on the basis of the period 1968-1997								
Rainfall	0.614**	0.558**	0.33	0.1	-0.13	0.14		
DI	-0.13	-0.17	-0.05	0.08	0.04	-0.09		

Notes:

Rainfall figures are derived from the nine stations within the study region.

DI figures are based on figures for Kaya, Ouahigouya and Tougouri stations.

Yields are based on data for Sanmatenga Province in which much of the Kaya region is located.

Empty cells have no data.

** significant at the 0.01 level

As in the earlier cases, there are variations across the various crops as to the correlation between rainfall and yields, with millet and sorghum having high, positive and significant coefficients between rainfall and crop yields, and low and positive (though insignificant) coefficients between DI and yields. This is similar to what was found in Mali and at this level, the level for which it was designed, it seems that the DI is not behaving what as expected. Correlations between rice yields and rainfall and DI rice are quite similar to that found in Mali. The fact that yield levels fluctuate enormously is reason for care.

As far as other crops are concerned, it can also be said that higher rainfall levels do not always mean higher yields. We plotted the data for groundnuts and it appears that the relationship between rainfall and yields for this crop has become ever more negative in the course of the last few decades. The following plot shows the result.

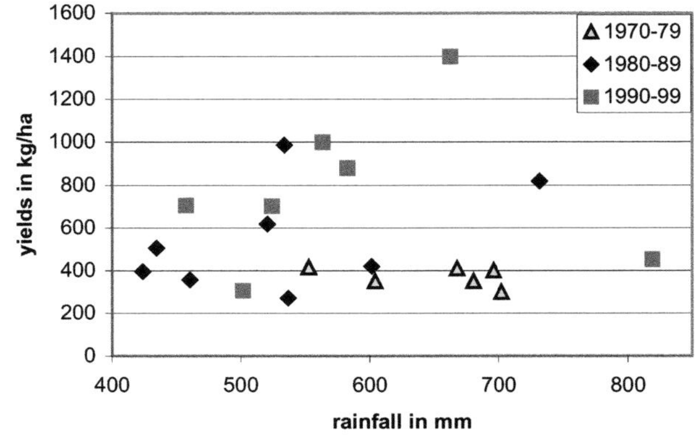

Figure 7.1. Relationship between rainfall in mm and groundnut yields in kg per ha, per decade, 1970-1999 *(See colour section, p. 461).*

The relationship changes in the course of time, indicating there is an intervening variable hidden in the dataset as presented in table 7.6. It may be caused by the selection of new varieties of groundnuts in response to favourable rainfall and high input levels. The lowest yields, occurring at low rainfall levels, are not below the level of the earlier varieties, while the highest yields occur with average rainfall levels and are very much above the highest yields achieved before. It may also be the case that we see the effect of adaptation to these average rainfall levels, with low yields when rainfall levels are very low and moisture is short, and high rainfall levels when damage through disease and waterlogging or flooding occurs. The graph also shows the resilience of the earlier system where yields of around 400 kg per hectare where achieved regardless of the rainfall level.

Table 7.6 Changing relationships over time between rainfall and groundnut yields

Period	Slope of the linear regression coefficient
1970 - 1979	-0.32
1980 - 1989	1.09
1990 - 1999	0.15

2.3.1 The household level analysis in Burkina Faso

For the household level analysis of yields and its independent variables, a dataset was used for the period 1993 to 1998, as collected by the monitoring and data collection service of the Ministry of Agriculture and Livestock Resources (MARA/DSAP), the Enquête Nationale sur l'Agriculture (ENSA) (Brons et al, 2000). This dataset was collected at household and plot level and is much more detailed though with a certain bias towards the purpose of its collection.

The following production function shows the result for the analysis of yields in kg/ha of millet and sorghum (as first crop). Information on the database and the regions are presented in Brons *et al.* (2000).

Table 7.7 Yields of millet and sorghum in kg per ha, Cobb-Douglas production functions, per region.

Complete model	**South**			**Centre**			**North**		
	Coeff.	*t-value*	*Sign.*	*Coeff.*	*t-value*	*Sign.*	*Coeff.*	*t-value*	*Sign.*
Area	1.07	13.7	***	1.17	14.7	***	1.23	18.4	***
Labour	0	0.1		-0.12	1.5		-0.16	2.4	***
Fertiliser	0.03	2.7	***	0.01	2.1	**	0.02	2	**
Rainfall	-0.1	0.3		0.06	0.3		0.39	2.9	***
Dummies									
Intercept	-3.19	1.7	*	-4.07	2.9	***	-6.54	7.5	***
Manure	0.1	2.2	**	0.11	2.6	***	0.01	0.3	
Manual soil preparation	-0.06	0.6		0.04	0.4		0.23	2.2	**
Animal soil preparation	0.07	0.9		0	0		-0.06	1.1	
Erosion control	-0.01	0.2		0.07	1.6		0.01	0.1	
Household plot	0.49	4.7	***	0.29	2.5	*	-0.11	0.7	
Cowpea intercrop	-0.02	0.3		-0.02	0.3		0.37	0.6	
Other intercrop	-0.1	0.9		-0.11	1.6		-0.04	0.5	
Plain	-0.1	1.5		-0.07	1.3		-0.05	0.9	
Slope	-0.15	1.8	*	-0.28	3.7	***	-0.12	1	
Near village	0.13	3	***	0.07	1.5		0.72	1.6	
Sorghum	0.16	3.9	***	0.15	3.9	***	0.09	2.1	**
Adjusted r-square	0.79			0.78			0.79		

Note:

The dummies are defined as opposed to no application of the relevant variable, except for Household plot (versus individually owned plot), Plain (versus basin-located plot), Slope (idem), Near village (versus at a distance) and Sorghum (versus Millet).

Household characteristics other than labour were insignificant and were left out of the model. Other variables at plot level are available but the data quality is limited and dummies were included to capture at least some of the explanatory power of these variables at this level.

The marginal productivity of land is the single most important and most significant variable in the model, with fertiliser application coming second. Labour appears only as a significant variable in the North, but with a negative sign. Probably, other and unobserved household characteristics are influential in explaining output variability from the labour coefficient. Rainfall appears only as a significant variable in the most northerly zone of the study while in the other zones it is not significantly different from zero.

The dummies show that the southern and central zones have quite similar significant scores on the dummy variables, with manure application and variables indicating access (both in terms of ownership and spatially) being important as well. Slope as opposed to basin location of fields is significant in these two zones as well, with the expected sign. Soil preparation in the north is a significant variable.

In conclusion one could say that at household level the area and fertiliser use are the most important factors in explaining yield differences, with rainfall being important and significant in the North only. Of the dummies at plot level, variables pertaining to access to land and manure (south and central) and manual soil preparation are significant and of the expected sign, though the coefficients are not always very high.

3. CONCLUSIONS FROM THE CASE STUDIES

It remained unclear for a long time why there should be no negative relationship between millet and sorghum and the DI in some cases. Together with the evidence from Mali and Ghana, we can now say that the DI, though developed for dryland regions and dryland crops, does not seem to predict yield levels very well. Also, it is now questionable whether the data on the Ghana Bolgatanga area is correct, when compared with a similar analysis (at the similar level of the region) in another region of Ghana and in two other countries. The DI coefficients for the other crops in Ghana are also not very high and often have the wrong sign. The data and findings at regional level for Mali are more comparable with those of Burkina Faso, with the exception of rice and cotton perhaps. For the Sahelian-Soudanian region as a whole there is no clear relationship between rainfall and yields for the two cash crops of rice and cotton. Within this zone, rice and cotton production tends to be an attractive option in the event of adequate rainfall conditions. This indicated that though the production and yields of these cash crops is insensitive to actual annual rainfall, it still needs higher average rainfall, harnessed through soil and water management. In areas where cultivation takes place in lower lying areas (the bas fonds), the highest rainfall levels cause damage and the crop is then more affected by flooding than by drought. Despite all this, the average rainfall remains a better and more significant predictor of yields, particularly of the dryland crops of millet and sorghum, than the DI.

Still, there are a number of remarkable explanations for some of the low correlations. The example of groundnut yields in Burkina Faso shows that the data, when split in decades, shows that considerable yield increases can be expected for rainfall levels around the average, either as a result of a process of adaptation of crops to the rainfall regimes, or as a result of soil selection for this crop and the addition of other inputs such as fertiliser, manure or labour.

The two Cobb-Douglas production functions, though different in implementation, can be compared when we look at the results for grains in the case of Mali, and Burkina Faso.

Chapter 8

POPULATION DYNAMICS

An important intervening variable

Ton Dietz and Els Veldhuizen

Abstract: Changes in population densities for the period 1960 – 2000 are presented in this chapter.

1. INTRODUCTION

Between 1960 and 2000, the population increase in the West-African drylands were among the world's highest, close to 3% per annum. In addition, a considerable section of the population migrated to the southern coastal states while the number of people seeking inter-continental migration, mainly to Europe, started to become considerable. Deteriorating situations in the arid and semi-arid areas of Senegal, Mali, Burkina Faso and Niger (due to rainfall decrease, land degradation, and violence) also resulted in a rapid intra-country migration southward and in a swelling of the big cities like Dakar, Bamako, Ouagadougou, Niamey and Kano. Map 8.1 shows a comparison for Africa as a whole between 1960 and 1994 (UNEP/GRID, 1998). West Africa shows a mixed picture. For the arid frontier zones in Mauritania, Mali and Niger, the data actually reveals a population decline. On the other hand, almost everywhere along the West African coasts, the population increased more than four-fold. Increases of between 250 and 400% are also visible in eastern Senegal, in a large area around Bamako, in Niamey and in mid-western Nigeria. However, populations in large areas of Burkina Faso and northern and eastern Nigeria did not grow so fast: here the population did not even double, as a result of massive out-migration towards the south and to countries abroad. Estimates for Burkina Faso suggest that close on half of the adult population born there moved, at least for part of the year, to coastal states like Ivory Coast and Ghana.

A.J. Dietz et al. (eds.),
The Impact of Climate Change on Drylands: With a Focus on West-Africa, 79–82.

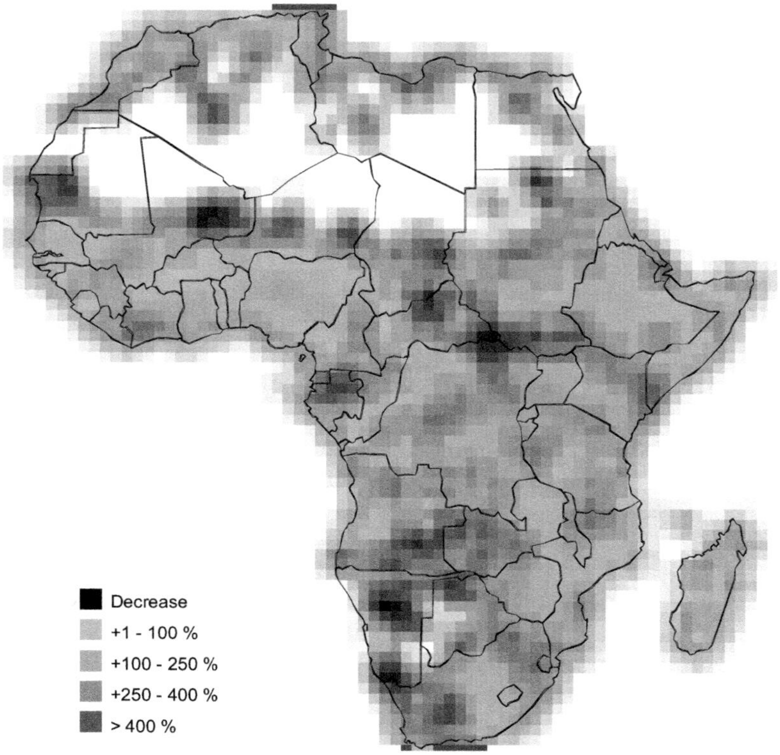

Map 8.1. Changes in population density (inh km^{-2}) between 1960 and 1994 (in %) *(See colour section, p. 462).*

A typology was made for the 1°×1° dryland areas in West Africa of population density and urbanisation for both 1960 and 1990/1994 (1990 for urbanisation; 1994 for population density). Table 8.1 shows the changes between the early 1960s and the early 1990s.

Further rapid population growth can be expected. For West African countries as a whole, the World Bank predicts an annual population growth of between 2.5% and 2.8% for 2000-2025 (World Bank, 1995), although recently long-term growth is thought to be lower because of increasing urbanisation and the impact of HIV/AIDS. The West African Long Term Perspective Study, a multi-donor endeavour to look into West Africa's near future, predicted that "West Africa's population will almost certainly double over the next thirty years" (see: Club du Sahel, 1995, p. 14). It was also predicted that there would be a major (further) population redistribution. "In the year 2020, the urban population in West Africa will most likely be over 60 percent of the total population, compared with 40 percent today. The urban population will be more than three times greater than in 1990. However, the rural population will also have increased by 40 percent" (Ibid.). Population growth and urbanisation will require massive investments in infrastructure and social services (water being a particularly problematic challenge) and a major intensification and commoditisation of food production. Poor world market perspectives for the area's current major export crops of cotton and groundnuts will also lead to a shift in favour of food crops and a further increase in the shortage of foreign exchange.

Table 8.1 Dryland West Africa: changes in population density and urbanisation, 1960 and 1990/1994.

	Semi arid cells					**Sub-humid cells**				
high density in 1960 and urban in 1960	2				2	1				1
High density in 1960 and rural in 1960	3	11			14		4			4
low density in 1960 and urban in 1960			1		1	1				1
low density in 1960 and rural in 1960	2	8		59	69	1	8	1	37	47
total in 1990/1994	7	19	1	59	86	3	12	1	37	53
	high density in 1994 and urban in 1990	high density in 1994 and rural in 1990	low density in 1994 and urban in 1990	low density in 1994 and rural in 1990	total in 1960	high density in 1994 and urban in 1990	high density in 1994 and rural in 1990	low density in 1994 and urban in 1990	low density in 1994 and rural in 1990	total in 1960

For seven regional 'cells' in the ICCD research project the following assessment and tentative prediction can be given for the demographic and food needs situation in 1990 and 2050 respectively: see Table 8.2. Food needs are based on average energy needs of grain equivalents of 225 kg a^{-1} per capita. We also add the calculated necessary yield requirements of grain (equivalents), in kg ha^{-1}, if all land were to be cultivated and if all food needs were to be provided by agricultural (grain) production in the regional cell itself. In many areas of dryland West Africa, the current actual land use for crop production purposes is less than 20% and the current yield per cultivated hectare less than 600 kilograms of sorghum or millet per hectare.In drought years the figures are much lower. Optimal harvests are in excess of 3000 kg ha^{-1} though. Theoretically, in all study regions, local grain production could be sufficient to feed the predicted population in 2050. In practice, however, crop land expansion, yield intensification, higher cash value agriculture in exchange for grains from elsewhere and non-agricultural income-earning activities in exchange for food from elsewhere will all be needed to prepare for a major challenge.

Table 8.2. Demographic change 1990-2050, predicted food needs and necessary yields in grain equivalents in 2050; seven major NRP-ICCD study regions.

1°×1° Cell (B = Burkina Faso, M = Mali, G = Ghana)	Inh km^{-2} 1990	Total inh. 1990 (mill.)	Predicted annual pop. growth 1990-2050 (%)	Predicted population density in 2050 (inh. km^{-2})	Predicted population in 2050 (in mill.)	Predicted food needs in 2050 (mill. kg grain equivalents)	Necessary yields in 2050 (kg ha^{-1})
Ouagadougou (B)	90	1.1	3.0	540	6.5	1460	1200
Kaya (B)	40	0.5	2.0	130	1.6	360	300
Gorom Gorom (B)	15	0.2	1.5	40	0.4	100	90
Bamako (M)	70	0.9	3.5	590	7.1	1600	1300
Koutiala (M)	20	0.3	3.3	120	1.5	330	300
Douentza (M)	10	0.1	2.0	30	0.4	90	70
Bolgatanga (G)	60	0.7	2.6	280	3.3	700	600

Chapter 9

DRIVING FORCES FOR CHANGES IN LAND USE

Johan Brons, Ruerd Ruben, Mohammed Toure and Boukary Ouedraogo

Abstract: In this chapter we apply a multivariate regression model to assess the effects of agro-climatic and socio-economic factors on land use and crop yields in Mali and Burkina Faso. We use 1965-1990 time-series data at national level on climate, prices, input use, infrastructure, and population density to explain land use and crop yields. The results show that infrastructure development and credit availability have, in particular, been stimuli which have enhanced cotton yields and expanded the cotton area. Farmers managed to increase cotton production against lower output prices mainly because costs were reduced by better crop management practices. Cereal and groundnut production benefited from spin-off effects at farm level, such as the availability of animal traction, fertiliser residual effects and improved market outlets. These findings confirm the importance of supportive agricultural policies that facilitate the optimal use of inputs and access to new technologies. Moreover, a sound production environment can offset the effects of adverse and deteriorating climate conditions. Further enhancement of food crop production will require additional public investments in infrastructure, markets and technological development.

1. INTRODUCTION

Changes in land use and resource allocation are caused by the interaction between a wide range of agro-climatic and socio-economic factors that influence regional ad farm-household options for crop choice, input use, labour allocation and marketing. In this chapter, we analyze the driving forces behind changes in land use and resource allocation in sub-Saharan Africa, considering the variable and changing agricultural production environment in which farmers face uncertainty about short-term and medium-term production conditions. Therefore, attention is given to (i) the analysis of the determining factors for

A.J. Dietz et al. (eds.),
The Impact of Climate Change on Drylands: With a Focus on West-Africa, 83–96.

resource use in agricultural production, (ii) the impact of population pressure and technical change on land use and factor intensity and iii) the differential response to rainfall variability for specific cropping activities.

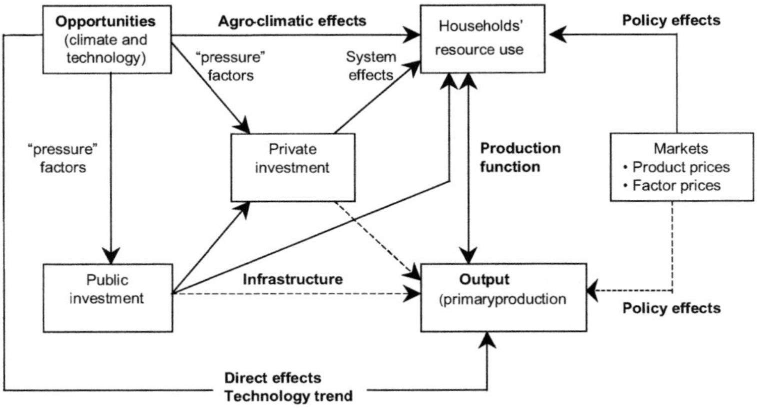

Figure 9.1. Major interactions between agro-climatic opportunities and agricultural output.

The adjustment of resource management strategies is considered to be highly dependent on prevailing conditions of factor and commodity markets. Favourable output prices offer incentives for intensification, but market and institutional constraints frequently limit farmers' capacities to mobilise the required resources for agricultural intensification. Figure 9.1 (adapted from Binswanger *et al.*, 1993) illustrates the relationships among agro-climatic endowments, public and private investment, prices and agricultural output. Public investment (in infrastructure) and farmers' investments (input use) can partly offset adverse weather or market conditions (Rosegrant *et al.*, 1998). In addition, technological change tends to reduce production costs. These relationships interdependently and simultaneously determine supply response in terms of the adjustments in cultivated area and yield level. If climate and market conditions are favourable, synergy effects can occur when public and private investments reinforce farmers' welfare.

The assessment of the impact of climatic variability on land use and food production enables the subsequent identification of different pathways and livelihood strategies for portfolio choice in West African drylands (see Chapter 16). Making use of farm-household village-level data, specific socio-economic and biophysical factors can be identified that influence decisions regarding choice of livelihood options, labour allocation and input use.

The analysis regarding temporal variability in land use at a regional scale in West Africa focuses on Mali and Burkina Faso, paying attention to the interaction of agro-climatic and socio-economic factors influencing farmers' supply response during the last three decades (Brons *et al.*, 1999). Different analyses are performed making use of statistical data obtained from national agencies (CMDT in Mali; ENSA and INERA in Burkina Faso), local research centres (CEDRES in Burkina Faso, IER in Mali) and from international organisations (FAO, World Bank, ICRISAT-SH). Analytical procedures are based on the statistical analysis of time series on rainfall variability (drought risk) and yields, and multiple regression to identify factors influencing area and yield response.

2. DRIVING FORCES OF CHANGES IN LAND USE IN WEST AFRICA

Variability in agricultural land use is related to natural processes and socio-economic forces. The relative importance of climate and demographic change, market development and the availability of productive and social infrastructure for adjustments in land use during a period of 25 years (1966-1990) has been analysed for Burkina Faso and Mali. Figures 9.2 – 9.4 provide an overview of the changes in land use, yields and production in cereal and cash crops in both countries and their relationship with rainfall patterns. As expected, cereal yields show more sensitivity to rainfall variability compared to cash crops. Moreover, short-term variations in crop yields are associated with rainfall, but long-term trends in yields generally improved.

Supply response models for changes in crop area and yields are applied to disentangle different patterns of agricultural development. The models rely on the Nerlovian supply response analysis technique, making use of multiple regression procedures. Harvested areas per rural inhabitant and crop yields (cereals, cotton, and groundnuts) are used as dependent variables. The Nerlovian supply response model compares different determinants as if they directly influence agricultural supply (Nerlove, 1958; Askari & Cummings, 1976; Sadoulet & de Janvry, 1995).

Despite its limited capacity to analyse features of simultaneity of determinants and long-term supply dynamics (Nerlove, 1979; Oyejide, 1990), the Nerlovian supply response model can be considered a standard approach to assessing supply response mechanisms and related policy issues (Rao, 1989; Fosu *et al.*, 1997). The basic Nerlovian supply response equation is as follows:

$$Q = \alpha_0 + \alpha_1 Q_{t-1} + \alpha_2 P_{t-1} + \alpha_3 Z_t + v_t$$

where Q represents production, P are prices, Z are structural (non-price) characteristics and t is time. The specification includes lagged variables that indicate a response to earlier events.

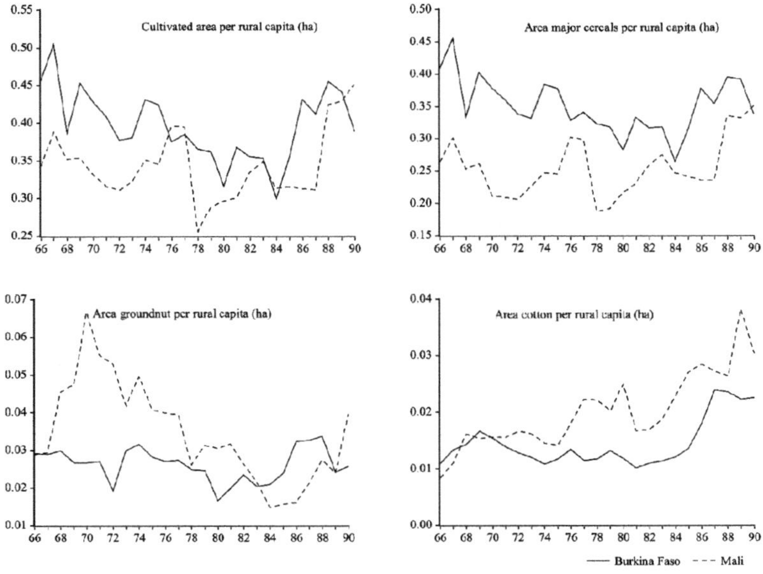

Figure 9.2. Area of cereals, cotton and groundnut (ha/inhabitant) for Burkina Faso and Mali.

The models are estimated on the basis of time-series data from FAO and World Bank. Climate data are from the Agrometeorological Group FAO-SDRN, NOAA and AGRHYMET. The data covers the period 1966-1990.

This model can be used to examine agricultural sector responses to changes in the agro-ecological and socio-economic environment. These responses can be specified in terms of adjustment of cultivated area per rural inhabitant (i.e. extensive growth) and crop yields (i.e. intensive growth). Extensive growth is usually associated with expansion of the agrarian frontier into less fertile lands, declining fallow periods and decreasing soil fertility. Intensive growth refers to yield enhancement by crop management practices that use more labour and capital per area unit. Extensive growth and intensive growth are not necessarily mutually exclusive.

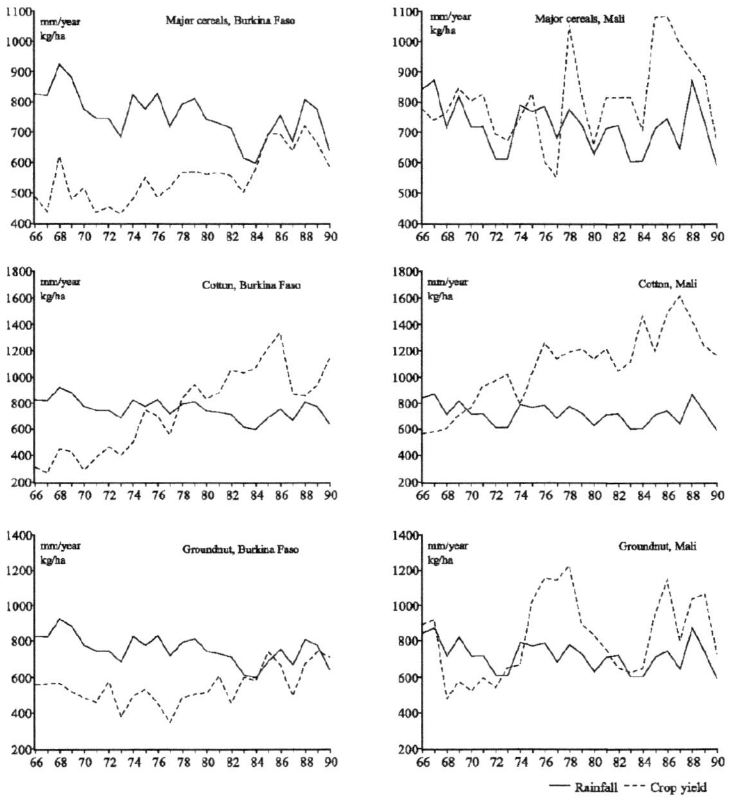

Figure 9.3. Rainfall and yield of major crops for Burkina Faso and Mali (1966 - 1990).

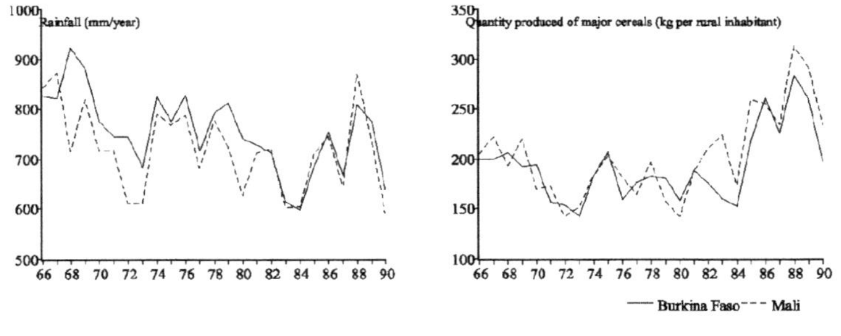

Figure 9.4. Total cereal production and rainfall for Burkina Faso and Mali (1966-1990).

Available time-series data at national level allows the use of proxies for the following explanatory variable categories: i) climate; ii) crop prices; iii) technical input-output relations; iv) infrastructure; and v) an autonomous technology. These factors are not truly exogenous but may be mutually related to each other. Production opportunities attract both public and private investments and induce farmers to invest in crop production. These supply determinants can reinforce each other either in space (i.e. within a specific region) or in specific sectors (i.e. within the cotton sector). In contrast, negative driving forces such as erratic and low precipitation and a weak infrastructure discourage agricultural investments.

To examine producers' responses to climate variability, two separate regression equations are specified for area and yield adjustments (Table 9.1). The first regression follows the Nerlovian supply response model in which the harvested area is a function of farmers' expectations and observations with respect to price and non price variables. Farmers' expectations concerning climate are included with a one-year lagged specification of yield, rainfall and temperature. Price expectations are taken into account by the previous year's sorghum and groundnut prices and the actual year's price of cotton (the price of cotton is announced before planting). Productive and social infrastructure is specified by tractor availability and cattle density (numbers rural inhabitant^{-1}), credit supply (FCFA rural inhabitant^{-1}), road density (km road km^{-2} arable land), and government expenses (FCFA capita^{-1}). Population density (inhabitants km^{-2} arable land) reflects an autonomous trend variable. Finally, a lagged specification of the dependent variable accounts for the possibility of a time-lag in the adjustment of crop area.

In the second regression equation, crop yield is a function of climate conditions during the actual year (rainfall and temperature), price expectations, factor availability, productive and social infrastructure, and an autonomous trend variable. In addition to the area response model, proxies are introduced for factor availability of labour (persons ha^{-1}), fertilisers (kg ha^{-1}), tractors and cattle (number ha^{-1}). A proxy for land use patterns is the share of cereals on the total harvested area. Infrastructure indicators are credit supply, road density and government expenses.

It is expected that climatic stress has an adverse effect on both crop area and yield, hence coefficients for rainfall and crop yield are expected to be positive, while the temperature coefficient is expected to be negative. Producers are supposed to respond positively to price increases, reflected by a positive price elasticity for the own crop and a negative elasticity for competing crop prices. However, commodity prices may have a limited or ambiguous effect on area and yield of food crops, since only 15% of the cereal output is marketed (Lecaillon *et al.*, 1987). In addition, illustrative examples of perverse supply reactions to increasing prices are known, particularly for subsistence food crops (Fosu *et al.*, 1997). Factor availability, infrastructure development as well as the trend estimator (changes in population density) are expected to have a positive effect on land use intensity and yield. Improved market access (i.e roads and means of transport) leading to reduced transportation costs can significantly enhance crop production (Heerink *et al.*, 1997).

Table 9.1.Supply response model for crop area and yield: variables in double log regression equations

Category	Area response model	Yield response model	(unit)
Dependent variable	Crop area (ha $person^{-1}$)[1)]	Crop yield (kg ha^{-1})	
Explanatory variables			
- Adjustment coefficient	Crop area $_{t-1}$ [2)]		(ha $person^{-1}$)
- Climate	Crop yield $_{t-1}$		(kg ha^{-1})
	Rainfall $_{t-1}$	Rainfall	(mm yr^{-1})
	Temperature $_{t-1}$	Temperature	(^{0}C)
- Price expectations	Sorghum price $_{t-1}$	Sorghum price $_{t-1}$	(FCFA kg^{-1}) [3)]
	Cotton price	Cotton price	(FCFA kg^{-1})
	Groundnut price $_{t-1}$	Groundnut price $_{t-1}$	(FCFA kg^{-1})
- Factor availability		Labour availability	(persons ha^{-1})
		Cereal share of	(%)
		harvested area	(kg ha^{-1})
		Fertiliser availability	(numbers ha^{-1})
		Tractor availability	(numbers ha^{-1})
		Cattle availability	
- Infrastructure development	Tractor availability		(numbers $person^{-1}$)
	Cattle density		(numbers $person^{-1}$)
	Credit supply	Credit supply	(FCFA $person^{-1}$)
	Road density	Road density	(km km^{-2})[4)]
	Government expenses	Government expenses	(FCFA $person^{-1}$)[5)]
- Trend estimator	Population density	Population density	(persons km^{-2})

Notes: 1) unless otherwise stated, persons refer to rural inhabitants; 2) the subscript t-1 denotes that the variable is one-year lagged specified; 3) prices are real prices (deflated by GDP index: 1987 = 100); 4) per km^2 of arable land; 5) FCFA per inhabitant for the total population.

The development of price and non-price variables is depicted in Figure 9.5 and 9.6 (annex). Real cereal prices increased only slightly until 1984, while cotton prices declined dramatically during this period (Lecaillon *et al.*, 1987). In Burkina Faso, cotton prices improved in relative terms in the mid 1980s. Over the whole period, however, the ratio cotton-sorghum price declined, basically due to the stronger tendency towards market liberalization in cereal markets. The availability of physical and social infrastructure and material inputs (tractors, fertilizers) shows a positive tendency in both countries. Tractor density and credit supply are substantially higher in Mali, but credit supply stagnated in the 1980s and even declined in Mali after 1984. Cattle density decreased in Mali after the drought years of 1971/72 and 1981/82.

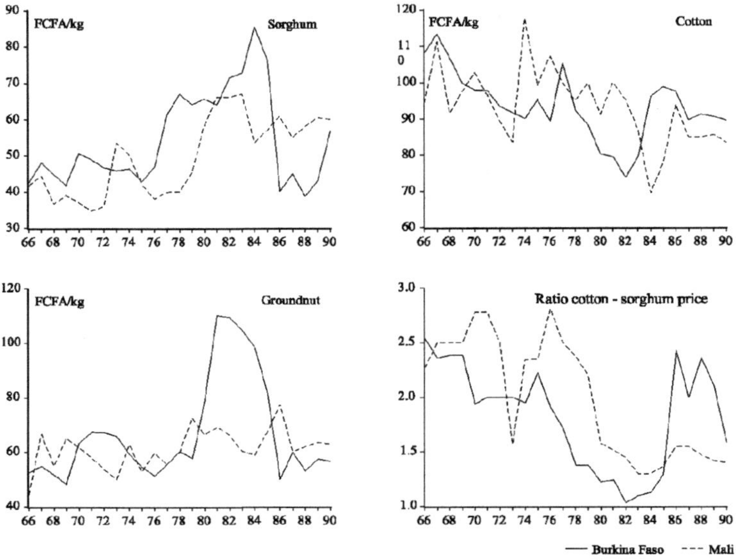

Figure 9.5. Official prices of major crops (FCFA/kg) and ratio cotton - sorghum price for Burkina Faso and Mali (1966 - 1990).

3. RESULTS

The model estimates for area response in Mali and Burkina Faso provide most consistent results for cash crops (cotton, groundnuts) but are less conclusive for food crops (see Table 9.2). The area adjustment coefficient for cotton is 0.63 and 0.69 respectively (significant at a 1% level for Burkina Faso and at 10% for Mali). For cereals and groundnuts, the model does not show a significant time-lag in area adjustment (g = 1) for Burkina Faso. For Mali, the cereal area adjustment coefficient is 0.52 (significant at a 5% level) and the groundnut area adjustment coefficient is 0.24 (significant at 1%). The cotton area equation shows a negative own price response for Burkina Faso (significant) as well as for Mali (not significant). This might indicate that the marginal costs of production have decreased and have permitted farmers to produce at lower prices. The own price elasticity of cereals is negative in Burkina Faso (not significant) and positive for Mali (significant at a 10% level). The positive own price elasticity for groundnuts for both countries is not significant. The substitution of groundnut for cotton in Mali was positively influenced by cotton prices. Climate variability does not appear to have much effect on harvested area, except for the lagged cotton yield specification for Mali. A positive coefficient indicates that farmers expand the cotton area in response to a good harvest in the previous year. For Burkina Faso, the coefficients of cattle density and road density are highly positive and significant. For Mali, the population density coefficient is positive and significant, indicating a trend of production increase that is not captured with the other variables.

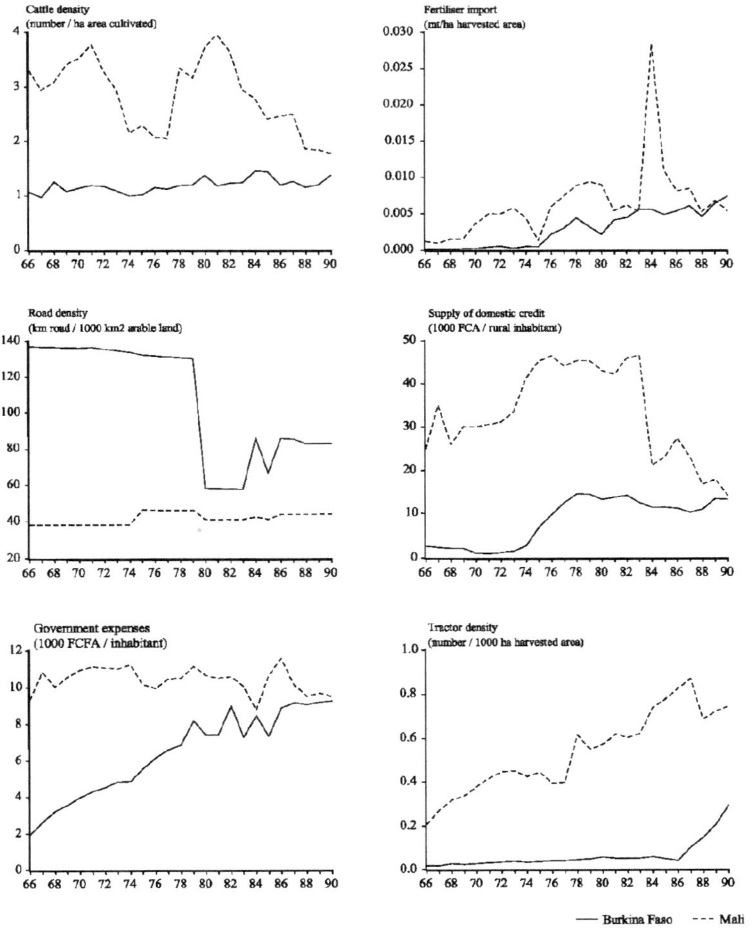

Figure 9.6. Productive and social infrastructure indicators for Burkina Faso and Mali (1966 - 1990).

Table 9.2. – Summary of outcomes of the reduced equations for area response.

	Burkina Faso			Mali		
	Cereals	Cotton	Groundnut	Cereals	Cotton	Groundnut
Area adjustment coefficient	1.00	+0.63 *** (3.84)	+0.78 ns (1.51)	+0.52 ** (2.94)	+0.69 * (1.73	+0.24*** (5.74)
Climate Yield lagged		+0.19 *** (4.42)				
Own price	-0.347** (-2.43)	-0.59 ** (-2.21)	-0.07 ns (-0.39)	+0.21 ns (1.55)	-0.44 ns (-1.40)	0.42 ns (0.98)
Cross Price						
Cotton						+0.89 ** (1.87)
Infrastructure						
Cattle density		+2.17 *** (6.44)				
Road density		+0.37 *** (4.40)				
Population density					+1.28 *** (2.06)	
Adjusted r-Squared	0.31	0.94	0.156	0.342	0.77	0.70
F-Statistics	6.20***	70.83***	3.12*	6.96***	26.57***	19.16***
LM F-statistic	1.58	0.13	0.54	0.00	0.23	0.10

Notes: 1) *** = 1% significance level; ** = 5% significance level; * = 10% significance level; 2) ns = not significant at a 10% level; 3) t-values are presented between parentheses; 4) an intercept is allowed for all equations.

The results of the yield response equation indicate that groundnuts in Burkina Faso and cereals in Mali are particularly responsive to the selected climate variables (Table 9.3). Coefficients of prices of the own crop are not significant in any of the equations. However, coefficients of competing crop prices are significant at a 10% level in the case of cotton and groundnut in the equation for Mali. Higher sorghum prices seem to increase competition for inputs, while higher groundnut prices enhance cotton yields. Important effects on yields are found for the variables related to input use. Credit supply is particularly relevant for improving cotton yields in Burkina Faso, and tractor availability stimulated cereals and cotton yields in Mali. In the latter case, the substitution of oxen by tractors had a significant positive yield effect. Cotton yields in Mali are also very sensitive to fertilizer use.

Table 9.3. Summary of outcomes of the reduced equations for yield response.

Yield response	Burkina Faso			Mali		
To:	Cereals	Cotton	Groundnut	Cereals	Cotton	Groundnut
Climate			+0.90 **	+0.91		
Rainfall			(2.25)	***		
				(3.70)		
Own price	+0.17 ns	-0.52 ns	+0.22 ns	-0.15 ns	+0.34 ns	-0.083 ns
	(1.12)	(-1.25)	(1.31)	(-0.90)	(1.66)	(-0.24)
Cross Price					-0.96 ***	
Sorghum					(4.81)	
					+0.99***	
					(3.98)	
Groundnut						
Factor use						
Labour					+0.66 ***	0.46 ***
					(3.88)	(1.99)
Cereal					+1.87 ***	
Percentage					(3.87)	
					+0.18 ***	
Fertiliser					(4.56)	
availability					-0.54 ***	
				+0.46	(-4.49)	
Oxen availability				***	+0.97 ***	
				(3.79)	(5.92)	
Tractor availability						
Population		+1.79 ***	+1.26 ***			
density		(4.32)	(3.12)			
Infrastructure		+0.20 ***				
Credit		(3.47)				
Technical	+0.61***					
Change (time)	(3.44)					
Adjusted r-squared	*0.37*	*0.87*	*0.39*	*0.49*	*0.90*	*0.14*
F-Statistic	*7.40 ****	*55.91****	*4.49***	*8.32****	*27.74****	*2.24*
DW statistic	*1.87*	*1.66*	*2.09*	*1.56*	*2.40*	*1.53*

Notes: *** = 1% significance level; ** = 5% significance level; * = 10% significance level; ns = not significant at a 10% level; t-values are presented between parentheses.
Source: Brons *et al.* (1999).

Increasing population density has a positive and significant impact on cotton and groundnut yields in Burkina Faso, due to the higher labour intensity of harvesting operations. In addition, the availability of rural labour improves yields in cotton and groundnut production in Mali. Labour scarcity is increasingly perceived as a limiting factor for agricultural intensification. Both cash crops exhibit a dominant autonomous technology development trend that enhances yields and input efficiency. The larger yield variability in cereals points to the absence of such a development trend in other cropping activities.

The combined results of the area and yield equations indicate that groundnut production in Burkina Faso and cereal production in Mali are, in particular, most responsive to climate variability. Own crop prices appear to be not significant, while coefficients of competing crop prices are significant at a 10% level only in the case of cotton and groundnut in the equation for Mali. Usual cross-price elasticities are registered for competing crops. Most input variables have the expected sign (except for oxen, due to the substitution by tractors) and are especially significant for cotton in Mali. Population density is a positive and significant factor in the case of cotton and groundnut for Burkina Faso. Moreover, the presence of a dominant autonomous technology development trend for these cash crops is confirmed. Due to large yield variability, such a trend is absent in other crops. Finally, infrastructure indicators are found to have a significant coefficient for credit supply in the cotton yield equation for Burkina Faso.

The findings also indicate that particularly variations in cotton area as well as yield are induced by infrastructure development, factor availability and an autonomous trend variable (related to population growth). Expansion of production has been realised in Mali as well as in Burkina Faso despite lower real prices of cotton. Farmers have been able to offset these negative determinants with an important reduction of production costs (crop variety choice, traction, fertiliser and insecticide application) that permitted them to produce against lower prices. In addition, technological progress might have neutralised the effects of climate deterioration, explaining the relative limited incidence of rainfall and temperature variables. The production of crops grown with traditional technologies (cereals and groundnut) is less responsive to changes in climatic, technical and socio-economic conditions. The results confirm that long-term changes in the quality of infrastructure, crop management practices and assets might have been at least as important than short-term variability in prices and climate for explaining variability in yields and cultivated area.

4. CONCLUSIONS

The importance of climate variability between the years as a determinant of crop choice is generally considered to reflect a rather indirect relationship. Prolonged periods of drought might, however, influence the availability of resources like seed material. Decisions on cultivated areas are clearly influenced by rainfall expectations (especially the timing of the start of the growing season), but other factors – like the availability of labour and traction – tend to be of greater importance. The role of prices and price expectations is difficult to register since marketing channels for staple crops were still highly institutionalised in the period under study. Consequently, price uncertainty was rather limited and input availability was determined on a contractual base.

Considering differences between agro-ecological zones, we can conclude that land availability and rainfall are most important factors for explaining yield differences especially in the more arid (northern) zone, while land and manure and fertiliser applications are more important for yields in the central and southern zones. Variations between villages and farmers may, however, be more important. Regions with rather similar rainfall levels, but with very different land use patterns show specific levels of adaptation by the local population while very different yield levels are achieved. In addition, relatively stable incomes are achieved during years with very different rainfall regimes, by making use of technological development based on local resources (mechanisation with limited land availability, non-input based land management change, etc.).

Crop choice, land use and input intensity are related to rainfall regimes and land endowments over the long term. While rainfall plays a prominent role in explaining variability of (cereal) yields and production, it is far less important as an underlying factor for differences in income, since a variety of other factors play an intermediate role. Aggregate production data suggests that mainly infrastructure development (roads and extension and credit institutions) stimulated the increase of cotton production, in particular, both through yield enhancement and area expansion. Since the input – output price ratio for cotton production deteriorated, the production increase was only made possible through a reduction in the production costs (crop management improvements such as variety development, fertiliser application, insect control, and mechanisation) and horizontal expansion (more farmers adopting similar technologies). Cereal production (in particular maize and rice) and groundnut cultivation benefited from spin-off effects at farm level (fertiliser residues, mechanisation, better market structure, etc.). Yet, the production of these crops also expanded mainly in a horizontal direction with more farmers cultivating larger areas.

As expected, climate variability proved to be particularly relevant for explaining differences in yield levels. Given the long time period, technological change and adjustments in market linkages (due to infrastructure investments) may have had a significant influence on improvements in yields. The consistent increase in crop yields throughout the whole period – maintained even under adverse weather conditions – points to the fact that farmers can adjust factor use as a compensatory device. This is especially the case with cash crops that register relatively higher supply response compared to food crops (perhaps with the exception of maize cultivated in rotation with cotton). Efforts to enhance agricultural yields should thus rely on an adequate mixture of public incentives and private investment (see Chapter 21).

It should be noted, however, that instead of average rainfall other variables, like the timing of rainfall, the first days of rainfall in relation to the rest of the agricultural year, or the moment when enough rainfall has fallen for the soil to be workable (depending on the soil) may be more important for explaining differences in yield levels. Local rainfall and yield data mostly show limited direct correlation (see also Chapter 7) and methodological problems (e.g. interpretation of the rainfall data for areas in between stations, yield measurement procedures, soil quality differences) strongly constrain the registration of such linear relationships.

A comparison of the supply response models for the different crops confirms the importance of supportive agricultural policies that facilitate optimal use of inputs and access to the innovations generated by technological progress. The reduction of production costs allows farmers to produce at lower prices. Particularly in the long term, technological progress supported by adequate infrastructure development is a more effective policy device than instruments of agricultural product pricing. Moreover, a sound production infrastructure can offset the negative effects of adverse risk-prone production conditions. Long term public investments do have a positive impact on agricultural production. However, these efforts tend to be strongly biased in favour of the cotton sector. Enhancement of the performance of food crops grown without modern technologies will thus require additional public investments in infrastructure, markets and technological development.

Chapter 10

COTTON AND CLIMATE CHANGE IN WEST AFRICA

Peter Ton

Abstract: As cotton has become the most important non-food cash crop in the West-African drylands, a tentative study was carried out into the possible impact of climate change on cotton cultivation and the vulnerability of the crop to increasing drought stress. Not much evidence could be found yet about a geographical movement of cotton production from northern to more southern areas.

1. INTRODUCTION

During the last twenty years, cotton production has become one of the most important economic activities in the West African drylands. In the late 1970s, cotton covered 900,000 hectares, of which 530,000 hectares in Nigeria. Total production was almost 500,000 metric tons of seed cotton and average yields were about 450 kg/ha. In the year 2001, the total cotton area had increased to 2.3 million hectares with total production increasing to 2.3 million metric tons of seed cotton (with yields more than doubling to 1,000 kg/ha). There was a major shift to Francophone West Africa. In 2001, Nigeria had 540,000 ha (and only 750 kg/ha), hardly any more than in 1979. Ghana's cotton production had increased from 12,000 ha to 52,000 ha. However, the cotton area in Mali had grown from 120,000 ha to 530,000 ha, in Burkina Faso from 72,000 ha to 346,000 ha, in Benin from 26,000 ha to 360,000 ha, in Togo from 16,000 ha to 150,000 ha and in Ivory Coast from 110,000 ha to 260,000 ha. Everywhere in Francophone West Africa average yields were above 1,000 kg/ha. In 2001, the total export value of cotton lint and seed for Mali was 83 million US$ and 65 million US$ for Burkina Faso. To put that in perspective: total (recorded) cattle exports were 35 million US$ for Mali and 8 million US$ for Burkina Faso, while total net cereal

A.J. Dietz et al. (eds.),
The Impact of Climate Change on Drylands: With a Focus on West-Africa, 97–115.

imports were worth 18 million US$ in Mali and 38 million US$ in Burkina Faso (data from http://www.fao.org; FAOSTAT AGRIC).

This chapter deals with the vulnerability and the adaptability of the cotton crop to possible climate change. It draws on an inventory of literature on cotton & climate change and on the author's working experiences in cotton growing areas in northern and central Benin and in eastern Senegal. Moreover, a number of cotton researchers (ICAC, the International Cotton Advisory Committee, based in Washington D.C. (USA) and CIRAD, the *Centre de Coopération Internationale en Recherche Agronomique pour le Développement*, based in Montpellier (France); CIRAD works on cotton research in almost all countries in West Africa) and cotton consultants were approached for references to relevant literature.

It should be noted that no specific publications on cotton & climate change were encountered, neither via the literature inventory, nor via the cotton researchers and consultants contacted. Several of them responded explicitly that they do not know of any publication which deals with cotton & climate change (Brüntrup, personal communication, 24-11-1999; Chaudhry, personal communication, 29-11-1999).

However, literature is available on cotton and water availability and on agricultural practices and water availability. In this paper we will limit the notion of climate change to the decreasing availability of water during the growing season. We will not, for example, go into the effects of climate change on rainfall patterns and atmospheric nutrient deposition (through Harmattan winds, lightning and rain), or into its effects on sunshine, air humidity, or cotton pest and predator population dynamics.

Sections 2 and 3 deal with cotton as a crop, and with the relationship between general and cotton-specific production practices and water availability. We will try to determine the main parameters that influence cotton yield variability. Literature on regional shifts in cotton production in West Africa is also limited. No publication was found that deals explicitly with these shifts. Only some scattered information was encountered about internal regional changes in cotton production within individual countries. Section 4 of this chapter deals with cotton production growth in West Africa and tries to explain why no major regional shifts in cotton production have apparently taken place.

2. COTTON AND WATER AVAILABILITY

The ICCD research proposal (1996) stated: "The problem of semi-arid livelihoods is not so much that rainfall is low (...)". The problem is that rainfall is so variable and erratic". This chapter deals with rainfall variability in the cotton growing areas in West Africa and with the vulnerability of cotton to variable rainfall.

Inter-annual rainfall variability is important in the cotton growing areas in West Africa. This is demonstrated in practice through the identification of agro-ecological zones. Beauval & Raymond (1992), for example, distinguish 11 agro-ecological zones in the small country of Benin, and then characterise the northern zones according to rainfall with variations between 700 and 900 mm for one zone, and 800-1200 mm, 900-1300 mm or 800-1400 mm for others. Data from specific sites provide an even more striking picture of inter-annual rainfall variability. Recent data from Koussanar (eastern Senegal), for example, indicates a rainfall pattern of 475 mm in 1995, 808 mm (+ 70%) in 1996, 577 mm (-29%) in 1997 and 713 mm (+ 24%) during 1998 (ENDA-Pronat, 1998 and 1999).

Intra-annual rainfall variability is high as well and may even be more important from the perspective of the farmer, i.e. in our case the cotton grower. During 1997, for example, the first rains in northern Benin started in early March, whereas they usually arrive in late April or early May. In 1999, however, the first rains in this area only started in June and this led to late sowing and to subsequently lower yields, at least in theory. When the rains really started, the amount of precipitation was such that lots of fields were inundated, leading to a serious setback in cotton crop growth. In other years, cotton growers in northern Benin sometimes witnessed drought periods of more than three weeks during July or August, which is generally considered to be too long for any annual crop to sustain without major drops in yield. Yet in other years or other regions, first rains may have fallen right on time and have been sustained over the months of July-September before stopping all of a sudden by mid-September instead of continuing as they usually do until mid-October. In that case, the cotton crop cannot mature adequately and average cotton lint quality will deteriorate as a result.

Coping mechanisms are available to deal with inter-annual and intra-annual rainfall variability, but the success of the strategies adopted by individual farmers can generally not be guaranteed beforehand, as we will see in section 3 while discussing general and cotton-specific cultural practices.

We will now deal with cotton and rainfall variability. There is plenty of literature on cotton-water relations, particularly in irrigated systems. For a detailed inventory of the principles of cotton water relations, see: Hearn (1995), The principles of cotton water relations and their application in management, and Lacape (1996), Effets de la sécheresse sur le cotonnier et amélioration génétique de son adaptation au déficit hydrique. Cotton production in West Africa is almost entirely rain-fed. Irrigated cotton is only found in the valley of the Senegal River (northern Senegal) and the Niger River (south-west Niger). Irrigation would technically be possible in other areas as well. However, according to Sément (1988), "except in the case of very good alluvial soils in extremely dry regions (less than 600 mm of rain a year on average), the extra yield obtainable with irrigation is insufficient to cover the costs incurred". Average yields obtained in West African rain-fed cotton production may go up to 1,300-1,400 kg of seed cotton/ha, but the overall average yield is about 900-1,000 kg/ha. Cotton yields have stagnated since the mid-1980s and even tended to decline during the late 1990s (Fok, 1999).

Cotton is a perennial crop which is now commonly grown as an annual crop mainly to curtail the development of pests and diseases. Cotton crop growth is thus indeterminate by nature. The end of the rainy season marks the end of the crop's growth (Lacape, 1996). Contrary to most cereal and leguminous crops, cotton is a deep-rooting crop. Its vertical taproot gives the crop access to water and nutrients in lower soil-layers. This makes the cotton plant a good rotational crop and one that is relatively tolerant to drought and to variable rainfall (Sément, 1988; Hearn, 1995). At the same time, however, cotton is highly sensitive to excesses of water (Lacape, 1996).

Cotton yield is determined by a large number of factors, e.g:

- climate (temperature, amount of sunshine, rainfall distribution);
- the physical and chemical characteristics of the soil; and
- the nature and extent of the plants' enemies (weeds, pests, diseases).

When all conditions are favourable, the cotton plant may produce more than three tons of seed cotton per hectare in tropical Africa. If, however, only one of the favourable conditions is lacking, yields are drastically reduced. The extent of yield reduction is dependent on the nature, the extent and the duration of the unfavourable conditions. If favourable conditions return before the season has progressed too far, the cotton plant may "recover", thanks to so-called compensatory growth (Sément, 1988; Lacape, 1996).

The first requirement for growing cotton is a climate offering conditions with regard to temperature, sunshine and soil moisture that favour a good crop, plus a marked dry season, which is essential both for the bolls to open out properly and for harvesting. The tropical savannah climate is therefore quite suitable from the point of view of temperature and sunshine. However, in transitional zones in which the climate borders on that of the moist tropical regions, the cotton plant has a longer production cycle because of lower temperatures and less sunshine (see Table 10.1). There, the risk of excess water will also be higher.

Overall, cotton requires a total of 105 to 125 days of sufficient soil moisture to grow. The water requirement of cotton plants varies considerably according to the stage of development and, at a certain stage, according to the intensity of sunshine and the humidity of the air. In tropical regions, 2 to 4 mm of water are needed daily at the beginning and the end of the growth period while, at the height of flowering, 5 to 7 mm are required daily according to climatic zone. Thus 500 to 700 mm of water are sufficient for the crop to develop fully. However, in practice rain-fed cotton can only be grown in regions where average annual rainfall is 700 mm or more, since inter-annual and intra-annual rainfall variability, and the amount of resulting run-off have to be taken into account(Sément, 1988). The CMDT (1996) refers to an agro-climatic model showing that cotton yield is strongly related to average annual rainfall for those production areas receiving less than 900-1,000 mm of rain. No reference is made, however, to any report or publication.

Table 10.1. Life-cycle of the cotton plant (prior to opening of the bolls).

Zone	Sowing to start of flowering	Ascending phase of flowering	Growth of bolls from peak flowering	Total period requiring watering	Probable end of rains	Sowing dates
Sahel/Sudan	55 days	25 days	25-30 days	105-110 days	20/9	2/6 - 7/6
Sudan	60 days	25-30 days	25-30 days	110-120 days	10/10	12/6 - 22/6
Guinea/Sudan	60-65 days	30 days	30 days	120-125 days	31/10	28/6 - 3/7

Source : Sément (1988), Cotton.

The rainfall regime in West Africa means that the amount of moisture in the soil is usually a limiting factor as regards production. This is the case because there is excessive moisture after heavy rain on soils which do not dry out particularly well, or - as is more often the case - because there is insufficient moisture as in the case in a very short or irregular rainy season (Sahel-Sudan or Sudan zone) (Sément, 1988).

There are a number of Cotton varieties, with different responses to climate conditions. For a detailed inventory of literature on cotton breeding and drought tolerance, see: Lacape (1996), Effets de la sécheresse sur le cotonnier et amélioration génétique de son adaptation au déficit hydrique. Cotton is a plant of the genus Gossypium. Four species are cultivated as a commercial crop. Cotton grown in West Africa is Gossypium hirsutum, or American upland cotton, which is an exotic crop to the region. The Gossypium hirsutum varieties were first introduced in West Africa from the 1950s onwards.

Unlike seeds for food crops, farmers in West Africa generally do not keep cottonseeds for sowing at the start of the next season. Each year, the national cotton research institutes and the cotton marketing boards distribute selected cottonseeds that originate from their own seed multiplication schemes. In some countries, like Benin, seeds are distributed for free to ensure homogeneity of varieties grown, which is important for ginning and marketing. However, in other countries, like Senegal, farmers receive cottonseeds on a credit loan, which is reimbursed at payment of the seed cotton. In francophone West Africa at least, there is no free market for cottonseeds, meaning that individual farmers are unable to choose from different varieties. The situation is probably different in the Anglophone countries in West Africa, where the cotton sectors are all privately run.

The productivity of a crop variety is to a large extent due to the excellent way it adapts to the environmental conditions (soil, climate, natural enemies, agricultural techniques) in which it has been bred. That is why varieties developed in Africa are rather different from the American varieties for instance, although they share the same origins. In tropical Africa, for example, selection has led to varieties with hairy leaves being able to withstand attacks by certain insects, or to varieties for which fruiting is spaced out over a longer period of time making the plant hardier (Sément, 1988). Moreover, the impact of diseases is generally negligible, as the varieties grown are mostly disease-resistant or disease-tolerant, against bacterial blight in particular (Gaborel, 1999). Still, cotton breeding in West Africa has principally focussed on improving ginning outturn and lint characteristics (Hau, 1995), rather than on performance at field level (Gaborel, 1999). Ginning outturn of 40-44 % is considerably higher than the usual 33-37 % common elsewhere in the world (ICAC, 1993). Lint characteristics are good (medium to long fibre, clear-white colour, clean, etc.), making West African lint popular on the international markets.

There is increased awareness that cotton varieties in West Africa should become more adapted to site-specific growing conditions. So-called "robust" varieties should be developed which are less vulnerable to adverse growing conditions, including rainfall variability. Gaborel (1999) states for example that: "The varieties grown are all similar to one another, making it uncertain whether they are always the most adapted to the enormous diversification of growing conditions which we currently witness." Vaissayre (personal communication, 21-3-1999), while explaining the dramatically low yields obtained in Senegal over 1998/99, claims that varieties grown were not well-adapted to local conditions, as the decreasing number of days of rain per season (about 60 over 1998) no longer fulfils the cotton plant's needs. According to Vaissayre, the varieties grown in West Africa were almost all bred in the coastal countries, under more humid conditions than those in which they are now grown.

Farmer-participatory breeding (PARAB, 1996; Sékloka *et al.*, 1999) may become an important tool for identification of more "robust" varieties that perform better in the field while continuing to match with the quality requirements of the international cotton markets. Rainfall variability and water availability will probably turn out to be important parameters for characterisation of the varieties identified this way.

3. PRODUCTION PRACTICES AND WATER AVAILABILITY

Van den Born *et al.* (1999) state: "Although models on the climatological constraints on crops illustrate that crop distributions are sensitive to climate change, this conclusion cannot be directly extrapolated to the agricultural sector in general. This sector is probably less vulnerable because its current management practices create many options for adaptive capabilities. Only under extreme, mostly dryland, conditions with marginal agriculture can the impacts of climate change on agriculture become a significant limiting factor".

This section deals with general and cotton-specific production practices and their impacts on water availability and identifies options for adapting current management practices in order to deal with possible climate change in West Africa.

Choice of land

The agricultural season starts off with the choice of fields. Normally, cotton would fit into a 3 to 4-year rotational cropping system. However, in a number of areas cotton has become so important that it exceeds 25-30 % of total cropping area. In northern Benin, for example, in 1996/97 cotton occupied 37 % of total cropping area on a provincial average, with some district averages even being as much as 50 %.

Where monoculture sets in, imbalances will rapidly appear in the soil's nutrient composition. The common practice of burning and the cotton-specific fertilisation, crop protection and crop residue management practices in West Africa (see below) further enhance depletion of the soil's organic matter content, leading to impoverishment of the soil structure, less retention of water and an increased risk of erosion. Proper crop rotation is an important element of integrated soil fertility and crop management practices and will also indirectly favour the availability of soil moisture.

Clearing of land

Parts of the fields which are used for cotton cultivation are fallow or bush lands. In the case of northern Benin, Katary (1998) concludes that total cotton production increases are not the result of yield improvement but rather of expansion of agricultural land, thus provoking deforestation and cultivation of fallow lands. Ton (1993) estimated that cotton could be held responsible in northern Benin for about one-third of all deforestation since the early-1980s.

Clearing of land is generally carried out in order to facilitate the use of animal traction, although animal traction is not common in southern cotton production areas like central Benin, where land is prepared manually. With the exception of a number of tree species, like karité (Vitellaria paradoxa), néré (Parkia biglobosa), baobab (Adansonia digitata) and Acacia and mango species (Hijkoop *et al.*, 1991), almost all vegetation is burnt, both on new and on old lands, so that erosion is increased, the upward movement of water reduced and deep percolation encouraged. The fact that significant amounts of organic matter are lost has a negative effect on the soil's retention of water during the rainy season. Mulching and composting of vegetative residues is not yet widespread in West Africa. These techniques could, however, be incorporated into strategies to cope with problems of water availability, through replenishment of the soil organic matter content.

Field design

Field design can make an important contribution to a more efficient use of water. First of all, existing small-scale variability within fields with respect to soils and crops may help to mitigate the effects of variability in nutrients and soil moisture (Brouwer & Bouma, 1997). Secondly, water infiltration may be increased through erosion control methods in and around fields, like stone ridges, live fences or vegetative strips etc. (Dupriez & De Leener, 1990). Erosion control measures are applied to some extent in West Africa, particularly in Mali (Afrique Agriculture, 1999b; Hijkoop *et al.*, 1991) and Cameroon (Béroud, 1999) to prevent the run-off of water and nutrients. Furthermore, fields could be designed in such a way that diversity is enhanced. Rather than sowing large areas with one single crop, one could opt for parallel production of different crops (for example, only 10 rows of one crop at a time), to reduce erosion and to increase populations of natural enemies of key pests. Strip cropping is very much favoured in organic agriculture, but is not yet applied on a large scale in West African agriculture.

Mixed cropping or inter-cropping can provide yet another tool for increasing the efficient use of water, particularly if the accompanying crop is a cover crop. Inter-cropping is discouraged by current state cotton policies and little research has been done on the subject (Elobu *et al.*, 1995). According to Sément (1988), "the problem with inter-cropping is that it is not particularly compatible with (…) the need for herbicidal and insecticidal treatments that are appropriate to only one type of plant". Note that pesticide residues may also lead to severe intoxication, as was sadly witnessed in Benin over the 1999/00 season when several dozens of people died due to cotton pesticide residues on maize and gombo plants that originated from conventional cotton fields. Inter-cropping can be an important tool in organic cotton production, where no synthetic pesticides are sprayed. Organic cotton production is a recent phenomenon in West Africa and is therefore experimental by definition (PAN-Africa, 1998; Ton, 1999; Myers & Stolton, 1999).

Special mention should be made of cotton grown as a "relay" crop in the southern production areas like central Benin. Here, the longer rainy season permits cotton to be planted in between maturing maize plants or after short-cycle leguminous crops, with only a limited need for additional soil preparation and a reduced risk of erosion.

Preparation of land

The preparation of land plays an important role as well in the availability of water. It is common practice in West African cotton production, and less so in cereal production, to break the entire field open through ploughing or ridging. This leads to, among other things, increased water retention , but it also increases the risk of erosion during the first weeks of the season when crops have not yet covered the soil adequately.

The desirability of ploughing and ridging is dependent on site-specific soil and climatic conditions (Munro, 1994). The agronomic effects of ridging can exemplify this. Ridging opens up the land to improve root development and suppress weeds, while at the same time increasing water retention. But ridging is also applied to prevent the roots from too much water and to enhance soil aeration; i.e. to limit the soil moisture content (Sément, 1988). This type of tillage may be questionable in dry areas with high rainfall intensity where it favours erosion, and in areas with relatively high rainfall where water evacuation may be the problem rather than water retention. Similar concerns apply to erosion control measures like contour ploughing. Inter-annual and intra-annual rainfall variability complicates decision-making on these issues even further. Where erosion is considered a problem, contour ploughing is generally being advocated. However, when farmers in northern Benin were asked why they ploughed along the slopes, their answer was that they preferred to get rid of rainwater, rather than risking inundation of their crops.

Preparation of land may thus be important to influence water availability. However, a trade-off will need to be sought by farmers between root development (tillage), sowing date (no tillage), water availability (tillage), water evacuation (no tillage), weed suppression (tillage), pest and disease control (tillage) and erosion control (no tillage), while taking account of the on-site soil characteristics and rainfall patterns.

Sowing

The sowing data is a crucial parameter for cotton production in West Africa. It should be noted that the first rains in West Africa are used to prepare and sow as much land as possible whether with food crops or cash crops. During the season, however, farmers may very well find out that the particular rainfall pattern is, for example, favourable to weed development and that the cultivated area is too large to weed properly. Part of the cropland will then be abandoned, resulting in lower average (cotton) yields (eastern Senegal, 1998). This is one of the factors that have hampered data analysis by the ICCD project with respect to the relation of (cotton) yield and rainfall variability. As the cotton crop needs a relatively long growth period, it is important to make full use of the rainy season to ensure maturity of the fibres before the dry season starts. The sowing date is a lesser problem in the southern cotton production areas, like central Benin, where the rainy season is spread out over a longer period of time and where the arrival of the dry season is less sudden (Sément, 1988). Also, early sowing is an effective pest control method in most cotton production areas in West Africa as it will help plants to escape from damage during its susceptible growth stages (Vaissayre, 1995).

However, at sowing time, food crops are usually given priority in West Africa over cash crops such as cotton, to ensure food availability for home consumption. Cotton yields will turn out to be lower as a result. The cotton sowing date could be speeded up if cotton were sown directly without prior tillage. Direct sowing is widespread in West African cereal production and the approach is now also being advocated for cotton in Cameroon (Afrique Agriculture, 1999a).

Furthermore, sowing and plant density may be adapted to improve water retention and moisture availability. Cotton plant density in West Africa is usually 40-60,000 plants/ha after thinning out. However, the CIRAD Cotton Programme is currently studying the introduction of "narrow row" and "ultra narrow row" cotton in West Africa (Hau & Vaissayre, 1999) which could -amongst others- lead to a better soil coverage and to subsequent advantages in water retention and moisture availability (Lacape, 1996). In turn, high densities generally favour insect pests and reduce the effectiveness of pesticide applications, especially in cases of high humidity and reduced insulation (in the southern production areas of West Africa) (Vaissayre, 1995).

Fertilisation

The fertilisation of crops is not common practice in West Africa, except in the case of cash crops like cotton. West African agriculture is still highly oriented towards slash and burn, with fallow lands providing natural regeneration of soils. However, this system cannot be sustained anymore (Van der Pol, 1992) due to demographic growth and expansion of cultivated land. Fallow and bush lands have come under pressure, both in quantity and in quality.

Cotton stands out positively as the only major crop that receives significant amounts of (synthetic) fertilisers. The synthetic fertilisers do restitute nutrients (N, P, K, and some S and B) that are extracted from the farming system through the export of seed cotton and burning of cotton crop residues. Additionally, cotton fertilisers have a high P-content, to help rectify the natural phosphate deficiency of West African soils.

However, the success of synthetic fertilisation depends on the on-site soil characteristics, product quality, application rates, rainfall patterns, etc. Synthetic fertilisers are prone to volatilisation and leaching but also to erosion. For example, over 1999 farmers in central Benin reported that fertilisers had been washed away by heavy rains (Le Matinal, 1999). Béroud (1999) of the French cotton company CFDT acknowledges: "(...) synthetic fertilisation slows down soil degradation, but it does not curtail it". Still, Cretenet *et al.* (1994) and Gaborel (1999) even report for the savannah areas with one rainy season that "a lack of potassium generally appears after about a decade of cultivation". Moreover, in southern production areas with two rainy seasons, the synthetic fertilisers in use are reported to enhance processes of acidification (Gaborel, 1999; Cretenet, personal communication, 2-9-1999) rather than slowing them down.

Synthetic fertilisers do not restitute any organic matter to the soil and they are only effective if the soil contains sufficient organic matter (Sément, 1988). Cretenet *et al.* (1994) and Gaborel (1999) conclude on the basis of long-lasting soil fertility research, that the soil organic matter content decreases rapidly in the northern savannah areas with one rainy season. Dakuo (1998) even states that the degradation of soil fertility, and of natural resources in general, has led to a shift of cotton production areas in each of the West African countries from the dry zones towards more humid ones.

The West African cotton sectors tend to neglect restitution of organic matter to the soil. The only major exception is Mali where the use of organic manure has become regular and significant (Béroud, 1999). In other countries research is being carried out on the subject (Berger, 1996; Dakuo, 1998; Gueye, 1998; Wennink, 1998) but very little effort is being made to promote the efficient use of local resources available for organic fertilisation, even though their socio-economic and ecological potential is high in some regions (Ton, 1998). Note that current state subsidies and credit loan systems also favour the use of expensive synthetic fertilisers without any organic matter, thereby discouraging the use of relatively cheap and locally-available resources of high organic matter content.

Organic fertilisation is therefore only applied to some extent and in some areas. It generally occurs "by default", for example through ashes of burnt vegetative residues, through manure from cattle roaming crop residues, or through temporary cattle camps installed by pastoralists. "Active" organic fertilisation (through manure, compost, cover crops, mulching, household residues, processing residues, etc.) is still rather exceptional and it is often complicated by its labour intensity, by a lack of transport, or by socio-cultural considerations. Cultivation of green manure crops, for example, runs contrary to the dominant culture of growing crops for food or cash obtained immediately at harvest. Moreover, local cultural habits may, for example, prevent the use of human excrements as fertilisers in agriculture.

Organic fertilisation of the cotton crop could contribute considerably to maintenance of the soil organic matter content in West Africa and consequently to better soil structure, increased water retention and reduced erosion.

Crop protection

The wetter the cultivation area, the richer the entomofauna. In tropical Africa, the cotton pest complex is limited to three main species of bollworms and two species of sucking pests (aphids, white fly) in the northern part (rainfall < 1.000 mm) but is much larger in the wet area (rainfall > 1.000 mm). Insect pests that affect cotton yields in West Africa significantly are: Heliothis/Helicoverpa armigera (American bollworm), Diparopsis watersi (Sudan bollworm), Spodoptera littoralis (armyworm), Earias spp. (spiny bollworm), and in the southern production areas Cryptophlebia leucotreta (false codling moth) and Pectinophora gossypiella (pink bollworm). Other cotton pest problems in West Africa include: Dysdercus spp. (cotton stainer), Aphis gossypii (cotton aphid), Bemisia tabaci (whitefly), and in the southern production areas Polyphagotarsonemus latus (broad mite) (Cauquil, 1993; Vodounnon & Silvie, 1997). As most of these insects have a number of different host plants, the natural vegetation preceding and surrounding cotton plays an essential role in insect abundance (Vaissayre, 1995). For a detailed description of the lifecycles, biology and ecology of individual cotton insect pests, see: Matthews & Tunstall (eds.) (1994), Insect pests on cotton.

Rain has a mechanical effect on minute forms of life such as aphids, eggs and the first instar of lepidopteran (bollworm) pests. Decreases in pests are often a consequence of heavy rainfall, especially in tropical conditions. Although raindrops have a mechanical action, the major effect of the crop water balance is the variation in the relative humidity within the plant canopy. Pest development is closely related to relative humidity (Vaissayre, 1995).

Generally speaking, pest incidence peaks when the cotton plant is affected by bad weather conditions (either too dry or too wet). A dry period following planting and emergence of the plant favours sucking pests. Attacks by bollworms are at their most intense during dry, sunny periods in the rainy season.

The beginning of the dry season is followed by outbreaks of aphids or white fly which cause soiling of the fibre in open bolls with honeydew (Vaissayre, 1995).

Insect pest management includes general cultural practices (choice of land, land preparation, plant density), choice of crop varieties, and certain cotton-specific cultural practices (sowing date, synthetic fertilisation, synthetic pesticide applications, crop residue management).

Synthetic pesticide applications are common practice in West Africa and they are generally felt to be obligatory (Sément, 1988; Matthews, 1989; Vaissayre, 1995). Successful synthetic pesticide use depends on the type of insect pest, product quality, number and dosage of sprayings, periodicity between sprayings, rainfall patterns, etc. However, pesticides pose serious threats to humans, animals and the environment in general. The long-term efficiency of the use of synthetic pesticides in West Africa is uncertain, as its environmental impacts have, as yet, hardly been assessed. There are no baseline-studies on the incidence of pest and natural enemy populations. Thus, the impact of pesticide sprays on natural enemy populations -and their contribution to subsequent pest resurgence- is unknown. The impact of synthetic pesticides on soil fertility is unknown as well. An assessment of the efficiency of synthetic pesticide use in West Africa has so far been limited to research on insect resistance to pesticides (Martin *et al.*, 1997; Vaissayre *et al.*, 1998). With respect to cotton & water availability, it should be noted that pesticides seriously damage (micro) biological activity in soils. As a result, the soil structure deteriorates, water retention is reduced and soils become more prone to erosion. During 1999, for example, farmers in central Benin witnessed the emergence of large numbers of earthworms, These then died after use of endosulfan pesticide sprays.

Cotton researchers in West Africa favour integrated pest management (IPM) strategies and they are experimenting with insect pest monitoring to reduce the number of pesticide sprays (Agoua *et al.*, 1997; Michel *et al.*, 1997; Ochou Ochou *et al.*, 1998). However, standardised calendar sprays (5-7 per season) are still the rule. Organic cotton projects (e.g. in Mali, Benin and Senegal) approach the pest and pesticide problems from a different perspective: the use of synthetic pesticides is prohibited and all efforts are supposed to be directed towards appropriate cultural practices and the creation of diversity on the farm, in order to prevent pest problems rather than to cure them.

Post-harvest crop residue management

Crop residues provide a habitat for over-wintering insects. Crop residues include: cotton stalks; stumps; loose seed cotton spilled around the fields; ginning residues and voluntarily emerging cotton seedlings (Munro, 1994). Post-harvest stalk destruction, which entails destroying the habitat of over-wintering pests or reducing available food for late season pests, remains one of the most powerful weapons against pests and insect-transmitted diseases, but has to be practised on an area-wide basis (Vaissayre, 1995). However, this standard recommendation has come under pressure recently (CTA, 1996), as it is common practice to leave the cotton stalks in the field until the next season and to lay cotton fields fallow without destruction of stalks. A so-called closed season may be desirable and effective, but it is apparently not feasible under West African growing conditions. Cotton stalks should rather be considered as valuable sources of organic matter for maintenance of soil fertility, but there are currently few initiatives to make use of cotton stalks for organic fertilisation.

Post-harvest cropland management

Cropland management in West Africa tends to be limited to the rainy season only. However, to facilitate ecological sustainability, cropland should be managed year-round. Mulching, manure application and composting could very well be carried out over the dry season, when the opportunity costs of labour are relatively low. This way, the need for investment in synthetic fertilisers could probably be significantly reduced (Ton, 1998). So-called improved fallow practices are another aspect of post-harvest cropland management. Hoefsloot *et al.* (1993) report on the opportunities and constraints for the "active" improvement of fields laid fallow.

Concluding remarks

We have seen that current cotton production practices do indeed leave many options for adaptation to climate change. Some of these options are already being applied in particular cotton growing areas in West Africa, while others are still the subject of research.

Overall, three options for the adaptation of current management practices can be distinguished: a) re-design (cotton) fields to favour the infiltration of water and to reduce erosion; b) optimise the recycling of vegetative residues and manure, in order to improve the organic matter content of soils, to increase water retention and to reduce erosion and c) evaluate and re-consider the cotton-specific fertilisation and crop protection practices in the light of land degradation in general and water availability in particular.

4. DYNAMICS OF COTTON PRODUCTION IN WEST AFRICA

In the previous sections, we have focussed on the relations between cotton and agricultural practices on the one hand and the availability of water on the other. In view of possible climate change in West Africa, we identified a number of options for increasing the adaptability to changing rainfall conditions of agriculture in general and cotton production in particular.

In this section we will look at the dynamics of cotton production in West Africa over the last decades, to see what factors may explain production growth and regional changes in cotton production in West Africa.

4.1 Cotton production growth in West Africa

Francophone West Africa has undergone enormous growth in cotton production over the last decades (Table 10.2). Both the cotton crop area and the production of seed cotton have more than quadrupled since the late-1970s. Cotton is now grown on more than 1.6 million hectares of land. Mali is by far the biggest cotton producer in francophone West Africa, followed by the Ivory Coast, Benin and Burkina Faso. Average yield went up during the early 1980s, stagnated in the mid-1980s and then decreased during the late 1990s. During 1995/96, seed cotton production in Anglophone West Africa was estimated at 265,000 tons for Nigeria (Madhavi, 1996) and 18,200 tons for Ghana (Esteulle, 1996). In 2001 it had increased to 402,000 tons in Nigeria and 44,000 tons in Ghana, but that was only 20% of total seed cotton production in West Africa. The Francophone countries had gained the upper hand and this has been the situation since the mid 1980s (http://www.fao.org).

Table 10.2. Cotton production in francophone West Africa (seasons 1978/79, 1988/89 and 1998/99).

Season	1978/79			1988/89					
Country	area (ha x 1,000)	seed cotton (Metric tons x 1000)	yield (kg/ha)	area (ha x 1000)	seed cotton (Metric tons x 1000)	yield (kg/ha)	area (ha x 1,000)	seed cotton (Metric tons x 1,000)	yield (kg/ha)
Benin	26	18	690	110	105	955	400	335	840
Burkina Faso	72	60	835	171	142	830	297	310	1045
Ivory Coast	107	115	1075	213	290	1360	270	350	1300
Mali	117	128	1095	190	251	1320	500	520	1040
Niger	9	4	445	7	6	860	n/a	n/a	n/a
Senegal	48	34	710	39	39	1000	45	12	270
Togo	16	14	875	81	87	1090	160	190	1190
Total	**395**	**373**	**945**	**811**	**920**	**1135**	**1672**	**1717**	**1025**

Source : MCD (1991) and APM-Afrique (1999b).

There are many explanations for this enormous cotton "boom" (World Bank (1988), MCD (1991), Deveze (1994) and Brüntrup (1997) and these can be divided into push and pull factors. Push factors include: the lack of markets for alternative crops whether food or cash crops, the lack of alternative sources of income, and the growing need for monetary revenues for consumption and production. Pull factors are: the creation of market outlets for cotton, favourable policies of national governments and the saving opportunities provided by the seed cotton purchase schemes in which payment is made in just one or two rounds.

The cotton "boom" was only able to materialise thanks to an increased number of cotton growers, demographic growth and a number of technological innovations: good quality seeds, synthetic fertilisers, synthetic pesticides and animal traction. Where seeds and synthetic inputs led to yield improvements, the increased number of growers, demographic growth and animal traction led first and foremost to the expansion of agricultural land. The successful introduction of animal traction explains a large part of the cotton production "boom", as it enabled farmers to expand the cotton crop area. However, at a growth rate of 3% per year, demographic growth alone would already explain for 86% of area and production growth over a 20-year period.

National cotton production programmes were set up in various countries in West Africa during the mid-1970s and the 1980s, with financial support from the French development co-operation organisations, the European Development Fund and the World Bank. The system put in place was similar all-over francophone West Africa and became known by its French name: l'approche filière. It encompassed the following elements: a guaranteed producer price set in advance; a stabilisation fund to ensure the minimum producer price; a cotton input supply system; a credit loan system to facilitate synthetic input use and animal traction; specialised cotton research; cotton extension services; renewal and construction of ginning mills; and a state monopoly on ginning and exports.

During the late 1990s, however, the filière approach came under severe criticism from the World Bank, whose aim is now liberalisation and subsequent privatisation of the state-led cotton sectors. Liberalisation and privatisation processes have been forced on all cotton-producing countries in West Africa, but at a varying pace and with mixed results so far (Afrique Agriculture, 1998; APM-Afrique, 1999a).

Regional shifts in cotton production in West Africa

Most cotton researchers agree that the integrated filière approach has played a decisive role in rural development in West Africa (See: Roupsard (1987), World Bank (1988), MCD (1991), Deveze (1994) and Brüntrup (1997). An export market for cotton has been created where no alternative outlets are available for other crops, the price guarantee reduced farmers' production risks, while state subsidies and the credit loan system proved effective in intensifying production through the use of synthetic inputs. Moreover, cotton became a vehicle for the introduction of agricultural innovations, like animal traction. From a political perspective, one should note that the rural cotton-producing areas have become much more integrated into national economies and societies than a couple of decades ago.

Regional differences in cotton production within countries can generally be explained by factors like: agro-ecological conditions, population density, access to markets for food crops, degree of agricultural intensification, political instability, etc.

There is no proof from the available literature that any major regional shifts have occurred (yet) within countries with respect to the volumes of seed cotton produced. Togo presents the only exception encountered. There, strong growth was reported for the central plateaux over the period 1992/93-1997/98, with stagnation of cotton production in the northern savannah region (Agossou, 1998). New regions became involved over the years and changes have occurred in the relative contribution of particular regions to national cotton production.

However, the available literature does not link this with production decreases in the traditional "cotton basins". In Benin, the central Zou province and the southern part of the (northern) Borgou province experienced an increase in their relative shares in national cotton production over the 1980s and 1990s, but production in the traditional "cotton basin" (northern Borgou) still almost doubled between 1991/92 and 1997/98! There are some indications that the most northern cotton-producing areas in Mali and Cameroon gradually lost, and more southern areas gained, but no literature was found with empirical evidence.

Production growth has apparently been quite general in the major cotton-producing countries in West Africa. This trend seems to reconfirm our thesis that national agricultural policies, supported by international donors, have been determinant for cotton production growth in the different countries and regions. A clear example is presented by Mali, where the formerly peanut-producing region of Kita was integrated into the CMDT structures over 1995 in order to promote cotton as a cash crop (CMDT, 1996). Admittedly, the first initiatives for cotton growing in Kita were taken by a local NGO, growing some 2,500 ha of cotton over 1994/95 (Sangaré & Seck, 1994), but production only really got going with the full involvement of the CMDT. Over 1997/98, the Kita region already grew 21,050 ha of cotton (Diabate, 1998).

5. CONCLUSIONS

The efficiency of water use is an important element of cotton growing and one of the limiting factors in rain-fed cotton production. This is particularly the case in the northern part of the ICCD research area, where average annual rainfall is < 700 mm per year. Drought-tolerant varieties may help to shorten the crop's life cycle and make full use of the rainy season. However, the cotton crop itself is already quite drought-tolerant thanks to the plant's morphology.

Factors other than water availability seem to be more important for cotton yields in those semi-arid and sub-humid zones of the ICCD research area where average annual rainfall conditions are > 700 mm. per year. Pest control is generally assumed to be the most important one, followed by soil fertility. Still, excess water will be detrimental to cotton yield, as well as prolonged periods of drought over the rainy season.

Cotton is particularly vulnerable to pests, due to its long lifecycle and the number of insects using cotton as (one of) their host plant(s). Manipulation of sowing time is important as a coping strategy to suppress damage by pests and diseases. Cultural practices like land preparation, proper fertilisation, plant density, weeding, and early harvesting are all other elements of integrated pest and crop management, but pest control in West African cotton production can generally still be equated with calendar sprays of synthetic pesticides.

Soil fertility is another limiting factor for West African cotton yields. Some nutrient deficiencies may occur naturally, but evidence of human-induced deficiencies is mounting. Cotton fertilisation is generally limited to the restitution of nutrients through synthetic fertilisers. However, synthetic cotton fertilisers in use are often unable to rectify the nutrient balance and there are even reports that they actually enhance soil impoverishment in the West African cotton belt.

Current cotton production practices leave many options for adaptation to climate change. Overall, we distinguished three options for adaptation of current management practices:

- re-design (cotton) fields to favour the infiltration of water and to reduce erosion.
- optimise the recycling of vegetative residues and manure in order to improve the organic matter content of soils, to increase water retention and to reduce erosion.
- evaluate and re-consider the cotton-specific fertilisation and crop protection practices in the light of land degradation in general and water availability in particular.

The first two options relate to what Hearn (1995) calls: "Soil surface management technology to retain rainfall and reduce run-off and soil evaporation". They will increase water availability, while at the same time favouring soil fertility management and pest control.

The third option is a result of the analysis that the synthetic inputs in use do indeed solve specific technical problems in cotton production in the short term. However, at the same time it poses serious threats to their users, and to the medium and long-term ecological sustainability of the cotton-based systems in West Africa.

Dakuo (1998) states that the degradation of soil fertility, and of natural resources in general, has led to a shift of cotton production in each of the West African countries from the dry zones towards more humid ones. This statement could not be proved by our literature research but the hypothesis is certainly worth studying.

Some authors suggest that cotton does relatively well on soils with low organic matter content and that synthetic fertilisers prolong the cultivation period, with enhanced soil degradation as a result. If this were true, cotton would do well on soils with a low water buffering capacity, which might thus discourage some of the options we identified for adaptation to climate change. Other authors, however, claim that soil organic matter deficiencies will lead to lower cotton yields. This would favour the adoption of the options to adapt to climate change, as well as the search for cropland in other -more humid- areas where the soil organic matter content is more favourable.

Still, no references were found to any major regional shifts in cotton production within individual countries, either in relation to soil fertilityor in relation to water availability.

Cotton production growth has been impressive in francophone West Africa over the last couple of decades. The very success of cotton production in West Africa is generally assumed to lie in the integrated character of the filière approach, adopted by almost all national governments with support from international donors.

Major threats to cotton production in West Africa are inherent in the uncertain outcomes of the liberalisation and privatisation processes and in the socio-economic and ecological sustainability of cotton growing. On the one hand, the institutional framework of the national cotton sectors will to a large extent determine the trust farmers have in the production and marketing systems put in place. On the other hand, land degradation poses a serious threat to the socio-economic sustainability of cotton as a crop and to the sustainability of the cotton-based production systems at large. Input credit loans currently make up for approximately 30 % of total seed cotton revenues and are likely to increase in the near future, in view of the reduction of state subsidies, increases in the world market price for oil and pesticide resistance built-up by cotton insect pests. Lower, and more variable rainfall could constitute an additional threat to what has become an economic lifeline for at least part of the West-African drylands.

Chapter 11

A LITERATURE SURVEY ABOUT RISK AND VULNERABILITY IN DRYLANDS, WITH A FOCUS ON THE SAHEL

Kees van der Geest and Ton Dietz

Abstract: While focusing on recent scientific literature about the Sahel, we present an overview of conceptual advances in understanding risk and vulnerability in dryland societies. The unreliability of rainfall and the seasonality of rainfall, agricultural activities and economic and social life as a whole have created the necessity to cope with vulnerability and stress. A central concept to understanding vulnerability is entitlement, but this is combined with insights from the empowerment approach, the political ecology approach, human ecology and political economy, creating a 'causal structure of vulnerability'. However, people's responses can be very different, based on different sensitivity and resilience. Incorporating concepts like insurance strategies, coping strategies and adaptation, a conceptual framework of farm household vulnerability is presented which can be used as a tool to study dryland societies like the ones in West Africa. Those who wish to predict climate change can learn from recent experiences in the region during adverse years, which experiences were the basis for most of the studies reviewed in this chapter.

1. INTRODUCTION

This book deals with the impact of unreliable rainfall, drought, seasonality of rainfall and climate change on rural people's food and livelihood security. It therefore touches on the current scientific debates on global climate change and its local and regional impact, the influence of climatic variability on rural people's livelihoods, the development of Early Warning Systems (EWS) against famine, agricultural intensification, livelihood diversification; migration and remittances; and the functioning of a 'moral economy'.

A.J. Dietz et al. (eds.),
The Impact of Climate Change on Drylands: With a Focus on West-Africa, 117–146.

The main focus of the research, however, is the debates on (1) rural people's vulnerability to hazards, like droughts and floods, (2) how rural people cope with such hazards and (3) how rural people adapt their livelihoods to changing conditions. In this chapter, a reconstruction of these debates will result in a categorisation of three different concepts of responses: insurance strategies, coping strategies and adaptive strategies (or adaptation). These concepts will be combined in a conceptual framework for studying 'farm household vulnerability and responses to normal opportunities and constraints, unusual events and changing conditions'.

We have avoided the philosophical debate about the role of human beings on Earth and their responsibilities toward other beings and 'nature' in general. We have investigated rural people's struggle to survive in a harsh environment without taking into account the well-being and survival of other creatures inhabiting the Earth. Virtually all research in this field has a clear anthropocentric (as opposed to eco-centric or bio-centric) perspective. In this field of research, nature is no longer seen as valuable in its own right (Sachs 1999: 58). Nature is seen instead as a set of resources that human beings seek to manage and utilise for their own survival and well-being. Escobar (1995) refers to this transformation as "the death of nature and the rise of environment."

In this chapter a start will be made to the debate about people's vulnerability to hazards in general and climate related hazards in particular. This will be followed by an outline of theory on rural people's strategies to offset risk and to pursue food and livelihood security in good times and in bad times. We will conclude this chapter with the aforementioned conceptual framework. The scientific debates about some contextually important issues, such as land degradation, agricultural intensification, livelihood diversification, migration and remittances and moral economy will not be dealt with in this chapter. The theory in this chapter is mostly focused on sub-Saharan West Africa and the Sahel in particular.

2. RAINFALL VARIABILITY: UNRELIABLE RAINFALL AND SEASONALITY

It is not the low *average* level of precipitation that makes an area drought-prone. It is rather the inter-annual *variability* of rainfall that causes dry years (Kemp 1994:42). This becomes clear when we take into account the difference between aridity and drought. Aridity results from a low average rainfall and is a permanent feature of a region's climate (*ibid*: 41). Drought, on the other hand, is a temporary deficiency of rainfall significantly below the normal or expected amount in a year, season, or month. The higher the rainfall variability, the higher the chance of receiving significantly under-average rainfall and thus the higher the risk of a meteorological drought that can evolve into an agricultural drought. When we talk of rainfall variability, we have to be more precise. Generally, there are three types of rainfall variability : spatial variability, inter-annual variability and intra-annual variability or seasonal concentration (Van Schaik & Reitsma 1992: 22-23).

Spatial variability concerns the differences in rainfall received between places, either structurally or proximately (in a given year). Spatial variability is high when major differences occur between places that are relatively near to each other. When two nearby villages are separated by a mountain range, one can expect structural differences in precipitation and thus high spatial variability. The village on the weather side will be wetter than the village on the lee side. In the absence of mountains, the amounts of rainfall can still vary greatly over short distances. This can result in different annual drought risks within a small area and this has consequences for the use of agro-climatic information to predict stress in agricultural production and for designing effective Early Warning Systems (EWS) against famine. High spatial variability has a positive side in terms of coping with food stress. If crops fail in one village due to drought, but neighbouring villages harvest well, part of the food gap can be filled by inter-village transfers (see Toulmin 1986: 65). Moreover, food prices are less likely to increase as sharply as in the case of a region-wide crop failure. This makes it easier for affected households to purchase food. There is a negative correlation between spatial variability and mean annual rainfall. "If mean annual rainfall is low, spatial variability tends to be high" (Foeken 1989: 9).

Inter-annual variability is the annual deviation from a long-term average, or the difference in rainfall between years. In statistical terms, inter-annual variability of annual rainfall is the standard deviation of annual rainfall divided by the average annual rainfall multiplied by 100%. The analysis of inter-annual variability is usually limited to a comparison of *total annual amounts* of rainfall in different years, while the year-to-year variation in the rainfall distribution is neglected (for example: the onset and offset of the rainy season, the occurrence of dry spells and excess rainfall, the number of rainy days, etc.; see e.g. Van Schaik and Reitsma 1992: 23). This is strange because it is the year-to-year variation in the distribution of rainfall that exposes rain-fed agriculturalists to uncertainty and risk. The analysis of inter-annual variability should therefore not only include the annual amounts of rainfall, but also the distribution of rainfall. Foeken (1989: 9) indeed highlights the importance of analysing the inter-annual variability of *monthly* rainfall.

There is a negative correlation between average annual rainfall and inter-annual variability of annual rainfall (Ruthenberg 1980: 22 and Foeken 1989: 9). In arid regions, inter-annual variability amounts to more than 50%, while in semi-arid regions the figure is usually around 30%. In sub-humid regions, it is less than 30% (Van Schaik & Reitsma 1992: 23). Higher average rainfall does not automatically mean lower inter-annual variability in total rainfall, however. In agricultural drought risk assessment, inter-annual rainfall variability is more important than average rainfall conditions (Kemp 1994:42).

Intra-annual variability or seasonal concentration refers to the distribution of rainfall *within* a year. It would be zero if every day - or month or whichever time unit is used in the analysis – were to witness exactly the same amount of rainfall (Foeken 1989: 7). In most of the semi-arid and sub-humid regions of sub-Saharan West Africa, the rainfall pattern is uni-modal, i.e. rainfall is concentrated in one wet season in which the rain-fed farming activities take place, leaving the dry season for other activities. This means that farmers can only harvest once a year, making the period to bridge between two harvests rather long and concentrating risk in one instead of two harvests. The months before the harvest are often difficult for farmers because food stocks run low and consumption has to be reduced while hard agricultural work has to be carried out (Dietz & Van Haastrecht 1997: 51). The seasonal concentration of rainfall gives rise to a seasonality in the agricultural cycle, in labour demands, food availability, food prices, the prices of consumer goods and labour, health, births, deaths (Dietz 1991: 86), celebrations and migration patterns.

In sub-Saharan West Africa, the length of the rainy season decreases as one moves from south to north. The rainy season coincides more or less with the summer in the northern hemisphere's temperate regions. This pattern of rainfall results from the annual north-south shift of the Inter Tropical Convergence Zone (ITCZ). This is the area where a dry, continental air mass and a moist, tropical air mass converge. The northward shift of the ITCZ brings moist and relatively unstable air from the ocean and causes precipitation. Droughts in West Africa have been associated with the failure of the ITCZ to penetrate northward as far as is usual (Kemp 1994: 48), but this phenomenon cannot explain all West African droughts (*ibid*: 66). It can certainly not explain the high spatial variability of droughts.

Without seasonal concentration, crop production would be impossible in many areas because an even distribution would mean that the monthly rainfall throughout the year would not be sufficient to sustain plant growth in any period (Van Schaik & Reitsma 1992: 23). Intra-annual variability can pose problems to farmers when it is so high that too much rain falls in a short period while rainfall during the rest of the year is insufficient for crops to fully develop. This is actually the case in some years. But again, we have to distinguish between inter-annual and intra-annual variability. The analysis of rainfall data in our Sahelian case studies suggests that the *average* seasonal concentration of rainfall does not pose problems for farmers. The problem lies in the fact that the distribution of rainfall varies from year to year. In other words: it is the inter-annual variation in the distribution of rainfall that causes agricultural stress. While the average seasonal concentration is favourable for rain-fed agriculture, there are years in which the rain is concentrated too much into one or two months and/or is interrupted by detrimental dry spells.

We have tried to indicate that the real problem for the farmer is the unreliability of rainfall, caused by inter-annual variability of both total amounts and distribution of rainfall. Erratic rainfall makes agricultural planning very difficult. Every year, before the farming season, decisions have to be made concerning crop mix, sowing moments, seed varieties, the location of fields, the application of manure and the seedbed type without knowing when the rains will start, how intensive the rains will be, how long the rains will continue and whether or not the rains will be interrupted by dry spells. This makes farming in areas with high inter-annual variability a risky enterprise (Van Schaik & Reitsma 1992: 25). Some farmers in the research area even compared farming with gambling or lottery staking. Similarly, Watts and Bohle (1993: 64) quote a Nigerian colonial officer who talks of the "annual lottery of the harvest". Apart from climatic risk however, he was referring to other factors that create uncertainty.

3. THEORIES ON VULNERABILITY

Unreliable rainfall poses production risks to farmers and other occupational groups and it has the potential to trigger off disasters. For many years, it was assumed that natural hazards - many of them associated with the climate and weather - *caused* natural disasters among human populations. It is now widely recognised that natural hazards do not necessarily lead to disasters (Cannon 1990: 1). A drought does not have to result in a famine. Two earthquakes with the same intensity in two different places can cause high mortality in one place and only small material damage in the other. When an area is affected by floods, for one family this can result in a tragedy from which it might take years to recover, while for a neighbouring family it might be a mere disturbance to daily life. Similar examples can be found in Blaikie *et al.* (1994: 47).

A natural hazard becomes a disaster when it hits vulnerable people (Blaikie *et al.* 1994: 22; Cannon 1990: 1). While the natural hazard acts as a *trigger event* for a disaster to occur, the underlying causes are to be found in people's vulnerability. These causes are often economic and political. Inequality is the basis of vulnerability (Ribot 1995: 121). It is not easy, however, to uncover the economic and political processes that make some people in a community vulnerable and others secure.

The concept of vulnerability needs further explanation. Vulnerability is often confused with poverty, but although poor people are usually more vulnerable than rich people, the two concepts are not the same. Vulnerability, to distinguish it from poverty, is "not lack or want, but defencelessness, insecurity and exposure to risks, shocks and stress" (Chambers 1989: 1). Vulnerability has an external side of exposure to risk and an internal side that consists of the inability to cope without damaging loss and the limited potential for recovery (Watts & Bohle 1993: 45). 'Loss' here can refer to becoming physically weak, economically impoverished, socially dependent, humiliated and/or psychologically harmed (Chambers 1995: 20). Vulnerability, and its antithesis security, are thus determined by the degree of risk exposure, coping capacity and recovery potential (Bohle *et al.* 1994: 39).

The above definition helps us to distinguish vulnerability from poverty. We talk of vulnerability as relative to a certain hazard (Blaikie *et al.* 1994: 59) and a certain consequence (Ribot *et al.* 1996: 16). People are vulnerable in different degrees to different hazards and consequences. Subsistence farmers are more vulnerable to food insecurity (consequence) caused by drought (hazard) than teachers. On the other hand, the latter group, , although generally less poor, may be more vulnerable to food insecurity triggered by hyperinflation because they rely more on the market for their food needs. Hence, hazards do not have to be natural. They can also be socio-economic and political. The difference between poverty and vulnerability lies in the external side of vulnerability: the exposure to risk. This is not to say that exposure to risk does not correlate with poverty. The poor are relatively more exposed than the wealthy because they often live in poor housing (exposing them to earthquakes and extreme weather events); because they often cultivate marginal lands (exposing them to agro-climatologic risk), etc. The internal side is more directly related to poverty. Inability to cope and recover is mainly caused by a lack of resources, alternatives and buffer capacity, associated with poverty.

Vulnerability is now a widely accepted concept in social science. Many scholars have written about it, although in very different ways and more often theoretically than empirically. The early theory on vulnerability has been developed to explain how famines have occurred (Watts and Bohle 1993: 47). The ultimate objective is to prevent future famines by 1) identifying which groups in society are vulnerable to different hazards on different moments; 2) developing Early Warning Systems against famine and 3) designing and implementing policy interventions that reduce vulnerability. According to Davies (1996: 38), however, "rarely are attempts made to monitor *how* people are vulnerable, *how* they are responding and hence what the most appropriate form of intervention might be." That is why conventional Early Warning Systems are rarely effective and capable of preventing famine. We should also bear in mind that vulnerability assessments are hypothetical and predictive (Blaikie *et al.* 1994: 59). Assumptions have to be made about which factors increase vulnerability and which factors create security. One can only 'prove' whether the assumptions were valid when the hazard strikes. What follows is a review of some important contributions to the theorisation of vulnerability.

4. THE ENTITLEMENT APPROACH OF VULNERABILITY

Sen's entitlement approach to hunger and famine is still very influential in vulnerability research. It argues that hunger and famines are often not caused by a decline in availability of food (i.e. production failure), but by a failure of people to manage food (i.e. exchange failures or entitlement failures). A person's entitlement is defined as the set of different commodity bundles (including food) that he can acquire by using his original bundle of ownership (his endowment) and the various alternative bundles he can generate by using this endowment (endowments include human resources such as the quality and quantity of labour and social claims over resources based on informal rights and networks; Swift 1989, in Bohle *et al.* 1994: 40). A person will go hungry if his entitlement set does not include a commodity bundle with enough food. Famines occur when large groups of people experience this type of entitlement failure (Sen 1987: 7-8). In normal years, entitlement to food and livelihood is gained through a combination of production (both primary and secondary), exchange (of cash, goods, services, sale of labour) and assets (including investments, stores and claims). These three categories can be labelled the endowments. Households with an adequate endowment portfolio are relatively secure. In times of food stress, coping strategies form an additional set of entitlements that are derived from the endowments (this formulation comes from Davies - 1996: 35-36 - who can be considered a representative of the *extended* entitlement approach). The original formulation by Sen (1981: 45-46, in Mortimore 1989: 88) was different. There, a distinction was made between a person's endowment (land, labour and other resources he owns) and what he can obtain in exchange for his endowment in a market economy: his exchange entitlement mapping.

The same accounts for adaptive strategies. Adaptation occurs when households have to respond to more permanent changes in their environment or to changes in the household's composition or entitlement base. To monitor food security, one should look at both sources of, and calls on, entitlements because both fluctuate over years and seasons. Calls on entitlements, or the expenditure side of entitlements, do not only arise from consumption. Investments and claims from other people or the state are also calls on entitlements (Davies 1996: 35-36). Consumption does not only concern food intake. Even the poorest households have essential non-food cash needs. In the 'two 80 percent rule', ultra-poverty is defined as those people eating less than 80 percent of dietary energy requirements, *despite spending more than 80 percent of their income on food* (Lipton 1986: 4). Most people will spend less on food and more on other needs. Sources of entitlements must also provide in these needs.

Food security is a sub-set within the pursuit of livelihood security. Contrary to what is often stated, poor households do not always pursue short-term food security. When they are faced with stress, they make a trade-off between satisfying immediate food needs and longer-term sustainability and survival. When the granary starts to become empty, a household can choose to sell a goat to buy food, but the household can also refrain from depleting certain assets if this endangers their *future* income opportunities. They can choose to go hungry in order to pursue future livelihood security (Swift 1993 and de Waal 1989, in Davies 1993: 60).

People are vulnerable when they face a high risk of entitlement deprivation. In early entitlements research, the most vulnerable people were considered those who were exposed to extreme market fluctuations and disturbances. The entitlement approach emphasises temporary shifts in entitlements and has been criticised to neglect the structural-historical processes that cause the unequal distribution of entitlements to resources. Another shortfall of this approach is the failure to explain what happens *after* a disaster, the recovery process (Watts and Bohle 1993: 47-48). According to Swift (1993: 4), the entitlement approach has neglected food production failures in favour of exchange failures or entitlement failures. By endeavouring to show that famines do not have to be caused by a decline in the availability of food, they have shifted emphasis away from production failures, while in fact there have also been many examples in which production failures and a decline in availability of food *did* cause famine. Crop failures still have the potential to cause serious food stress among cultivators (directly, as a production failure) and/or among market-dependent households because of skyrocketing food prices (indirectly, as an exchange failure). In the same vein, Leach *et al.* (1999: 232) warn against an excessive polarization of the distinction between *availability of* and *command over* food because in practice the two phenomena are often interconnected. More recent entitlement research, or extended entitlement research (e.g. Davies 1996) has incorporated these criticisms.

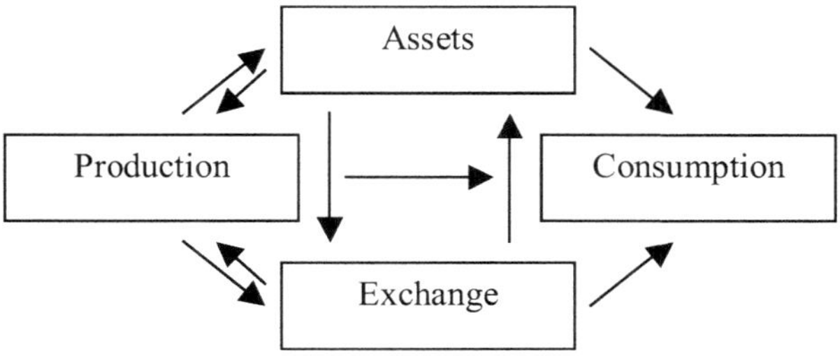

Figure 11.1. The role of assets and exchange as a buffer between production and consumption. (Source: Adapted from Swift (1989: 11))

Swift (1989) has developed a relatively simple model of four factors that determine immediate, short-term vulnerability (see figure 11.1). He acknowledges that other, underlying factors explain *structural* vulnerability. In this model, production (failures) and consumption are mediated by exchange (failures) and by *assets*. 'Exchange' concerns a household's position and participation in labour and commodity markets. Assets are defined in a broad sense to include investments, stores and claims. Assets form a buffer between production, exchange and consumption. Assets are created when a surplus in production and exchange is – willingly or unwillingly – used to invest, to build up stores or to put resources in the community. In times of dearth, these buffers can be converted into productive inputs or into food for consumption, either directly or through selling, buying, bartering and/or inter-household transfers.

5. "THE CAUSAL STRUCTURE OF VULNERABILITY"

The empirical analysis of vulnerability at household level is a relatively unexplored field. According to Watts and Bohle (1993: 45) "vulnerability as a concept does not rest on a well developed theory; neither is it associated with widely accepted indicators or methods of measurement." In their article, they endeavour to narrow the theoretical gap in vulnerability analysis by bringing together different approaches. This effort is an important contribution to the theoretical debate about vulnerability (see Ribot 1995). They do not, however, succeed in providing methods of measurement, especially at household level.

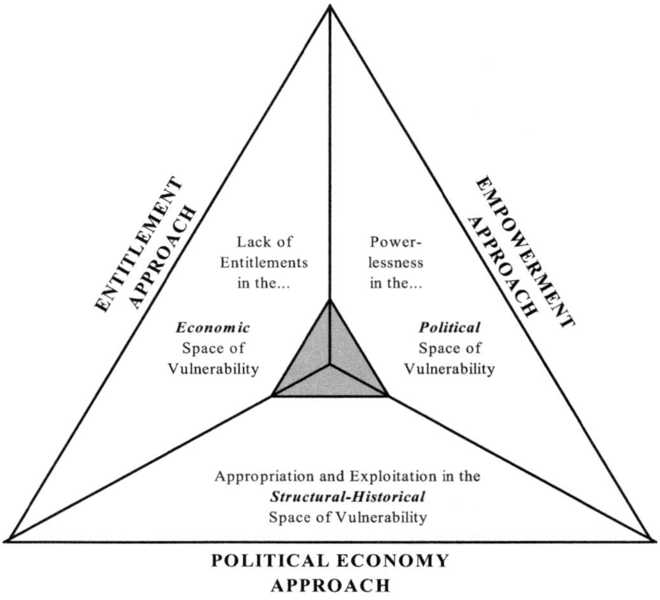

Figure 11.2. The causal structure of vulnerability.
Source: Adapted from Watts and Bohle (1993: 53, 54)

Watts and Bohle (1993) present what they call "the space of vulnerability: a causal structure of hunger and famine." They distinguish and review three complementary approaches to vulnerability that together form an 'analytical triangle': the entitlement approach, the empowerment approach and the political economy approach (figure 11.2). The entitlement approach has already been outlined above. The *empowerment* approach emphasises that limited command over food results from limited rights and power in three political domains: the domestic domain, referring to intra-household politics, the domain of work, referring to production politics and the public-civil sphere, referring to state politics (*ibid*: 49-51). The *political economy* approach uses a class perspective to provide a historical explanation of the structural patterns of entitlement and empowerment in a society. Commercialisation, proletarianisation and marginalisation are processes that increase inequality and vulnerability through the appropriation of surplus from direct producers (*ibid*: 51-52). Although there is some overlap between the approaches, each emphasises its own causality. Vulnerability is caused respectively by:

1. Lack of entitlements or command over food in the economic space of vulnerability;
2. Powerlessness in the political space of vulnerability and

3. Appropriation and exploitation in the structural-historical space of vulnerability.

The space of vulnerability is the intersection where these three causal powers determine risk exposure, coping capacity and recovery potential. The three bundles of causality are not mutually exclusive: they exist simultaneously and reinforce each other. Their relative weights can explain the distribution of food insecurity and security between different regions and social groups in the real world. Vulnerable groups in society are (1) the resource poor and those vulnerable to market disturbances, (2) the powerless and (3) the exploited. Vulnerable regions are (1) the marginal regions, (2) the peripheral/dependent regions and (3) the crisis-prone regions (*ibid*: 52-57).

Watts and Bohle (1993: 57-62) also present five historical case studies from South Asia and sub-Saharan Africa in which they attempt to link the *empirical* differences in the space of vulnerability to their model. They differentiate into class, livelihood system and gender and trace changes in type and degree of vulnerability over different historical periods. They also differentiate between livelihood system vulnerability in different years (crisis vs. normal) and seasons (slack vs. peak). Such an analysis can give very interesting insights into the evolution of vulnerability of different social groups over time, but their case studies draw on secondary sources: on researches that were not designed to analyse vulnerability empirically. Consequently, social groups are assigned different types and degrees of vulnerability in different epochs, based on interpretations of historical processes. Therefore, we see no improvement in terms of methods of measurement (we should mention, though, that Watts and Bohle do not claim that they *did* develop a method of vulnerability measurement. That was not their objective: "Each of the following five cases are necessarily sketchy but our intent is to trace, comparatively, the broad contours of vulnerability across space and time"; Watts and Bohle 1993: 57).

Empirical analysis of the political and structural-historical space of vulnerability differs fundamentally from an empirical analysis of people's entitlements to food and livelihood in the economic space of vulnerability. According to Adger (1999: 253) entitlements to resources are also difficult to measure because of their temporal and seasonal dimensions and intra-household transactions. In his assessment of vulnerability to climate variability and change in Coastal Vietnam, Adger uses income as a proxy for poverty, and poverty as a proxy for entitlement to resources. Adger's study will be briefly dealt with at the end of this section. According to Davies (1996), vulnerability analysis requires a careful dis-aggregation of poverty and a detailed insight into the way people gain access to food, both in normal years and in stress years. She further emphasises the importance of seasonal variation in access to resources across occupational groups. Contrary to Adger, she therefore does not use proxies (like poverty) for vulnerability.

Powerlessness and exploitation in the political and structural-historical space do not cause vulnerability *directly*. They cause some people to have a limited set of entitlements that in turn produces direct food and livelihood vulnerability. The three spaces of vulnerability do not work simultaneously: two spaces relate to underlying causes, while the limited set of entitlements causes immediate vulnerability. The latter type of vulnerability can be measured, but the underlying causes of vulnerability can only be recognised and described. It should be noted that a lack of entitlements is not only caused by powerlessness and exploitation. The entitlement approach has its own cluster of causes (explaining *temporary* shifts in vulnerability) to which these two clusters of more *structural* causes can be added.

If the objective of vulnerability analysis is to design effective early warning systems against famine and disruption of livelihood systems, it should focus on entitlements to food and livelihood, taking account of the changes over time in sources of, and calls on, entitlements. It should be borne in mind that a target group's entitlements to food in a given year (or *proximate* food vulnerability) can only be fully understood if more is known about medium to long-term changes in the baseline of entitlements.

If, on the other hand, *structural* reduction of vulnerability is the objective of a research project, then the conjunctural aspects of entitlements, though they have to be understood, can be given less attention with the primary focus being on the processes that can explain why some groups in society have a more limited entitlements set than others. Following Ribot's (1995: 120) line of thought: "empowerment is the ability to shape the political economy that in turn shapes entitlements." Combining the two policy objectives is most desirable and perhaps an imperative. If combined, vulnerability should be analysed by looking at both its immediate and its structural causes.

6. VULNERABILITY TO UNRELIABLE RAINFALL, SEASONALITY AND CLIMATE CHANGE

In this section, we will narrow the discussion on vulnerability by focusing on rural people's vulnerability to unreliable rainfall, seasonality, and climate change. Obviously, rural people face many other additional sources of risk. The emphasis lies, however, on unreliable rainfall because the purpose of the ICCD research project is to assess people's vulnerability and responses to climatic variability in order to inform policy interventions related to climate change. Moreover, unreliable rainfall is one of the principal sources or *the* principle source of risk for rain-fed agriculturalists in the Sahel. As a *hazard*, unreliable rainfall can either be a situation of insufficient rainfall or excess rainfall. Such a hazard will trigger food and livelihood stress when vulnerable people are affected. In many regions, unreliable rainfall is also a 'normal' characteristic of the natural environment and so is the seasonality of rainfall. In that sense, these climatic phenomena should figure, with other natural characteristics like low soil fertility, among the *structural causes* of vulnerability.

We started the section on vulnerability by stating that natural hazards do not *cause* disaster. Hazards become disasters when they hit vulnerable people. They act as trigger events. Natural factors do not come to the fore in the vulnerability theory, as outlined in "the causal structure of vulnerability". Social, economic and political factors act together to cause limited entitlements and therefore vulnerability. This notion combats physical determinism in studying disasters (Ribot 1995: 120). It might, however, introduce some kind of social or political-economic determinism (see Blaikie *et al.* 1994: 12) that overlooks the importance of the natural environment for rural people's livelihoods. Natural factors not only act as trigger events. In agricultural settings, natural factors largely determine people's entitlements to food and livelihood in 'normal' years, their prospects for creating a surplus and their ability to accumulate assets that reduce vulnerability. In rural areas, people's own food production is often still the major source of entitlements to food, although this is rapidly changing as a result of livelihood diversification and de-agrarianisation (see Bryceson 1997). Rural people who live in areas that are endowed with high quality natural resources and a favourable climate have a more reliable set of entitlements than people in risky environments with poor soils and limited wild natural resources. To summarise, natural factors can act as trigger events as well as the causes of vulnerability. To explain why some communities live in areas with low quality natural resources (marginal areas), underlying factors in the social, political-economic and cultural domain will have to be considered.

Ribot (1996: 16) argues that environmental (including climatic) variability and change should be incorporated in the social framework of vulnerability. "Vulnerability occurs at a junction of *physical* (our emphasis), social and political-economic processes and events. Hence, complete climate impact analyses must include this multi-causal perspective, placing climate as one causal agent among many". Reintegrating natural or environmental variables in the causal structure of vulnerability requires a careful distinction between natural factors as *causes* of vulnerability and natural factors that act as *trigger events*.

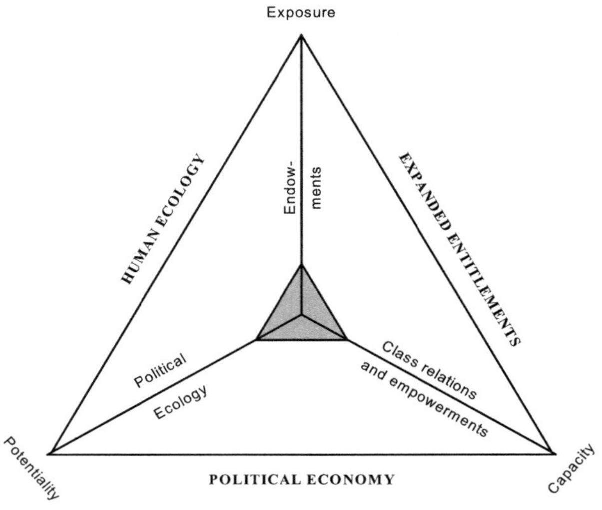

Figure 11.3. The causal structure of vulnerability after integration of the 'human ecology' approach. Source: Bohle *et al.* (1994: 39)

Land degradation as a long-term process and 'normal' climatic variability in semi-arid regions are not trigger events. They are natural factors that make people or regions more vulnerable because they put a structural constraint on farmers' productive entitlements and their capacity to accumulate assets.

In an article about climate change and social vulnerability, Bohle *et al.* (1994) present an adjusted causal structure of vulnerability (see figure 11.3). Following Dreze and Sen's (1989) incorporation of 'totality of rights' in the entitlement approach, the 'empowerment' and the 'entitlement' approaches are grouped together under 'expanded entitlements'. The 'human ecology' approach of vulnerability is now added to the analytical triangle. Human ecology refers to the relationship between *nature* and *society*. It focuses on understanding both the risk environment that vulnerable groups confront and the quality of their resource endowments, including their *natural* resource endowment. This adjusted model is more suitable for studying rural people's vulnerability to unreliable rainfall, seasonality and climate change than the one presented above. Bohle *et al.* (1994: 42) position vulnerable livelihood groups in their model according to the three causal powers that determine vulnerability. Refugees, cultivators, pastoralists, urban poor and wage labourers obviously face different types of vulnerability. Subsistence farmers are positioned in the human ecology space of vulnerability because they are vulnerable to climatic perturbations and because their livelihoods depend to a large extent on natural resources. A degraded or resource-poor natural environment exacerbates farmers' vulnerability.

Their vulnerability to food entitlement decline results from the low productivity of their livelihood system, rather than from a situation of exploitation or powerlessness. Within livelihood groups, certain social groups (women, the elderly, new settlers, etc.) can be additionally vulnerable because of limited rights and powerlessness (Bohle *et al.* 1994: 42).

Adger (1999: 251) has tried to assess vulnerability to climatic variability and change in coastal Vietnam. He distinguishes between *individual* and *collective* vulnerability. Individual vulnerability is determined by "access to resources and the diversity of income sources, as well as by social status of individuals or households within a community". Collective vulnerability of a nation, region or community is determined by "institutional and market structures, such as the prevalence of informal and formal social security and insurance, and by infrastructure and income." An additional characteristic of collective vulnerability is high inequality as regards access to resources. Adger's collective vulnerability is similar to what Cannon (1990: 5) has called the *social-protection* element of vulnerability. It concerns the level of 'preparedness' of the state and civil society to reduce the impact of a hazard. Adger uses poverty indices and the proportion of income dependent on risky (climate related) resources as quantitative indicators of individual vulnerability. GDP per capita and income inequality are used as quantitative indicators (proxies) for collective vulnerability. In his study, the quantifiable factor that is related to climate is the degree to which household income activities are directly dependent on the climate. Qualitative data was also gathered.

Adger found that some changes in the macro-economic and institutional environments (e.g. liberalisation, increased income equality and erosion of collective measures to protect against coastal storms) increased vulnerability. The rolling back of the state had had an ambivalent impact on vulnerability. It had reduced individual vulnerability due to higher incomes from commercial crops. It had, however, increased collective vulnerability because it had undermined existing institutional security-nets (Adger 1999: 266-267).

Adger's distinction between individual and collective vulnerability is very valuable because it conceptually separates *internal*, household-related variables and *external*, area-related or community-related variables (see also de Bruijn & van Dijk 1998: 1 and Dietz 1992: 39). The natural environment, the economic environment, the socio-cultural environment and the politico-institutional environment together determine the collective vulnerability or security of a certain area or community. In comparative vulnerability research between, say, agro-ecological zones or between central and peripheral regions, it is useful to distinguish between individual and collective vulnerability. There are secure environments and risk-prone environments. People living in politically marginalized areas with infertile soils and a virtual absence of alternative income opportunities are *collectively* vulnerable because these characteristics of the local environment affect everybody. Obviously, this does not mean that all the people in that area face equal vulnerability. The extent to which people are affected when a hazard strikes also depends on their individual vulnerability. Some people in a region or community may even benefit from the vulnerabilities of others in times of stress. There is a differential distribution of individual, household and livelihood system vulnerability *within* the area. This distinction between individual and collective vulnerability has its parallel in terms of risk. *Idiosyncratic* risks affect specific individuals or households (at the micro level). Examples are illness, cattle theft or loss of property and shelter in a fire. *Covariate* risks affect a whole village or region (at the meso-level). Examples are droughts, earthquakes and plagues (see Baas *et al.* 2001 and Nijzink 1999).

7. LINKING VULNERABILITY AND RESPONSES

As we mentioned earlier, we have not found many empirical studies on vulnerability concerning the Sahel. There is an important exception, however: Davies' (1996) research in Sahelian Mali. Davies' study can be considered a detailed extension of the entitlement approach. It is an important contribution because it combines a very strong theoretic framework with extensive data gathering. It also pays a lot of attention to the historical processes that are at the root of present-day vulnerability. Davies links responses to food entitlement decline with different types and degrees of vulnerability and she traces changes in livelihoods over the past three decades, following the Sahelian droughts of the 1970s and 1980s.

According to Davies (1996: 22-23), the notion that disasters are not caused by hazards, but by the underlying causes of vulnerability can result in a static approach to vulnerability. It does not consider how the productive capacities of livelihood systems vary from year to year and from season to season. In Davies' terminology: it looks at *structural* vulnerability (more or less permanent) while neglecting *proximate* vulnerability (changing from year to year). Similar distinctions can be found elsewhere: Kates and Millman (1990) write about "underlying processes" and "immediate causes" and Bohle *et al.* (1994) talk of "long-term structural baseline" and "short-term conjunctural condition".

Davies further distinguishes between *livelihood system* vulnerability and what Swift (1989) called *differential* vulnerability. Both can be either proximate or structural. Differential vulnerability concerns differences in vulnerability between households within livelihood systems and between individuals within households. It results from bio-physical characteristics (e.g. children under five, crippled people), type of household (female-headed, high dependency ratios), status in the community, wealth, etc. Livelihood system vulnerability results from the structural and/or proximate productive (in)capacity of livelihood systems. Conventional Early Warning Systems often monitored the structural differential vulnerability (Davies 1996: 23). Davies' study, on the contrary, focuses primarily on livelihood system vulnerability. Secondly, she analyses differential vulnerability to explain differences in vulnerability between individuals and households *within* livelihood systems. Table 11.1 shows the different types of vulnerability. An individual or household can fall into more than one of these categories. The more categories there are, the more intense the vulnerability is. A sick child of poor Sahelian farmers in a drought year, for example, falls in all the combinations of categories. Structural differential vulnerability arises from the fact that the individual is a child and the household poor, proximate differential vulnerability because the child is sick; structural livelihood system vulnerability because Sahelian farmers are structurally vulnerable and proximate livelihood system vulnerability because in a drought year the productive capacity of the livelihood system is even lower than usual.

Table 11.1. Nature and level of vulnerability.

Nature of vulnerability	Level of vulnerability
Structural vulnerability	Differential vulnerability
Proximate vulnerability	Livelihood System vulnerability

Source: Davies 1996: 30.

By looking at people's vulnerability from the point of view of the livelihood systems they are part of and by studying how people in different livelihood systems gain access to food in different seasons of good years and bad years, one can identify why and how different groups of people face shortages at a particular moment. With this understanding, Early Warning Systems against famine and policy interventions to reduce vulnerability become more viable.

To understand how livelihood systems become more vulnerable - or more secure - over time, we have to take account of two dimensions of vulnerability: *sensitivity* and *resilience*. Sensitivity concerns the intensity with which shocks are experienced. Resilience is the capacity to bounce back to a normal state after a crisis (see figure 11.4). Secure livelihoods are characterised by low sensitivity and high resilience. Vulnerable livelihood systems are highly sensitive and not very resilient. In highly sensitive livelihood systems, negative shocks easily cause food insecurity. In livelihood systems that are characterised by low sensitivity, drawing on available buffers can easily absorb the impact of negative shocks.

Households in resilient livelihood systems are able to bounce back to the level of livelihood security of before the shock, while in livelihood systems characterised by low resilience, it will take a long time before households recover from a shock. Some household will not fully recover and their livelihoods become more vulnerable. People will try to increase the resilience and reduce the sensitivity of their livelihoods. The livelihood systems that were analysed by Davies had experienced a reverse trend, however. In the past three decades, they became more sensitive, less resilient and thus more vulnerable (Davies 1996: 25-29).

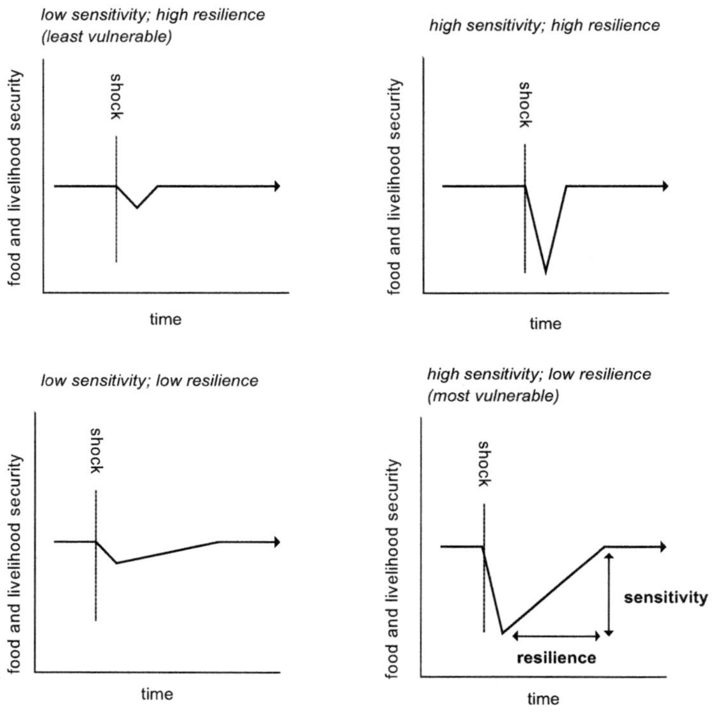

Figure 11.4. Sensitivity and resilience. Source: Adapted from Davies (1996: 27-28).

In Davies' framework, households in secure livelihood systems are able to meet food needs in most years through their primary production. Secure crop cultivators, for example, will be able to harvest at least a year's food supply. Income from secondary activities can be used to accumulate assets and to meet non-food consumption needs. In an exceptionally bad year, when primary production does not meet food needs, secure households can cope with this transitory food shortage by shifting emphasis to secondary activities, by engaging in tertiary activities, by drawing on assets and/or by temporary changes in consumption patterns. The following year, *if* conditions are better, secure households will be able to recover without a fundamental shift in their entitlement base.

In structurally vulnerable livelihood systems, on the other hand, households are not able to meet food needs through primary production in most years. Their food insecurity is chronic. They face an annual food-gap that has to be filled by engaging in secondary and tertiary activities. They have a limited capacity to accumulate, even in good years. In an exceptionally bad year, the prospects for coping are limited because their asset base is small and they already depend on secondary and tertiary activities in normal years (Davies 1996: 43). The following example will illustrate this difference between secure and vulnerable households.

In secure livelihood systems, seasonal labour migration, as an *unusual* secondary activity, can be an effective strategy to cope with food stress in an exceptionally bad year. It can supplement own food production, reduce household consumption needs and thus fill the food gap. In vulnerable livelihood systems, on the other hand, young men have to engage in seasonal labour migration every year to fill the food gap. Consequently, in an exceptionally bad year seasonal labour migration alone cannot offset stress. Among vulnerable households, seasonal labour migration is no longer a *coping* strategy. It has become permanently integrated into the normal cycle of activities. This is what Davies calls *adaptation*: permanent changes in livelihoods. "Proximate security is the ability to cope, whereas proximate vulnerability is the necessity for constant adaptation" (Davies 1996: 29). In the face of a particularly adverse event, vulnerable households have to tap additional sources of entitlement and/or they have to reduce calls on entitlements. Such additional responses are often erosive, i.e. they endanger the future livelihood security. In such situations, vulnerable households do not *cope*. They become more vulnerable in the face of a new cycle of stress (Davies 1996: 55-59).

In this section, we have used some insights from Davies' study in an attempt to link the concept of vulnerability with the concepts of coping and adapting. These latter concepts will be further elaborated in section 11.9. First, however, a third concept of responses has to be introduced, namely *insurance* strategies. In the last section of this chapter, we will bring the concepts of vulnerability, insurance, coping and adapting together in a *conceptual framework* for studying 'farm household vulnerability and responses to normal constraints, unusual events and changing conditions'. The three concepts of responses (insurance, coping and adapting) together form the overall livelihood strategies of households.

8. INSURANCE STRATEGIES AND COPING STRATEGIES

Besides coping and adapting, a third concept of responses concerns what people do to *avoid* food and livelihood stress in the - nearby or distant - *future*, that is before an adverse event hits them. This concept can be labelled 'insurance strategies'. In Davies' study, insurance strategies receive less attention than coping strategies and adaptation. Davies quotes Corbett's (1988: 1107) "sequential uptake of coping strategies" in which the first stage of coping is called insurance mechanisms. While for Corbett insurance *mechanisms* are an early stage of household coping strategies, Davies separates insurance *strategies* and coping strategies. She defines insurance strategies as "those activities undertaken to reduce the likelihood of failure of primary production" (Davies 1996: 47-48). In the case of crop cultivators, this would concern those activities undertaken to avoid - partial - crop failure. When presenting her empirical findings of insurance, coping and adaptation, Davies (1996: 239-246) seems to have abandoned this narrow definition of insurance strategies. Insurance strategies are now adopted more generally to offset potential risk of entitlement failure in the different entitlement bases (production, assets, exchange). This typically includes farmers' risk avoidance in cropping strategies, but it can also include agricultural intensification and extensification ("Agricultural intensification is increased average inputs of labour or capital on a smallholding (..) for the purpose of increasing the value of output per hectare"; Tiffen 1994: 29. Agricultural extensification is expansion of the acreage under cultivation, which is measured by increased farm size per capita). When farmers are confronted with declining soil fertility and deteriorating climatic conditions and the likelihood of not harvesting enough to meet consumption needs increases, they have to take countermeasures. The difference between insurance strategies on the one hand and insurance mechanisms on the other is that insurance strategies are adopted *before* an adverse event occurs and insurance mechanisms 'start working' when people are confronted with the impact of such an event. The former are preventive and the latter are curative.

Coping can be defined as "a short-term response to an immediate and unhabitual decline in access to food" (Davies 1993: 60). "Coping strategies are employed once the principal source of production has failed to meet expected levels, when insurance strategies have failed or are failing and producers literally have to cope until the next harvest" (*ibid*: 65). Similar distinctions between insurance and coping are made by Ellis (1998: 13) who speaks of 'ex-ante risk management' and 'ex-post coping with crisis' and by Dietz and Van Haastrecht (1997: 53-54) who distinguish 'preventive' coping strategies and 'curative' coping strategies.

What Corbett (1988: 1107) regards as insurance *mechanisms* are coping strategies that do not affect future sources of food and livelihood. Typical examples are the disposal of non-productive assets, the collection of wild foods, reliance on inter-household transfers and seasonal labour migration. In fact, even these actions affect future income sources. If you sell a bicycle or your goats, you cannot sell them again. What is meant here is that these actions do not seriously affect the productive capacity of the household. According to Corbett, insurance mechanisms are often employed to cope with *predictive* and non-severe risks. The difference between Davies and Corbett becomes clear here. For Davies, coping strategies are responses to *unhabitual* events, while in Corbett's terminology, people also cope with *predictive* and non-severe risks. The period over which these insurance mechanisms (as coping strategies) will be adequate depends on the extent to which the household has anticipated crisis during good years.

It is this *anticipation* or preparation of insurance mechanisms or coping strategies that we have added to Davies' (1996: 47-48) definition of insurance strategies ("those activities undertaken to reduce the likelihood of failure of primary production"). Strategies *could* in general be defined as "systematic or purposeful behaviour, using all available means to reach a long-term goal" (Dietz 1992: 37). In coping strategies research, however, the term 'strategy' is used to indicate that people have different *options*. They can, and indeed have to, make choices in the pursuit of food and livelihood security. In other words, they have 'room to manoeuvre'. This seems logical, but in a lot of research, it has not always been acknowledged (*ibid*). Corbett herself (1988: 1100), while reviewing some case studies of famine in South Asia, states that "farmers living in a drought-prone area will develop *self-insurance strategies* to minimise risks to their food security and livelihoods." However, she does not elaborate on the distinction between insurance mechanisms and insurance strategies and sometimes she uses the two interchangeably. Insurance strategies are thus defined by us as those activities undertaken to avoid future livelihood stress and food shortages. It should include those activities undertaken to reduce the likelihood of future entitlement failure altogether, rather than a failure of primary production alone. Investing in food stores, livestock, saleable assets, human resources and social networks are insurance strategies. Livelihood diversification is an insurance strategy because it enhances a household's portfolio of options to deal with crises. 'Playing the market' (buying and selling when prices are favourable) is an insurance strategy against exchange entitlement failures, etc. In this variety of insurance strategies, a division can be made between insurance strategies that are meant to:

1. Avoid the risk of primary production failure;
2. Diversify the sources of food and livelihood;
3. Create a buffer against future food and livelihood stress and

4. Offset seasonal shortages.

In the case of subsistence farmers, the first category of insurance strategies determines whether or not a household will be self-sufficient in its food production in a given year. The second category determines to what extent households are dependent on primary production. Well-prepared farm households can fall back on secondary and tertiary activities when primary production fails. The third category partly determines the success of coping strategies in times of crisis. When risk has not successfully been avoided and people are faced with food stress, they will start to depend on the buffer they have created in better years. Such a buffer not only consists of tangible assets but also includes social networks or social support mechanisms. A strong social network is an important asset for people who have to cope with food stress. The construction of buffers in good years is an insurance strategy. The depletion of buffers in bad years is an insurance mechanism or a coping strategy. This system of investment in, and exploitation of, buffers is relatively straightforward in the case of food stores and livestock. Investment in, and the utilisation of, social networks is, on the other hand, more complicated. Investment in human resources, like education, is an insurance strategy that does not so much follow this pattern of accumulation and depletion. Investment in education is a long-term strategy that often - but not always - continues in bad years. As a long-term insurance strategy, investing in formal education can be very rewarding, for example when it enables a son or daughter to find a secure, formal sector income, and when that means that the son or daughter becomes a dependable source of remittances.

In general, however, people try to increase their buffer capacity in good years. In crisis years, insurance mechanisms function as an early stage of coping behaviour. Some early coping strategies do not depend on insurance strategies and do not show the same pattern of accumulation and depletion, however. Examples are seasonal labour migration, collection of wild foods and reduction of consumption. It could, however, be argued that these activities also require a certain anticipation, preparation and/or experience to be successful. Seasonal labour migrants use networks and contacts to find relatively rewarding employment and cheap accommodation. If parents do not transfer knowledge of wild plants, new generations will not be able to effectively adopt wild food collection as a coping strategy. Similarly, it could even be argued that people can only reduce consumption levels during peak agricultural labour when they have steeled their bodies and souls against hardships. This requires a certain preparation or training, too.

It is confusing to call these early coping strategies insurance mechanisms as well, as Corbett (1988) does. It should also be noted that many households not only insure against food shortages in bad years, but also against expected, normal food shortages in the lean season (the fourth category of insurance strategies). There is an intra-annual cycle of seasonal insurance strategies and seasonal coping strategies, especially in households with a vulnerable livelihood.

When people are confronted with a certain hazard, production and exchange failures can sometimes successfully be avoided through risk-avoidance in primary production and through livelihood diversification. When these measures are not adequate, people will have to cope with the resultant food and livelihood stress. If it is not a very severe crisis, most people will be able to cope by drawing on the buffers that they have created, by finding additional sources of entitlement to food and/or by altering consumption patterns, without jeopardising future livelihood security. These coping strategies can be labelled 'non-erosive'. If the crisis is more severe, for example when an area is hit by drought in several subsequent years, or when several hazards strike simultaneously, the set of non-erosive coping strategies will soon be exhausted and people will have to take more drastic actions to combat the crisis. These actions can seriously affect people's future livelihood security and these 'coping strategies' can be labelled 'erosive' (De Waal 1989, in Davies 1996: 54). The label 'erosive coping strategy' contains a contradiction in terms, however. 'To cope' literally means: to deal successfully with something difficult: to manage. When a certain response to entitlement decline jeopardises a household's future food and livelihood security, this household is not 'coping'.

9. COPING STRATEGIES AND ADAPTATION

In the 1960s and 1970s, poor people were often depicted in social science literature as passive victims who were economically exploited and politically marginalized. In this view, they themselves could not do much to improve their lot or to ward against disaster. In the 1980s, it was realised that even very poor people have different livelihood *options* (Dietz *et al.* 1992: 37). Research started to focus on how some people managed to overcome extreme difficulties associated with recurrent drought and other stresses, while other people did not. Answers to this difficult question where found both in differences in vulnerability and in coping strategies. Many questions remain unanswered, however. In the case of coping with drought, Webb and Reardon (1992: 230) argue that most studies have tried to identify *general patterns* of coping rather than differentiating between agro-ecological zones, villages and types of household. In the 1990s, scientists have endeavoured to fill this gap in our understanding of how different types of households deal with stress.

It was often assumed that coping strategies show a sequential uptake and that increased knowledge about the sequence of uptake could inform Early Warning Systems designed to avoid famine. There are several caveats in the monitoring of coping strategies for early warning, however. Sequential uptake suggests that there are discrete stages of responses to food deficits. Each response (or cluster of responses) is adopted and exhausted before the household moves on to the next response. In reality this is not the case, as Devereux (1993: 54) argues, because different responses do not have the same 'time relevance'. Coping strategies can involve discrete, 'only once' events (e.g. distress migration), a series of discrete events (e.g. animal sales), or continuous processes (e.g. rationing of consumption). Although there may be a certain order in people's responses to stress, it should be realised that different responses occur simultaneously, as parallel processes rather than sequential events. Besides this practical critique on the 'sequential uptake approach', Davies (1993: 61) argues that coping strategies are too often seen as an "inherently good thing". Her concern centres on four points:

1. 'Coping strategies' is often used as a catch-all term for anything people do over and above primary productive activities." (..)

2. Focusing on coping strategies in situations of food stress can imply that people *do* cope and thus that food insecurity is a transitory phenomenon. (..)

3. While coping strategies may be useful in the short-term (..), they may be bad for longer-term development. Implicit in coping strategies is that the entire working life of subsistence producers is taken up in acquiring food, enabling people to stand still, but preventing them from moving ahead. A focus on coping strategies also hides the (increasing) need of rural producers to develop livelihood strategies, which will provide for greater numbers of people in the future.

4. Coping strategies are not necessarily economically and environmentally sustainable. (Davies 1993: 61).

Davies argues that people's coping strategies are not cast in stone: that they change over time. By focusing on coping strategies in the conventional way, structural changes in people's livelihoods and worsening conditions might be overlooked. To preserve the usefulness of the concept, both in monitoring proximate food insecurity and in strengthening people's capacity to avoid disaster, coping strategies have to be defined more narrowly and distinguished from adaptation. In Davies' definition, coping strategies concern people's short-term responses to *unhabitual* food decline. Genuine coping strategies are abandoned once the worst stress is over and households start to recover

(Davies 1993: 62). Whether a certain response should be labelled as 'coping' or 'adapting' depends on the intensity, timing, effectiveness and sustainability of the response and most of all on the reason why the household adopts this particular response (the motivation). This differs per household, per livelihood system, per region and over time.

Many farm households in dryland West Africa are confronted with food shortages almost every year. This usually occurs in the lean season, prior to the new harvest, when grain stores are running low. As mentioned above, people also have insurance and 'coping' strategies to deal with these predictive, seasonal shortages. In Davies' view, these livelihood systems are structurally vulnerable and the strategies to fill the annual food gap are not coping strategies, but adaptive strategies because they have become permanent features of livelihoods. Adaptation concerns permanent changes in the mix of ways in which food is acquired, irrespective of the year in question (Davies 1993: 60). It is a bit confusing, however, to apply the label 'adaptive strategy' or 'adaptation' to responses that people have already been adopting for (many) years to fill the food gap. Adaptation implies change. If we compare two static periods (the 'present' and the 'past'), it is clear that permanent changes in livelihoods are adaptations. In an ongoing, dynamic analysis of the ways in which people gain access to food and livelihood, an adaptive strategy in one year becomes part of the overall livelihood strategy in the following years.

Longhurst (1986: 27-33) distinguishes *seasonal* coping strategies and *famine* coping strategies and Campbell and Trechter (1982) distinguish between coping with expected and coping with *un*expected food shortages. It is acknowledged that the two are closely linked and that it is sometimes difficult to draw the line between these two categories. For crop cultivators, there is a continuum between a good harvest and a total crop failure. Given the same agro-climatic conditions, some households may face a non-severe seasonal shortage while other households in the same area may face real hunger conditions. Moreover, famine coping strategies are often an intensification of seasonal coping strategies. Similarly, Dietz (1991: 87) argues that: "a normal hunger season during a year with average rainfall and a severe hunger season as a result of drought are not two distinct categories. In practice it is a difference between few and many households coping with a crisis situation". He also argues that when a crisis is very severe it is confusing to speak of 'coping' strategies. 'Survival' strategies would be a more appropriate term (*ibid*: 88). Indeed, one can question whether we should speak of 'coping' when a household sells all its livestock at low prices to buy grains, while at the same time eating tree leaves, betrothing a daughter, going hungry and/or endangering next year's harvest with 'hunger trips' for temporary low-yielding wage work during the farming season. People lose. If not their lives, then at least (part of) their means of livelihoods, making them destitute. If coping means *successfully* dealing with difficulties, it is quite clear that people in these situations do not cope.

In Davies' study in Mali, vulnerable households are those that are not able to meet a year's food demand by primary production. Unlike most scholars, Davies does not regard livelihood diversification as a positive development *per se* because it is often a *forced* adaptation to deteriorated conditions. As Davies (1996) argues: "Diversification in the Sahel has followed a pattern of change that makes people more vulnerable. Activities that in the past were only carried out in periods of stress (as coping strategies) are now pursued every year, limiting the possibilities of coping in the next cycle of stress. They have become part of the normal livelihood strategies." In the Sahelian zone of Burkina Faso, Reardon *et al.* (1988: 1065) found that "almost all households rely to a certain extent on food purchases" and incomes are highly diversified in order to "insulate food consumption from broad swings in the local cereal sector", caused by climatic variability. They found that most households were production deficient. Nevertheless, the majority were food secure. They relied for more than 75% on non-cropping income, and because income opportunities were multi-sectoral they showed low levels of covariant risk. It seems that these households have quite *successfully* adapted to high levels of inter-annual rainfall variability, mostly through diversification.

In a review article about rural livelihood diversification, Ellis (1998: 2-3) summarises some of the conflicting findings in diversification research. Sometimes livelihood diversification is found to be a "deliberate household strategy" and sometimes it is an "involuntary response to crisis", depending on location and the economic status of households involved. Rural livelihood diversification is defined here as "the process by which rural households construct a diverse portfolio of activities and social support capabilities in their struggle for survival and in order to improve their standards of living" (*ibid*: 4).

To avoid confusion, it is necessary at this point to define adaptation or adaptive strategies. In Mortimore's (1989: 3) important study on adapting to drought in northern Nigeria, "adaptation is understood as a sequential process in which solutions to problems in turn become a part of the next problem." Mortimore does not use the concept of coping strategies because in the livelihood systems he studied, uncertainty was the norm rather than an aberration. Responses to drought in his study are short-term adaptations, whereas in the theory outlined above adaptation was a longer-term or even (semi-) permanent phenomenon. A definition of livelihood adaptation, provided by Davies and Hossain (1997: 5) takes into account that adaptation - like livelihood diversification - can be both positive and negative and distinguishes adaptation more explicitly from coping:

> ... livelihood adaptation is the dynamic process of constant changes to livelihoods which either enhance existing security and wealth or try to reduce vulnerability and poverty. Positive adaptation is by choice, can be reversed if fortunes change, and usually leads to increased security and sometimes wealth. It is concerned with risk reduction and is likely to involve an intensification of existing livelihood strategies or a diversification into neighbouring livelihood systems. (..) Negative adaptation is of necessity, tends to be irreversible and frequently fails to contribute to a lasting reduction in vulnerability. It occurs when the poor are forced to adapt their livelihoods because they can no longer cope with short-term shocks and need to alter fundamentally the ways in which they subsist. (Davies and Hossain 1997: 5).

In Davies' research area, adaptation occurred when coping strategies became permanently incorporated in the normal cycle of activities (Davies 1996: 35). However, as becomes clear from her definition of adaptation, this is not the only way in which adaptation can occur. People can also structurally improve their livelihood security, for example by changing from hoe-farming to plough-farming, by starting a dry season garden, by buying a sewing machine to become a dry season tailor, by geographic spread of social networks, etc. This type of adaptation can more generally be called 'livelihood strategies' (see Dietz *et al.* 1992: 38). Some – but not all – of these strategies are indeed intensifications of earlier coping strategies but if they help people to accumulate and diversify more it also makes their livelihoods more secure. Even though these activities can no longer be used as genuine coping strategies, a necessity to cope will occur less frequently. If these adaptive strategies result in, or are accompanied by, a drastic decline in own food production, households will become less vulnerable to climatic stress, but more vulnerable to market perturbations. Research at household level should be able to identify how the balance between primary productive capacity and livelihood diversity evolves over time.

Many adaptive strategies involve more efficient time and labour management, especially in areas with a marked seasonality in agricultural activities. Instead of 'idling' between one year's harvest and next year's land preparation, farm households engage in other income generating activities. This is not necessarily an improvement in the *quality of life* for everybody. An intensification of livelihood strategies will decrease leisure time and rest.

10. CONCEPTUAL FRAMEWORK OF FARM HOUSEHOLD VULNERABILITY

In this section, we will present a conceptual framework (see figure 11.5) that seeks to combine the theory on vulnerability and responses to stress as outlined in this chapter. The conceptual framework is an iterative model with an annual cycle. The units of analysis are individual farm households. For non-farm households and for farm households in areas with two growing seasons, similar models can be designed with shorter cycles of production and consumption. The model takes into account that part of the farm households in the research area have to deal with *seasonal* food shortages even in normal years: that a good number of farm households are not self-sufficient in their food production. Their entitlements to food through primary production are inadequate. To supplement their subsistence production, these farm households adopt *seasonal* coping strategies.

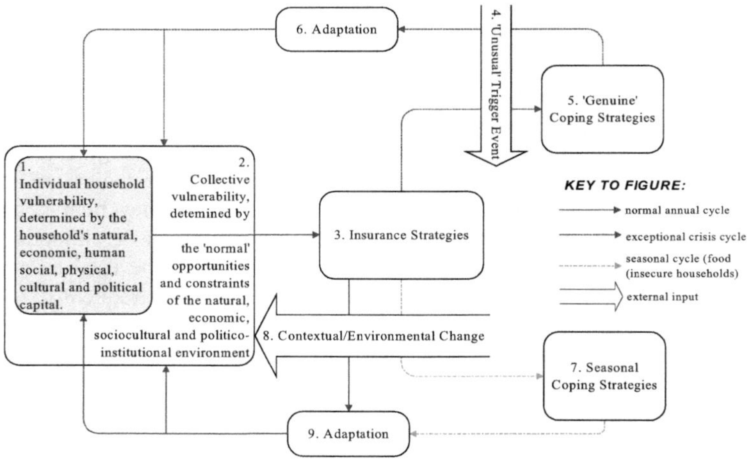

Figure 11.5. Conceptual framework: Farm household vulnerability and responses to normal opportunities and constraints, unusual events and changing conditions *(See colour section, p. 463).*

In years of unusual stress (e.g. in drought years), the cycle is broadened to include 'genuine' coping strategies in the narrow definition of Davies. Some trigger events or sources of food and livelihood stress concern idiosyncratic risk while other events concern covariate risk. In a given year, some households may be confronted with an unusual stress whereas other households will only have to deal with normal constraints.

The model assumes that it is possible to distinguish between normal opportunities and constraints, unusual events and gradually changing conditions and that it is possible to distinguish between seasonal coping strategies in 'normal' years and genuine coping strategies in 'exceptional' years. Davies' study shows that this is very difficult, but not impossible.

The conceptual framework is dynamic in the sense that the outcome of responses in one year determines the household's point of departure in the next year. In theory, it is possible to follow and quantify a household's livelihood situation throughout its own history and thus 'fill in' the model for a specific household. A more feasible possibility is to use the model to reconstruct a household's or an individual's livelihood history from the past to the present (to reconstruct its 'pathway'; see chapter 20). The conceptual framework also takes into account that the adopted strategies of the farmers in the research area have an impact on their environment. For an elaboration of the feedback of human activity to 'the environment', see Leach *et al.* (1999: 219):

> ... environments are constantly transforming and emerging as the outcome of dynamic and variable ecological processes and disturbance events, in constant interaction with human use. In other words, environmental conditions at any given time can be seen as the product of both ecological and social history. (..). Seen in this way, the environment provides a setting for social action but is also a product of such action. (..). As present practices build on the legacy of past ones, so the causality of environmental change may need to be seen as cumulative, sequential or path-dependent. (Leach *et al.* 1999: 219).

We will briefly explain how the model works. Block 1 contains the characteristics or the 'capitals' of individual households that determine their individual vulnerability at a given moment. Block 2 shows the different types of environment in which these households live. The opportunities and constraints of the environment determine vulnerability at the area or community level. Block 1 and 2 together determine the household's food and livelihood situation in a given year and how the household can insure (block 3) against unusual trigger events (block 4). If such an event occurs, the household characteristics, the opportunities and constraints of the environment and the adopted insurance strategies together determine how a household can cope (block 5). When coping strategies become recurrent and part of a household's normal livelihood, the household is adapting to modified conditions (block 6). But this is not the only way in which adaptation occurs. Adaptation is a constant process. People can also adapt their livelihoods as a response to improved opportunities. Adaptation can be forced or voluntary, but even when it is forced, adaptation itself does not make people more vulnerable. When households adapt their livelihoods to deteriorated conditions, they seek to minimise the impact of this deterioration. Sometimes adaptation is reversible. Sometimes it is not.

In this framework, adaptation concerns both structural changes in livelihoods and short-term shifts in the entitlement base. The former aspect of adaptation is derived from Davies' definition of adaptation and the latter aspect echoes Mortimore's approach of adaptation. In the latter sense, adaptation is not really an active strategy. It functions instead like a *balance* between a household's entitlements to food and livelihood (or resource base) in a given year and their strategies to cope with seasonal (block 7) or unusual (block 5) stresses.

An extremely simplified example can illustrate the mechanism: A farm household with five household members owns six goats and two pigs and it has two bags of millet in store. In the year under investigation, the rains are not very good and the household harvests seven bags of millet, which is not enough to feed the family until the next harvest. They need ten bags of millet to secure their food needs. To cope with the food gap, they eat their millet,which they had stored; they send a son on seasonal labour migration and they sell one pig and two goats. Five young goats are born. With the revenue of the animal sales, the household head buys four bags of millet. With the off-farm income of the son, they cover the non-food expenses. At the start of the following farming season, the household will have three bags of millet in store, nine goats and one pig, which is more than they had a year before. It is not a fundamental or permanent change in the household's livelihood, however. It merely concerns the balance of a year's production and consumption. If the balance is positive in several subsequent years, the household becomes more secure because they increase their buffer capacity. They may decide to invest the accumulated surplus in productive assets, like bullocks and a plough. This can be considered a fundamental change in their livelihood. If the balance is negative in one isolated year, the resource base decreases and the household will become slightly more vulnerable. No fundamental or permanent change in the household's livelihood occurs, however. If the balance is negative in several subsequent years, or if the balance is extremely negative in one year, the household's security endowment portfolio will substantially weaken and the household will become more vulnerable. If, in this state of vulnerability, the household is confronted with a particular adverse event, the household may be forced to sell its land and migrate to an urban centre. This would be a fundamental and possibly permanent change in the household's livelihood situation.

PART B

CASES FROM THE SUB-HUMID AND SEMI-ARID ZONES IN THE WEST AFRICAN SAHEL

Chapter 12

CLIMATE AND LIVELIHOOD CHANGE IN NORTH EAST GHANA

Ton Dietz, David Millar, Saa Dittoh, Francis Obeng and Edward Ofori-Sarpong

Abstract: The case study area of North-east Ghana consists of a densely populated zone in Upper East Region and a less densely populated area in Northern Region. It was selected as an example of a rural area with sub-humid conditions, a relatively high average population density and relatively severe land degradation. A major part of the chapter describes the lack of consistency between rainfall data and crop yield data and tries to explain farmers' behaviour as a constant adaptation to the rainfall situation, embedded in a volatile socio-economic environment. Farmers acknowledge climate change, and they, government institutions and non-governmental agencies have responded with more emphasis on water provision and land management improvements and with more emphasis on sector and geographical differentiation, including much more reliance on migration and remittance income.

1. INTRODUCTION

Northern Ghana is a sub-humid area, combining areas with high population densities and high reported levels of land degradation with scarcely populated areas, which have low levels of land degradation. It consists of three administrative Regions: Upper East Region (densely populated, around Bolgatanga and Bawku in particular), Upper West Region (pockets of dense population - around Wa, Nandom and Lawra - amidst low population densities) and Northern Region (mainly low population densities, with the exception of the area of the major town of Ghana's northern area: Tamale). The ICCD research was concentrated in the Bolgatanga area, but later extended to include the Nandom area in Upper West (as the start of a proposal to develop a Climate Change Preparedness Programme in Northern Ghana, financed by the University

A.J. Dietz et al. (eds.),
The Impact of Climate Change on Drylands: With a Focus on West-Africa, 149–172.

of Amsterdam). In the absence of useful longitudinal data at village or household levels, it was decided to organise two expert meetings (workshops) in collaboration with the University of Development Studies at Tamale, the University of Ghana at Legon and a local NGO, CECIK.

Most of the ICCD research activities took place in the so-called Bolgatanga cell, an area between 10° and 11° North and 0° and 1° West. It covers the eastern part of Upper East Region and the north-eastern part of Northern Region. Around 1960, the cell had an average population density of less than 50 inhabitants per square kilometre (although by that time, parts of the northern area already had densities far beyond that). On average, though, the density could still be regarded as 'low' compared to other drylands in the tropics. Currently, the Bolgatanga cell has between 0.7 and 0.8 million inhabitants, which means an average population density of between 60 and 70 inhabitants per square kilometre. This is high in relative terms. The part of the cell which is located in the Upper East Region has a very high population density with an average of 200 persons per square kilometre.

2. CROPS AND LAND USE DYNAMICS IN NORTHERN GHANA

Crops that are relevant in the northern parts of Ghana include maize, sorghum, millet, rice, groundnuts and cotton. Most of the crop (harvest area) data recorded by the FAO for these crops for Ghana as a whole can be attributed to the northern areas.

Maize has almost always been the most important grain crop of Ghana, in terms of hectarage (although more important in the centre-north areas and not in the upper-north areas). The maize area increased from between 200,000 and 300,000 ha in the 1960s to a level between 600,000 and 700,000 ha in the late 1990s. The year 1984 was an absolute peak year, with 720,000 ha. The years 1965 and 1978 were the lowest with less than 200,000 ha.

Sorghum and millet are the most important crops for the upper north areas. The sorghum area increased with ups and downs from 150,000 ha in the 1960s to more than 300,000 ha in the late 1990s. The area of millet production increased from 100,000 ha in the early 1960s to 180,000 ha in the late 1990s, but with higher figures in between (more than 240,000 ha in 1970, 1979 and 1989).

The rice (paddy) area increased from 25,000 ha in the early 1960s, to more than 105,000 ha in the late 1990s. This can be regarded as a steep increase. However, rice hectarage showed extreme fluctuations: up to 130,000 ha in 1977 and down to 40,000 ha again in 1983. Part of the area is irrigated.

The area of groundnuts production was 60,000 ha in the early 1960s. This dropped to half this hectarage in the mid 1960s and increased to fluctuating levels around 90,000 ha in the 1970s, 120,000 ha in the 1980s and more than 160,000 ha in the late 1990s. The seed cotton production area has always been much less, ranging from almost zero in 1960 to 25,000 ha in 1970, down again to almost zero in 1978, but increasing considerably until the 1990s (up to 50,000 ha).

Looking at the 'northern crops' as a whole we observe a steep increase in total area: from less than 600,000 ha in 1960 to more than 1.4 million ha in the late 1990s. In the total Ghanaian area of arable crops the 'northern crops' increased its share from one-third to half of the land use importance. There are probably two causes. The first is that arable land use in the north increased, following an impressive increase in the rural population. However, 'northern crops' also steadily 'moved south'.

Using estimated area and estimated production data, the FAO data also suggests changes (and fluctuations) in yield levels. These are shown in Table 12.1.

Table 12.1. Ghana, 'northern crops', 1960-1998, yield levels (in kg ha^{-1}).

Crop	Av yield 1960-1998	Lowest Yield (yr)	Highest Yield (yr)	Standard deviation yields (and %/av)	Yield level 1960s	Yield level 1970s	Yield level 1980s	Yield level 1990s
Maize	1155	430 ('83)	1648 ('98)	258 (22)	1200	1100	1000	1400
Sorghum	753	482 ('83)	1124 ('96)	161 (21)	600	700	700	900
Millet	665	477 ('65)	1020 ('96)	131 (20)	600	600	700	800
Rice	1234	590 ('82)	2075 ('98)	438 (35)	1100	900	1200	1800
Groundnuts	1033	670 ('91)	1697 ('80)	248 (24)	900	1200	1200	800
seed cotton	696	305 ('70)	1014 ('85)	208 (30)	500	600	800	no data

Source: FAO data; compiled by Maaike Snel and Jacoline Plomp, supervised by Marcel Put, in March 1999.

The FAO data suggests a number of interesting conclusions about yield developments:

- For all grains, the 1990s seem to be 'breakthrough years' with sudden and considerable increases in yield levels. There is no evidence of this improvement for groundnuts (decrease) and for cotton (no data yet);
- For maize and rice, crops were poorer in the 1970s than the 1960s, with stagnating millet levels, some improvement for sorghum and vastly improved yield levels for the cash crops groundnuts and cotton ;

- During the 1980s there was a further deterioration for maize, stagnation for sorghum and groundnuts and improved levels for millet and cotton;
- In terms of average yields of the grain crops for the period as a whole, rice leads, followed by maize, sorghum and millet. It is interesting to note that maize had better average yields than rice in the 1960s and 1970s;
- Looking at the standard deviation of annual crop yield data and comparing this with average yield figures, it is evident that millet are the least risky crop (in terms of yield fluctuations), closely followed by sorghum and, surprisingly, by maize. Groundnuts also exhibit quite comparable fluctuation levels. Both rice and cotton are 'gamble crops', with quite extreme fluctuations.

Increased hectarage and increased yield levels for all major grains have resulted in a considerable increase in grain production. In the early 1960s, total grain production reached 410 m. kg, or 60 kg cap^{-1} for Ghana as a whole. In the late 1990s, the total production had more than quadrupled, to 1790 m. kg, or 105 kg cap^{-1} for Ghana as a whole, despite a 250% increase in the number of people since 1960 (from 7 to 20 million inhabitants). This very positive result can mainly be attributed to the 1990s, when both yield levels dramatically improved and hectarages increased.

The FAO yield data allow a preliminary analysis of relatively bad years in productivity terms: all the years with a lower yield than the previous year are referred to as 'bad years'. A really bad year is when all six crops have lower yields than the previous year. A year in which all crops show improved yields compared to the previous year can be regarded as a 'good year' or at least (much) better than the previous year. The results are shown in Table 12.2. There have been three extreme years , with 1975 being the worst year , when all six crops had lower yields compared to 1974. However, 1980 and 1982 were problematic as well. In 1980 only groundnuts and in 1982 only cotton had better results than in 1979 resp. 1981. All other crops fared worse. The analysis also shows that the 1990s were much better than previous decades in terms of short-term yield deterioration.

Sorghum and millet have always been most important for the food security situation in the upper north area. In the 1960-98 period there were nine years in which both the sorghum and millet yields were lower than the previous year. This is an indication of food security problems. The years in question were 1965, 1968, 1975, 1980, 1982 and 1983, 1990, 1994 and 1997.

The northern part of Ghana can also be regarded as the most important livestock area of the country. Livestock production trends can also be found by using the FAO database. Total (commercial) meat production steadily improved in the 1960s, from a level of 60,000 metric tonnes to about 80,000 metric tonnes. After 1976 there was a significant increase to a level of 140,000 metric tonnes in 1984 and afterwards this level was maintained. However, the recorded beef and veal production in Ghana shows a slightly downward trend from the 1960s until now, although it can also be said that the period 1974-78 showed a tremendous downfall, after which production improved again to the current level which is still slightly below the high 1970-74 level of 22,000 metric tonnes. Both goat meat and mutton and lamb production improved steadily from 3,000 metric tonnes each in the early 1960s to 6,000 metric tonnes each at present. Other meat sources are chicken, guinea fowl and pigs.

Table 12.2. Yield data of six 'northern crops'* compared with previous years (1962-1998).

Decades Crops with lower yields than previous year	1960s	1970s	1980s	1990s**
all 6 crops	-	1975	-	-
5 crops	-	-	1980, 1982	-
4 crops	-	-	-	-
3 crops	1962, 1965, 1967, 1968, 1969	1970, 1976	1983, 1989	1990, 1991, 1994
2 crops	1963	1972, 1974, 1977	1981, 1984, 1988	1997
1 crop	1964, 1966	1971, 1973, 1978	1986, 1987	1992, 1996
no crops	-	1979	1985	1993, 1995, 1998

* Maize, sorghum, millet, rice, groundnuts, cotton.

** Cotton yield data is missing for the period 1993-1998, so data is provided on only five crops.

Source: calculations, using FAO data compiled by Maaike Snel and Jacoline Plomp, supervised by Marcel Put, in March 1999.

Commercial milk production increased from a level of 10,000 metric tonnes in the early 1960s to 24,000 metric tonnes currently. There was a steady increase with the exception of a severe crisis between 1974 and 1980. The Ghanaian production of meat and milk combined remained rather stable per capita, at a level of 10 kg cap^{-1} per annum.

Table 12.3 is a summary of the production data. If we examine the production variability (standard deviation divided by the mean for the period as a whole) it becomes clear that beef and veal production was the most stable livestock food, closely followed by goat meat, mutton/lamb and milk. Total meat production showed the largest fluctuations (because of chicken, fowls and pigs).

Table 12.3. Livestock production in Ghana, 1961-1998 (in metric tonnes).

Product	Average Production	Minimum (yr)	Maximum (yr)	Standard Deviation (and SD/av)	Production level 1960s	1970s	1980s	1990s
Meat Total	107,055	61,284 ('62)	143,923 ('98)	31,610 (30%)	70,000	100,000	120,000	140,000
Beef & Veal	19,213	12,995 ('78)	23,000 ('69)	2,661 (14%)	22,000	17,000	18,000	20,000
Goat Meat	4,295	3,040 ('65)	5,938 ('98)	821 (19%)	3,000	4,000	4,000	5,000
Mutton & Lamb	4,910	3,102 ('62)	6,545 ('98)	1,102 (22%)	3,000	5,000	5,000	6,000
Milk	18,621	9,750 ('61)	23,920 ('95)	4,052 (22%)	14,000	16,000	19,000	23,000

Source: FAO data; compiled by Maaike Snel and Jacoline Plomp, supervised by Marcel Put, in March 1999.

Food security not only means the capacity to feed the population with food that is produced in the country itself. Food security can also be facilitated by food import. Food imports can come through food aid or trade. In Ghana the food aid component of food imports has mostly been small. According to food aid data for the period after 1970, the average cereal imports were about 80,000 metric tonnes. There were peaks in 1977-80, 1983-85, 1987, 1991 (an absolute peak of 200,000 metric tonnes) and 1992-93. Cereal aid mainly consisted of wheat, wheat flour and rice, although coarse grains were included as well. After 1993 cereal aid gradually came to an end.

The US dollar value of total agricultural imports in Ghana has risen steeply: from a level of 50 m$ in the early 1960s to 350 m$ currently. The increase mainly started in 1986. However, registered livestock imports decreased considerablyfrom a level of 120,000 annual cattle imports in 1961 to almost zero after 1975 ($ value: from about 8 m$ to less than 1 m$). The import of goats decreased from 130,000 per year to almost zero after 1977 (value dropped from 1.5 m$ to less than 0.2 m$) and the import of sheep from 100,000 to less than 10,000 after 1978 (value dropped from 1.2 m$ to 0.6 m$). Most animals used to come from Burkina Faso, but after the 1974 drought the livestock trade petered out (at least the registered trade). Nowadays most agricultural imports consist of food grains, but there is no FAO data.

Between 1960 and 1998, the consumption of food in Ghana as a whole (per capita) changed as regards composition, with grains becoming a much more important element of the average diet and hence much more important for the 'northern crops' (and for grain imports):

- maize consumption increased from a level of 20-25 kg cap^{-1} in the early 1960s to 35-40 in the 1990s, with initial peaks in around 1970 and after 1984;

- rice consumption is on the increase, from 10 kg cap^{-1} until 1990 to between 20 and 30 kg cap^{-1} in the 1990s;
- millet consumption was rather stable, with 8 kg cap^{-1} (peaks in the 1970s); sorghum consumption increased (from 9 kg cap^{-1} in the early 1960s to 13 kg cap^{-1} nowadays, after rather low levels of 6-7 kg cap^{-1} in the 1980s;
- groundnut consumption first increased from a level of 2 kg cap^{-1} to between 4 and 6 between 1970 and 1990 and then dropped to a level of between 2 and 3 kg cap^{-1} in the 1990s
- meat consumption was quite stable at 10 kg cap-1 (but less for beef, veal, goat, and mutton), while milk consumption deteriorated (between 6 and 12 kg cap-1 before 1978 and between 2 and 6 kg cap-1 afterwards).

3. AN ANALYSIS OF CLIMATE DATA FOR THE BOLGATANGA AREA

According to UNESCO's aridity assessment (based on data for the period between 1930 and 1960), the whole Bolgatanga cell belongs to the sub-humid zone. However, it can be deduced from rainfall data for the years between 1960 and 1997 that the period 1982-1986 (or maybe even 1974-1988), in particular, had a semi-arid climate. Noteworthy drought risk years were 1962-63, 1967, 1970, 1977, 1981, 1984, 1990, and probably 1995 and 1997.

Ghanaian climatologist Prof. Ofori Sarpong carried out a survey for all data that exists on rainfall in northern Ghana as a whole, for the period 1900-1993. The average rainfall for this area (which is considerably bigger than the Bolgatanga cell and which includes rainfall stations to the more humid south as well) for the century as a whole was slightly higher than 1200 mm. The five-year moving average (see Figure 12.1) shows interesting and quite substantial fluctuations. The century started with a bad rainfall situation between 1900 and 1915, with droughts in 1904 and 1912. The late 1910s were good, with an all-time annual peak of close to 2000 mm in the year 1917, followed by drought again in 1918-1920, The period 1920-1935 was more or less average, followed by a dry period between 1935 and 1945. The period 1945 to 1975 was very good on the whole, with the exception of 1957-1960 (1960 being extremely bad). After 1975, the rainfall situation deteriorated a lot, reaching averages of 200 mm below the level for the century as a whole. However, the situation improved somewhat after 1985.

The climate is influenced by the convergence of two air masses, the boundary of which is called the Inter-Tropical Convergence Zone. The one air mass is continental and dry and is associated with the Azores High, which extends over the Sahara and gives rise to the north-easterly Trade Winds of 'harmattan'. In the dry season almost all of Ghana comes under the influence of these winds. The other is the moist tropical air mass, which originates from the South-Atlantic anticyclone and is associated with the moist south-easterly winds, which bring copious rainfall to Ghana during the rainy season. The boundary fluctuates north-south. In the Bolgatanga cell, annual rainfall averages about 850-1100 mm in the northern part and 1000-1200 mm in the southern part. Figure 12.2 shows the 5-year moving average for Bawku in the driest part of the research area. Rainfall is concentrated in the period from April to October, during which 95% of the precipitation occurs. The Monsoon rains reach their peak levels in August with about 1/4th of the annual rainfall. Rainfall intensity can be high during this period, with rainstorms causing severe erosion of unprotected soils.

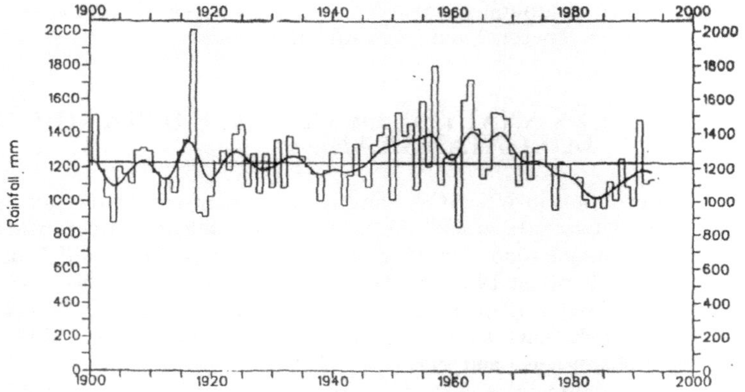

Figure 12.1. Rainfall in northern Ghana, 1900-1993.

Temperatures are high throughout the year, averaging about 28°C (with a range between 26-32 °C). An analysis of temperature trends in northern Ghana shows a gradual increase (up to 1 °C) during the 20th Century. During the dry season from November to April day and night temperatures are high, up to 46 °C during the day in the shade (with the exception of night temperatures in December and January, which are relatively low, e.g. 15 °C). High temperatures, dry conditions and (harmattan) wind encourage bush fires. Relative humidity fluctuates considerably, from less than 35% during the dry season to more than 70% during the rainy season. Diurnal fluctuation is substantial as well, with the highest humidity recorded in the morning, when it often exceeds 90% from July to September, while the lowest level is recorded in the late afternoon. The length of the growing period for rain-fed crops is more than 80 days in the southern part and less than 60 days in the north-eastern part.

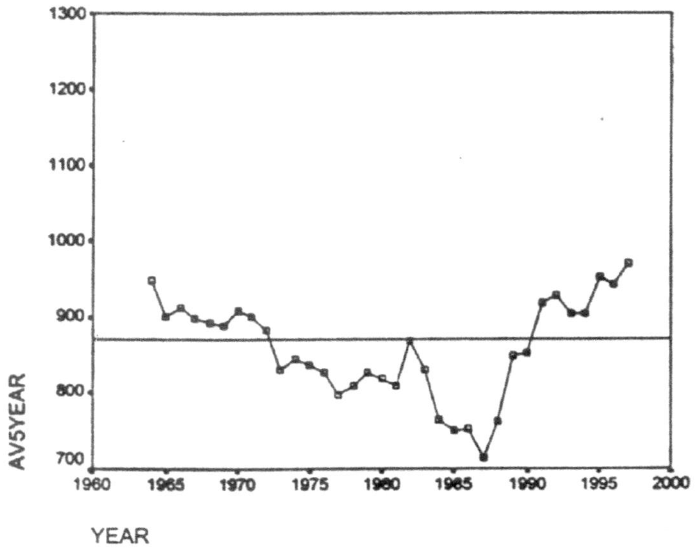

Figure 12.2. Rainfall in Bawku, 1960-1997, 5-year moving average.

Long-term climate data to the present day exists for three relevant stations: Gambaga in the central part of the area, Bawku, just outside the north-eastern corner of the cell and Navrongo, just outside the north-western corner of the cell. Data from FAO's Agrhymet source is slightly different from data provided by ISRIC because reference years differ. For Gambaga the first source can be used to calculate an average annual rainfall (based on 1960-1997) of 966 mm. ISRIC gives 950 mm. In the rainy season (defined as five months, March-September) averages are 855 mm resp. 813 mm. August is clearly the wettest month, followed by September and July. For Navrongo, the FAO and ISRIC data gives more or less the same results: 966 mm on average and 840 mm in the rainy season (however: in the case of Navrongo, eight years are missing in the FAO dataset). For Bawku we only have FAO data, showing an annual average of 871 mm, of which 733 mm in the rainy season. Gambaga and Navrongo receive more rain compared to Bawku. There are also other rainfall stations in the area (e.g. Zuarungu, Bolgatanga, Nakong, Paga, Sandema, Wiaga, Binduri, Garu, Kugri, Vea), but for those centres only a few years are available and the Ghana Meteorological Services Dept. does not regard this data as very reliable. There was a period in which they engaged schools to carry out rainfall measurements, but during school holidays results became very unreliable (according to a meteorologist working in the area it was not always clear if they measured water or urine…). For Navrongo and Bawku (and for a rainfall station at the Vea Dam site, in the north-western part of the research area) we received data for the years 1994-1997 from the Ghana Meteorological Services Department at Bolgatanga.

Annual rainfall totals can be quite different from year to year. In Gambaga the lowest rainfall measured during the 1960-1993 period was 731 mm (in 1962 and in 1977) and the highest rainfall was 1222 mm (in 1991). In Navrongo, the extremes were 776 mm (1984) and 1272 mm (1973) and in Bawku 644 mm (1983) and 1118 mm (1989). If we combine the data for these three stations (although there is no data for some of the years in the case of Navrongo) we are left with an assessment of the rainfall differences between the years for the region as a whole. For 1994-97 we combine the data for Navrongo, Bawku and Vea. Low rainfall years (or worse: periods) were 1961-62, 1964-65, 1972, 1977, 1981, 1983-87, 1990, 1992-93 and 1995 (see Table 12.4).

Table 12.4. Relatively good and bad rainfall years in Upper East Region, 1960-1997.

mm	Year
<700	1977
700-799	1983, 1984, 1985, 1990
800-899	1961, 1962, 1964, 1965, 1972, 1981, 1986, 1987, 1992, 1993, 1995
900-999	1966, 1967, 1970, 1971, 1973, 1974, 1975, 1976, 1978, 1980, 1982, 1988, 1997
1000-1099	1960, 1968, 1969, 1979, 1994, 1996
1100-1199	1963, 1989, 1991

Is there a rainfall trend in the area? The combined rainfall data for the research area as a whole shows an upward trend in the 1960s (a five-year average figure that moves from about 940 mm to more than 1000 mm in 1969) and a more or less continuous downward trend from 1000 mm in 1969 to slightly above 800 mm in 1985. In the second half of the 1980s the trend improves again and does so rather rapidly (to an average of 975 mm in 1990), after which year the trend moves slightly downwards again towards a level of 930 mm in the mid 1990s. The lowest point was reached in the 1981-'87 period, after which the 'Sahelian crisis' seems to have been over in many other parts of western Africa as well. Data for the period 1985-1997 certainly does not suggest a further deterioration of the rainfall situation.

Annual rainfall data is not the most relevant data to use for an analysis of agricultural (crop cultivation) risks. Data for the growth season is much more important and especially if it takes into account whether there are dangerous dry spells during the rainy season. For the ICCD project, a 'Drought Index' was developed with a six-point scale, from no (0) to extreme (5) drought risks for the cultivation of sorghum and millet. During the period 1960-'93 there were 13 years with no drought risks (index 0) in Gambaga, 12 years with very light drought risks (Drought Index 1), 7 years with light drought risks (Drought Index 2) and (surprisingly) 2 years with extreme drought risks (Drought Index

5). These two extreme years were 1962 and 1970. However, in 1962, the drought risk for Navrongo and Bawku was only very light (Drought Index 1) and light (Drought Index 2) in those two centres in 1970. The average drought risk assessment for Gambaga was 1.1, which is, on average, very light. In Navrongo the number of years with no drought risk (Drought Index 0) was much lower: only five, while there were 11 years with very light drought risks and 9 with light drought risks. There was one year (1967) with a moderate drought risk (Drought Index 3). On average the drought risk was only slightly above the one for Gambaga: 1.2 (but based on only 26 years of data). In Bawku the average drought risks were considerably higher: 1.8. Here there was only one year without drought risks, 16 years with a very light drought risk, 16 years with a light drought risk while one year (1980) had a moderate drought risk. Within the study region in most years the drought risk figures do not show a consistent pattern. If we consider an index of 2 or higher as a signal of drought problems (in relative terms) there are only six years in which all three rainfall stations show the same relatively problematic situation: for 1963, 1970, probably 1977, 1981, 1984, and 1990. Of those years only 1977, 1984 and 1990 also had a relatively low annual rainfall total.

4. DETAILED ANALYSIS OF THE RAINFALL SITUATION 1987-1997 AND OF THE VARIATION IN AGRICULTURAL PRODUCTIVITY: HOW TO EXPLAIN THE LACK OF CONSISTENCY?

For the years 1987-1997 'official' data exists on the agricultural production per hectare in Upper East Region, for five crops: millet, sorghum, groundnuts, maize and rice. We also have rainfall data, both annual and monthly, for 11 rainfall stations in this region and 2 in the nearby parts of Northern Region which enables us to calculate the drought risk assessment per station per year and for the area as a whole (although not far all stations data are complete).

The period covered starts after the bad rainfall period of the late 1970s and early 1980s. The years 1987-1994 are generally quite good, with 1990-91 a bit less so in Upper East (but not in Northern in 1991; rainfall there was quite bad in 1987 and part of 1988 , as well as 1992-93). In Upper East the period 1995-1997 is worse though, with Drought Risk Assessment figures of 2.6 for 1996 and 2.5 for 1997, in contrast to an average for the period of 1.4. On an annual basis, the rainfall situation was not at its worst during these two years though. The average rainfall annual total was lowest in 1990 (808 mm, with a Drought Index of 1.7). It was highest in 1989 (1158 mm) with the lowest average Drought Index (0.3)

Table 12.5. Bolgatanga study area, rainfall in the rainy season, 1996, in mm

Rainfall Station	May	June	July	August	September	Drought Risk Assessment
Upper East Region						
Bawku	139	149	71	257	261	3
Binduri	175	117	71	117	71	3
Garu	126	170	98	316	231	1
Kugri	82	175	70	258	73	4
Zuarungu	64	nd	nd	578	298	nd
Vea Dam	141	196	73	278	245	3
Navrongo	195	207	108	301	207	1
Paga	51	87	nd	nd	nd	nd
Wiaga	140	212	129	344	314	1
Sandema	71	124	nd	323	232	nd
Nakong	29	98	149	36	130	5
Northern Region						
Gambaga	141	196	73	278	245	3
Walewale	135	98	67	325	210	5 (?)
Average	115	152	91	284	210	2.9

Looking at the regional details, the large resemblance in some years is striking (e.g. 1987-1989 with Drought Index figures either 0 or 1), while in other years the location was a determining factor: in 1996 and 1997 the Drought Index was 1 in the best areas, but the Drought Index suggests an absolute crop failure (Drought Index 5) in some other parts, like Nakong in 1996 and Paga in 1997. The overall situation in 1996, (see table 12.5) was one of good rainfall in June (in some places even starting in May), a rather dangerous dry spell in most of the area in July and abundant rainfall in August, continuing in most areas into September. In Nakong - in the extreme north-western part of Upper East Region - the season started late, July was exceptionally good, but August exceptionally bad and most farmers will have experienced severe difficulties. In places nearby, like Navrongo and Wiaga, the situation was not bad at all though. In Kugri, in the extreme eastern part of UER, the month of July posed a problem, but thanks to surplus rainfall in June, the soil was probably moist enough for the crops to survive. The situation was the same in Walewale, in Northern Region. Here, surplus rainfall in June was probably insufficient , so farmers were faced with a high risk of crop failure, especially those farmers with their fields in uphill and sandy locations. In 1997 (see table 12.6), the overall rainfall situation was not very promising. In a place like Paga it was even very bad during all months.

Table 12.6. Bolgatanga study area, rainfall in the rainy season, 1997, in mm.

Rainfall Station	May	June	July	August	September	Drought Risk Assessment
Upper East Region						
Bawku	105	236	115	216	201	1
Binduri	82	187	99	169	210	1
Garu	125	238	92	233	186	1
Kugri	124	228	85	204	145	3
Zuarungu	101	231	76	153	225	3
Vea Dam	99	194	103	128	209	1
Navrongo	156	204	91	195	179	3
Paga	62	31	30	70	62	5
Wiaga	68	123	76	125	151	3
Sandema	89	159	73	148	131	3
Nakong	52	140	134	80	173	3
Northern Region						
Gambaga	99	194	103	128	209	1
Walewale	139	91	173	116	116	?
Average	101	174	96	151	169	2.3

Acknowledgement: rainfall data was provided by the Meteorological Services Departments in Bolgatanga and Tamale; drought risk assessment calculations were carried out by Dr Marcel Put, based on a computer model developed by Dr Sjoerd de Vos.

Tables 12.7 and 12.8 show the variability of crop harvests for five crops in the same period for Upper East Region as a whole and link the yield indexes of these crops to the Drought Index data. If the Drought Index were a good 'predictor' of yield levels, there would be a negative correlation between yield index data and drought risk assessment data. The correlation that is found, though, is completely spurious for all crops individually and for the five crops combined. On the basis of Drought Index data, the year 1996 could be expected to be the worst performing year. Instead, it was almost the best year as far as yield levels (kg ha^{-1}) were concerned, with the exception of groundnuts. The year 1997 was almost as bad in Drought Index terms, and it was indeed the case that the yield situation was poor (with rice as an exception). The best years in yield terms for individual crops were all years with a relatively high Drought Index: 1995 for millet and rice (Drought Index 1.7), 1996 for sorghum (Drought Index 2.6), and 1991 for maize (Drought Index 1.7). Only groundnuts had its best performance in a year with a low Drought Index (1987: Drought Index 0.7).

For millet and sorghum, which provide the bulk of food grains in Upper East Region, there have been three years during this period when yield levels for both crops dropped compared to the previous year, namely in 1990, 1994 and 1997. If we compare this with the data for (northern) Ghana as a whole, the same years were mentioned. The data on the 'bad years' in the last decade seems to be consistent. For Upper

East particularly, the situation in 1997 must have been rather dramatic: compared to the yield level in 1996 the millet yield was 44% less and the sorghum yield 34% less.

Table 12.7. Harvest estimates (index figures) in Upper East Region, 1987-1997.

Year / Crop	Millet	Sorghum	Groundnuts	Maize	Rice	Five crops
1987	76	77	140	88	63	89
1988	82	116	83	92	74	89
1989	101	110	87	64	120	96
1990	70	92	74	88	40	73
1991	72	82	99	153	95	100
1992	70	84	98	118	102	94
1993	142	99	113	112	98	113
1994	101	86	104	100	120	102
1995	158	113	111	115	144	128
1996	145	145	96	106	132	125
1997	81	96	98	59	108	88
Average in kg ha^{-1} (index = 100)	740	830	820	850	1670	982 (unweighted)

Source: Ministry of Food and Agriculture UER annual reports for the kg ha^{-1} estimates, based on expert assessment by MOFA field-level staff at district level, converted to an estimate at regional level by the Head of the MOFA at UER level. The staff making these judgements were changed continuously.

Table 12.8. Harvest indexes compared with Drought Risk Assessments in Upper East Region for five crops, based on data for 1987-1997.

Crop/ Drought Index	0.3	0.7	0.7	0.9	0.9	1.2	1.7	1.7	1.7	2.5	2.6
Year	1989	1994	1987	1988	1992	1993	1995	1990	1991	1997	1996
Millet	101	76	101	82	70	142	158	70	72	81	145
Sorghum	110	77	86	116	84	99	113	92	82	96	145
Groundnuts	87	104	140	83	98	113	111	74	99	98	96
Maize	64	88	100	92	118	112	115	88	153	59	106
Rice	120	63	120	74	102	98	144	40	95	108	132
All five	96	89	102	89	94	113	128	73	100	88	125

We may conclude that we are confronted with a situation that is difficult to explain (also see chapter 7): the expectations about the role the Drought Index could play as a tool in crop performance assessments appear to be illusions. What is even worse, if Drought Index data is so problematic when predicting harvest levels of crops, and especially for crops for which the Drought Index has been developed - in this case millet and sorghum -, it cannot function as a tool in harvest scenarios following climate change either.

So it is wise to brainstorm about the various reasons why the correlations between yield levels and Drought Index data are so spurious.

A workshop in Ghana in March 1999 (Bolgatanga) discussed the various reasons at length. A training session of CERES PhD students in the Netherlands did the same on 30 March 30 1999 (Hilversum). We acknowledge their contributions to this section.

- The reliability of the rainfall data can be regarded as rather questionable. Bad recording and mistakes made while copying data from one level to the next can cause problems of interpretation. Changes in personnel can cause gaps in reliability and care. Higher-level 'corrections' of lower-level data can sometimes be very confusing;
- The rainfall data are point data. We have tried to use as many stations as possible to arrive at a 'regional figure'. However, the large intra-regional variations and the fact that rainfall often comes in localised rainstorms, makes any 'up-scaling' of point data to area data questionable. The location of the rainfall stations might not give a representative overview of the most important agricultural production areas either. Some productive areas might not be easily accessible for agricultural staff during the harvest months, due to the road conditions and lack of transport;
- The yield data can be regarded as very questionable as well. In Chapter 7 a condensed comparative analysis was presented of statistical relationships between rainfall and crop yields for a number of areas, including Bolgatanga. Here, strong doubts were expressed about the yield data for this particular area. Crop acreage, kg ha^{-1} assessments and total production figures are based on 'expert' assessments (by local low-level personnel of the MOFA) at district headquarters, later up-scaled to the level of the Region as a whole. Agriculture is mainly subsistence oriented. It is unclear whether 'eating from the field' in the pre-harvest period is included, whether volume assessments are based on 'wet' or 'dry' harvests not whether yield assessments are based on representative samples or just a guess from the office. In a situation where not all planted land is also harvested, and where mixed cropping is the norm, mistakes can easily be made and confusion reigns. Later, higher-level civil servants arrive at a 'regional' figure. "(Some) Garbage In, (More) Garbage Out". Some civil servants tend to increase the yield data, either because their reference farmers tend to be the rich and more

successful ones, or because they want to 'prove' that their extension and other efforts have been successful. During perceived drought situations the opposite can also happen: harvests are underreported and the food crisis exaggerated so as to impress on national decision makers the urgent need for relief food;

- Changing yield data for particular crops might not refer to responses to the rainfall situation, but to genetic improvements in particular varieties or crops. It might also refer to changes in fertilisation, use of pesticides, irrigation, water harvesting and drainage and farm management techniques (as a result of changes in extension or otherwise) as well as reflect gradual changes in soil fertility, which can be considerable over an 11-year period;
- The Drought Index might be faulty, or the cut-off points between classes wrong. The Drought Index method itself is based on monthly rainfall data. For soil moisture assessments (relevant for the jump from Drought Index 3 to Drought Index 4 and from Drought Index 4 to Drought Index 5) more detailed analysis would in fact be required based on weekly or 10-day averages and specifying for different soil and landscape types. In the Drought Index method the rainfall data are taken from real measurements, but the temperature data and evapo-transpiration data are estimated and 'static'. It is probable though that actual temperatures are higher (and more fluctuating) during droughts and hence actual evapo-transpiration is higher as well, resulting in even more drought stress for plants;
- The Drought Index method does not take into account that plant moisture needs to vary over the growing cycle of plants, with the highest moisture needed somewhere in the middle: the impact of a dry spell in July can thus even be more dramatic than the Drought Index model predicts;
- The yield variability due to nature's whims is, of course, not only a result of droughts. Excess rainfall at the wrong time can also be quite devastating. Harvests rotting in the field, excessive weeds, pests (e.g. fungi in maize) and plant diseases can cause havoc. For quite a number of millet and even sorghum varieties that are used locally a lower rainfall and higher Drought Index (e.g. between 2 and 3) could even be a more optimal situation than a Drought Index between 0 and 2;
- The most important factor that explains the differences between Drought Index scores and yield levels is probably the preventive, coping and adaptive strategies of the farmers:

- Many farmers combine a rather large variety of crops and crop varieties and various rounds of seeding to spread risks and harvest moments. Many farmers start with fast-maturing millet, followed by other millet and sorghum and then some maize and beans, often in the same fields and with overlapping farm management cycles. During a drought they give up on more risky crops as far as drought stress is concerned and concentrate their efforts on crops and varieties which are more likely to succeed;
- Many farmers combine various locations, with different soils, exposure to the sun, slope, and water conditions. In many parts of the area farmers combine 'compound farms' and 'bush farms' with different, but varying, time and care investments, depending on the quality of the season; especially on compound farms, near the homestead, additional care (adding water, better weeding, manure application) can result in good harvests even if the rainfall situation is quite bad;
- There is probably an inverted U-shaped curve of care (and labour investments): when the rainfall situation is good, farmers tend to be 'lazy' (that is: they prefer leisure or other types of work to hard agricultural labour); when the rainfall situation begins to be tricky, farmers increase their care (e.g. apply more manure; re-sowing if needed) and labour input (more weeding); when the rainfall situation deteriorates further farmers tend to give up and spend time on non-agricultural alternatives. A perception of drought can also alter labour patterns (e.g. more child labour). Each farmer will have different behavioural patterns in this respect, so the combined effort is a result of highly varying farm management practices. The overall result is that yields during good years will not necessarily be better compared to yields during drought years, except for very dry years when most farmers give up. In the research area these extreme conditions hardly occurred though;
- After a good year, with relatively abundant harvest and a good food stock, farmers probably tend to 'relax' their effort; after a bad year farmers tend to increase their acreage, favour less risky crops/varieties and spend more time in their expanded fields. However, they often have to divide their efforts over larger areas and fields that are further apart; during and after a bad year their 'effort capacity' can also be undermined by a higher disease occurrence and higher labour investments in activities outside home agriculture (e.g. 'hunger trips');

- After a very bad year there might be a seed problem. Many farmers have to buy, barter or 'beg' seeds, sometimes with unknown characteristics, which creates a more insecure situation, as well as a higher 'experimentation attitude'. After a very bad year, food donations and other institutional interventions can change farmers' attitudes to farming and farm effort.

"In a dry year there will be a good harvest. Formerly the young did not understand, but now they do." (old farmer in Bongo).

5. ARE THERE INDICATIONS OF CLIMATIC CHANGE IN THE STUDY AREA?

It is evident that a comparison between the rainfall situation in the middle of the 20th century with the period 1970-1990 reveals a major climate deterioration, but that after the late 1980s the situation improved again, until the 1997 drought which was generally seen as problematic, but not causing a major crisis. The farmers who were interviewed generally saw a lot of evidence of long-term climatic change. They observed a change in the natural vegetation and in the relative importance of some trees (e.g. the gradual disappearance of the economically important *dawadawa* tree). They also observed lower water reliability and a shift in the planting season. In the past, many farmers started planting as early as in April, or even late March, but many have now changed to May or even June. The early maturing millet varieties are on the increase, late millet is disappearing and in the mix of white and red sorghum the more drought-tolerant red varieties have become more important. People say that the traditional signs of the start of the rainy season are no longer reliable, such as the behaviour of birds and ants, the changing of the winds, the coming of new leaves, the water tables in wells, the harvest of *dawadawa* trees. People are confused nowadays. They also regard the rainy season as more uncertain and shorter than before. People who used to rely on riverbed cultivation after the rainy season are nowadays regularly confronted with dry riverbeds that are 'dead'. Farmers also observe a growing importance of sheep and goats and a diminishing emphasis on cattle, which are confronted with an increasing 'feed stress' unlike sheep and especially goats. On the other hand, cotton is observed to be on the increase, especially in the southern part of the study area, near Langbensi.

People report that the number of 'swimming pools' have greatly decreased (especially pools along rivers have disappeared) and that also many running streams become stagnant water pools much earlier on in the season than before. As a result, the water quality from those sources becomes worse and there are more mosquitoes breeding. Some water sources, which used to be good, have now become salty. In addition, chemical pollution (pesticides; mining chemicals) has increased. However, due to greatly improved modern drinking water supplies, the average quality of drinking water has improved a lot, certainly for the urban population. Hand-dug wells and bore-holes provided by some NGOs helped a lot in this respect. It is interesting to note that in urban areas all over northern Ghana, water is rapidly commercialising, with water vendors selling water to those without their own decent supply.

6. WHAT WERE THE DRIVING FORCES OF LAND USE CHANGE?

A number of geographical trends in land use in the last decades are worth mentioning:

- a shift towards the valley (Volta) and marshy (Nasia) lands, facilitated by lower health risks (eradication of tsetse fly) and changes in farm technology (oxen ploughs);
- the development of irrigated agriculture in a number of 'schemes', and with it the expansion of commercial rice and vegetable cultivation. From the mid 1960s onwards the valleys of the major and minor tributaries of the Nasia river, as well as a number of small-scale irrigation areas (e.g. Vea) have become the main areas of rice production in Northern Ghana. During the 1960s many small water dams have been constructed. According to representatives from the Ministry of Food and Agriculture there were more than 200 or even 250 in Upper East Region alone (An IFAD-funded project counted 256 of them). An average dam can irrigate 10 hectares of land, mainly for tomatoes, onions or other vegetables, or even rice. According to a natural resource management expert in the area the Region needs at least 450 additional dams. Most of the existing small dams silted up during the 1970s and '80s. Recently a start has been made on rehabilitating these dams and encouraging farmers to use them for dry-season cultivation. In 1999, 44 of these dams had been rehabilitated and their use in areas of intensive cultivation of rice and horticultural produce has indeed increased considerably. People have come to recognise their importance and a much greater effort is being made to maintain and preserve them;
- many (parts of) so-called bush farms were converted to compound farms (due to population growth and establishment of many new farm households by the new generation), resulting in intensification and better care. Many farmers still try to maintain multi-location farming by

combining compound with bush farming and - if possible - a stake in either irrigated farms or in niches with low drought chances (soils with high water retention capacity; marshy areas).

- the growing importance of cotton cultivation could be seen as part of a 'southern shift' of the cotton belt and one of the consequences of climate change. What used to be 'normal' in southern Burkina Faso and Mali now becomes 'normal' in northern Ghana. However, it is too easy to see this as straightforward 'proof' of climate change. The increasing activities of cotton factories in the area, the recent attention it has received from government agencies and 'public opinion', and the improved access to cotton inputs are all factors which can also be seen as important;
- many farmers have been trying to cultivate more land during the last decade. Where possible (that is, not in Bongo and in the most densely populated parts of the Gambaga area) holdings have been expanding. This can be regarded as a response by farmers to higher risks, but it can also be seen as a move towards more commercial agriculture and a response to generally more favourable agricultural market conditions for Ghana after the economic crisis of the 1970s and '80s had ended.
- In a recent study (Dietz et al. 2002) on the impact of two Dutch-sponsored NGOs a.o dealing with natural resource management in North-eastern Ghana (the Catholic Diocese of Navrongo Bolgatanga, and the Presbyterian Church of Ghana) it became clear that a lot of effort had recently gone into improving land management. Many farmers had indeed adopted soil and water conservation methods and on-farm tree planting. In four research villages, which had been these NGOs' target villages , the percentages of survey households which had adopted 'new' farm management methods were as follows: grass bunding: 88%, on-farm tree planting: 83%, ploughing along the contour: 83%, involved in gully control: 83%, ploughing with bullocks: 80%, using compost 78%, using collected cow dung on their fields: 78%, dry-season garden: 31%.

7. WHAT ARE THE CHANGES IN COPING STRATEGIES WITH REGARD TO DROUGHT?

It is clear that people are changing the composition of livelihood portfolios by relying more on non-agricultural sources of income, by adding more market-oriented agricultural activities (cotton, onions, tomatoes) and by changing their food production strategies to more drought-adapted varieties on the one hand and to less water-stressed fields on the other (where they produce vegetables and rice). Animal husbandry is changing as well with relatively more emphasis on goats, pigs and fowls. Many old people told the researchers that they fear that the body of indigenous knowledge that enabled them to survive stress situations in the past (including the traditional institutions of land chiefs (*tindanas*), soothsayers, rainmakers and medicine men) is disappearing. "If nothing is done to re-discover this heritage", says Grandfather Akkare Adongo, "the next severe stress will find us all in our graves, for we cannot cope with it".

What is likely to happen in the event of (further) climate deterioration? If the temperature increases by on average one or two degrees and rainfall deteriorates by 20%, the area will clearly become semi-arid (P/ETP < 0.45) with rainfall between 700 and 800 mm. A probably shorter and less reliable rainy season will result in an average drought risk situation between 2 and 3, with a higher chance of drought risk situations of 4 and 5. Depending on market price developments for animals, groundnuts, onions, tomatoes and cotton in the southern parts of Ghana a possible strategy for farmers may be to rely less on autarchic food production at household level and to try and diversify agricultural and non-agricultural risks. The southern move of the cotton belt could present a major challenge, with secure marketing arrangements and price levels as an important prerequisite for gaining farmer's trust. With cotton and groundnuts as dryland crops in non-valley fields and rice, tomatoes and other vegetables in valley/riverine areas, it might well be tha case that there is a shift in the area away from a strong reliance on the millet-sorghum-legumes complex. With more emphasis on higher-yielding, but drought-tolerant millet and sorghum varieties, the most densely populated areas could be assisted in a necessary intensification process (higher average yields per hectare). The farming experiences and practices of the many immigrant farmers from Burkina Faso (who are often from the same ethnic macro-group of Mossi-related cultural backgrounds) are bound to reveal interesting possibilities.

8. HOW TO MITIGATE THE CLIMATE CHANGE RISKS?

It is evident that increased chances of low harvests from time to time will mean that more farm households will be confronted with inadequate cereals from their own fields to get them to the next harvest. Old traditions might be revived to try and store harvests from excess years and this means that areas where people are still able to expand their fields (within Upper East Region the areas in between the Voltas and areas towards the west and south; and in Northern Region considerable areas in the Nasia and Volta valleys) should be supported in producing and storing more cereals. For the densely populated areas (like Bongo, Bawku, Gambaga) market-led forms of intensification might be a solution, supported by increased forms of irrigation (water dams) and water harvesting techniques. In these areas it is also obvious to us that agricultural depopulation should be encouraged. The natural increase of the population should be lowered, or it can be done by out-migration, more remittances and more non-farm income. It would probably be a wise decision to encourage more children to go to secondary schools, especially in the most densely populated areas, as a preparation for either out-migration or non-farm activities in the area itself. The recent approach to selling food aid instead of giving it out for free or as 'food for work' should be encouraged, with exception of arrangements for old and sick people. Traders perform key roles in a more commercially oriented food acquisition system and the infrastructure for trade should be safeguarded. Food aid arrangements can easily undermine existing systems of commercial exchange of food.

- Which research activities are required most urgently in to be better prepared for climate change? During a workshop in Bolgatanga, in 1999, the following suggestions were made for further studies related to climate change observation and mitigation:
- Is it indeed true that indicator species in the natural vegetation show that 'normal' species are disappearing and that more 'northern' (Burkina) species are becoming more important? What commercial/useful species of the zone 200-300 kilometres more to the north could be developed more rapidly in the Bolgatanga region?
- Is it true that hydrological features are of a lower quality than before (less water in the Volta and other rivers; seasonal rivers drying up more rapidly; groundwater tables going down; dams drying up)?
- How important is the influx of immigration from Burkina Faso and, if engaged in farming, how do their farm practices differ from those of the indigenous population? Are farm practices of immigrants from the north more appropriate for drought conditions? Do immigrants from the north generally get lower quality land?
- If micro-differences in terrain are indeed so important for farmer's strategies, it would be good to start in-depth research over a number of years in which a few farmers with diversified plots are monitored with regard to their actual land use practices, their harvests and the rainfall conditions.

- If farmers tend to drift to riverine and (ex-)marshy areas, what does this mean in terms of crop risks (e.g. floods; crop diseases) and human health risks (water-borne diseases)?
- More research is needed to explain the differences in yields between farmers and to explain the seeming incompatibility of rainfall (drought risk) data and harvest data over years.
- Is it indeed true that 'wealthy' farmers are those who never experience food shortages using their own fields or is there a category of farmers who specialise in more lucrative commercial crops at the expense of autarchic food security at the farm level? How risky is this strategy in terms of food stability and income variability?
- It is important to find out if land tenure changes result in land use changes and changes in land management practices; do 'private owners' take more risks? Do they experience a higher yield variability between bad rainfall years and good rainfall years?
- Is it indeed true that the non-agricultural element in livelihood portfolios is increasing and that this is becoming an important element of food stability and food security at the household level?
- If it is true that migrant (remittance) income becomes more important, how important is labour migration for which types of households? Is the image of multi-location, intra-lineage assistance in dire times a correct one, or could lineage network analysis indicate a more restrictive mutual support system?
- Is it true that state and NGO food aid kills initiative and severely threatens existing mutual assistance patterns?
- Is there a changing cost-benefit situation with regard to wood/charcoal versus other energy sources? What are the experiences of forest rehabilitation (e.g. in shrine and grove areas; on hillsides; on watersheds)?

The survey research revealed a number of issues which should be covered in a more in-depth follow-up study, if such were to take place:

- the role of sacred water bodies in environmental preservation; their role in biodiversity;
- the institutional structures and organisation to regulate water use during stress situations;
- the order of use of 'strategies' during stress and the order of loss as a result of stress;
- Regenerative efforts after one stress period to pre-empt subsequent losses;
- monitoring change e.g. the use of resource mapping to empirically ascertain change.

In general it is recommended that monitoring studies be started (e.g. by using the same villages) to find easy early warning indicators of stress by combining efforts of the Ministry of Energy, Ministry of Food and Agriculture, Ministry of Lands and Forests, the Meteorological Service and strategic persons in the region.

Chapter 13

PATHWAYS TO MITIGATE CLIMATE VARIABILITY AND CLIMATE CHANGE IN MALI: THE DISTRICTS OF DOUENTZA AND KOUTIALA COMPARED

Han van Dijk, Mirjam de Bruijn and Wouter van Beek

Abstract: Two Malian case studies are compared, both representing dryland areas with relatively low population densities and relatively low levels of land degradation. Douentza is in the semi-arid-to arid zone, where crop cultivation is very risky and pastoralism a more 'natural' way of making use of the environment. However, recently crop cultivation has been expanding rapidly and has partially recovered from the droughts of the 1970s and 1980s. Crop cultivation now provides a livelihood for impoverished former pastoralists (like FulBe, and Tamachek), but also for groups who have always been cultivators (like the Dogon). Many people originating from this area have extended their geographical network and can be found in areas much further to the South. Many retain their relations with their areas of origin, though. Koutiala is an area in the sub-humid zone and in a region which benefited from the Malian cotton boom economy of the 1980s and 1990s. Not only did the local Minyanka prosper, the expanding economy also provided a livelihood for many immigrants who had escaped the drought conditions of the northern area.

1. INTRODUCTION

Climate change is not a phenomenon that can be instantly observed. It manifests itself in small, gradual changes in temperature, evaporation and rainfall figures. However, in the long term, climate change can have a tremendous effect, for example when the growing of a certain crop is no longer possible. In semi-arid environments, the variations in agro-ecological conditions in time and space are so great that small changes are hard to detect. Climate change, therefore, tends to manifest itself as an increase in extreme events such as excessive rainfall or, at the

A.J. Dietz et al. (eds.),
The Impact of Climate Change on Drylands: With a Focus on West-Africa, 173–206.

opposite end of the scale, meteorological droughts. Policy makers need to be informed about the likely responses of the local population to such events in order to be able to implement measures to accommodate the consequences of any change in climate. This can only be done by a detailed analysis of local-level situations.

Over the past decades the Sahel has gone through one such phase of change which indeed seemed to be characterised by a noticeable increase in extreme events. However, it is not only the climate that is an unreliable partner in this area. Market prices and social institutions also respond to rainfall fluctuations, world markets, political unrest and so on. An analysis of a whole set of contextual factors is therefore required. Risk and uncertainty are the most basic conditions for decision-making. This chapter considers the ways in which local farmers and herdsmen deal with climate variability and other uncertain contextual factors. The analysis is based on a comparison of two rural districts in Mali: Koutiala in the semi-humid south of Mali at the 12th parallel, and the district of Douentza south of the 15th parallel, with its semi-arid climate.

In the first section, the outline of the research is discussed, followed by a short introduction to the research and the two research areas. The third section offers an analysis of the pathways of local farmers and herdsmen in Koutiala and Douentza. In the final section, some conclusions are drawn with respect to the methodology used and to the pathways that have been developed over time by the various population groups in the study areas.

2. THE TWO STUDY AREAS IN A COMPARATIVE PERSPECTIVE

2.1 The Research

Research in the two areas consisted of a survey of the available literature supplemented with field research focused on specific groups of actors. In the Douentza District, previous research on the FulBe (or Fulani) herdsmen (De Bruijn & Van Dijk 1995) was extended with additional research on a number of other population groups such as the Dogon group of cultivators (Maas 2001, Brandts 2001); the *riimaybe*, the former slaves of the FulBe (Griep 2001) and the various population groups of the district capital Douentza (Zondag 2000), as well as the trading Tamasheq from outside the area (Rutgers Van Der Loeff 2001). Specific topics such as agricultural expansion, the migration of FulBe pastoralists (De Bruijn & Van Dijk 1999) and the dynamics of the herbal and woody vegetation (De Boer 2001) were investigated.[2]

It was assumed that more information on the Koutiala area would be available in the form of research reports resulting from the long-term engagement of researchers with the Malian Cotton Company and agricultural research in the area. Most of this research was conducted either at regional and village level (Hijkoop & Van der Poel 1989, Bosma *et al.* 1996) or focused on simulation models (Struijf-Bontkes 1999, Sissoko 1998) and did not deal with farming strategies in detail. In the end, two studies were made of two population groups, the Minyanka (Nikiéma 1999) and the immigrating FulBe (Van Steenbrugge 2001). The immigrating Dogon were covered while undertaking a study on agricultural expansion. Other basic data were obtained from non-ICCD research such as that conducted by Jonckers (1987, 1995), Degnbol (1999), Benjaminsen (2001) and Mosely (2001).

Both study areas are located in the Republic of Mali. Mali is a landlocked country inhabited by around 11 million people of whom 80% live in rural areas. Estimates of population growth vary between 2% and 3.2% a year according to the source (IOB 2000: 121). More than half of Mali's 1,230,190 km^2 belongs to the Sahara and around 25% of the country can be used for agricultural purposes. Agriculture and animal husbandry in the Sahel and Sudan zone are subject to high risks, especially in the Sahel. The country's natural soil fertility is generally low and soils tend to be old and weathered. Levels of organic matter are depleted because of low levels of biomass production due to insufficient rainfall and high year-round temperatures causing an excessive rate of decomposition of organic matter.

Mali is one of the poorest countries in the world both in terms of economic production and human development. The agricultural sector accounts for most of its export earnings with cotton being the most important export item, accounting for almost half of all exports. It is followed by livestock, which is estimated to account for a further quarter of total export earnings. Average income in 2000 was $240 (World Bank 2002: 233). However, income distribution is extremely uneven, with 69% of the population living below the poverty line in 1998. Poverty is concentrated in the rural areas where more than 80% of the population

[2] This research was carried out in collaboration with Wageningen University and Research Centre (Dept of Agronomy and Tropical Nature Conservation and Ecology of Vertebrates), Utrecht University (Dept of Cultural Anthropology), and the Amsterdam Research Centre for Global Issues and Development Studies (University of Amsterdam).

was classified as poor in 1994. In the cities, only 24% lived below the poverty line (IOB 2000: 124-5).

The study areas can be said to fall in the extremes of the poverty-wealth continuum. The district of Douentza is located in the Mopti region, the poorest in Mali in terms of economic and human development. Child mortality is high (200 per thousand in the first year of life and over 350 for all under-5s) and life expectancy is less than 40 years, probably even around 30-35 years in the most rural areas (Hill 1991: 171-73, 178). These estimates are based on demographic research in Central Mali. For the whole of Mali, child mortality of the under-5s is estimated at 223 per 1000 and life expectancy at birth at 43 years in 1999 (World Bank 2002).

Koutiala is located in the Sikasso region and is the heart of the cotton-growing area. It is the wealthiest in terms of economic development and in its degree of urbanisation with two major cities, Koutiala and Sikasso (Staatz *et al.* 1989). The cash incomes of farming households are far above the national average (Degnbol 1999: 118) and most farming families own considerable assets (Assets owned per capita for rural households in the Siwaa area, southeast of Koutiala varied between $96 per capita in the poorest category to $358 per capita for the wealthiest households; Mosely 2001: 138). However, if expenditures are considered, the region ranks as the second poorest after the Mopti region. Neither health provisioning nor educational levels are above average (Degnbol 1999: 120).

2.2 Koutiala

Koutiala District is located between 12° and 13° N and between 5° and 6° W (see map 13.1). The climate is Sudanic with rainfall concentrated in the months of June-October. Total annual precipitation amounts on average to 900 mm and varies between 15-20%. Average temperatures are high, ranging from 22°C in the cold season in January-February to over 30°C during the hot season in May. Total rainfall declined in the 1970s and 1980s but recovered somewhat during the second half of the 1990s (see Figures 13.1a & 13.1b).

The population density in the district is not (yet) very high. As of 1992, more than 300,000 people were living in an area of 8.740 km^2, resulting in densities of around 40 inhabitants per km^2. Population growth is far above the national average of 2% due to the influx of poor migrants from the north. The district is mainly inhabited by Minyanka, with minorities of other ethnic groups such as the FulBe and Dogon. Some of the FulBe and all the Dogon are relative newcomers to the area and migrated there in the wake of the droughts of the 1970s and 1980s.

However, almost all the cultivable land has been put into production. Around the district capital of Koutiala, agricultural occupation has almost reached its maximum, whereas in the more remote areas more land is available and fallow cycles are still followed, though allowing land to lie fallow for long periods is no longer feasible (Hijkoop & Van Der Poel 1989: 17). In densely populated areas, most of the land is under permanent cultivation (Hijkoop & Van Der Poel 1989).

Map 13.1. Overview of Central Mali and the study areas.

Altitude varies between 280-400 metres above sea level and relief levels are gentle. Geologically, most of the district belongs to the Koutiala Plateau, consisting of low hills with gently undulating slopes. On the hills, the soils are characterised by hardpans, with shallow soils and large amounts of gravel in the subsoil. Lower down, the soils gradually change from loamy-sandy soils that are poorly drained and highly susceptible to erosion, via sandy-loamy soils to heavy more fertile clayey soils in the valleys. The natural vegetation consists of forest to tree savannah or shrub savannah.

On higher land, the main constraint on use is erosion, and in the valley bottom poor drainage imposes its own limitations. Usage is adapted to the soil qualities with the degraded soils mainly used as pasture area. Vegetation consists of annuals, trees and shrubs. Different soils are used for the cultivation of cereals: millet is grown in the driest and most shallow soils, and maize and sorghum in wetter soils and soils which are poorly drained. Cotton is cultivated near the valley bottoms where gardens, tree yards and dry-season pastures can also be found.

Farming systems

The district's rural economy is founded on two pillars: the cultivation of cereals as a subsistence crop and the growing of cotton for cash. Millet, maize and sorghum account for more than two-thirds of the area cultivated and cotton makes up around 20%. These figures may vary from one area to another with a share of up to 50% for cotton in some individual households. The rest is used for gardens, pulses and peanuts. Most farming units combine the cultivation of cereals and cotton. There are generally two types of farming systems , one predominantly oriented towards cereal cultivation in combination with cotton for cash, with livestock production for the provision of traction animals, manure and capital outlays for meagre years. The other system concentrates on transhumant livestock production with supplementary cereal production. In terms of numbers, the first system is by far the most dominant and is operated by sedentary farmers, the Minyanka, who are the main ethnic group in the area, and some Bamana. The livestock-based farming system is in the hands of the FulBe, who have moved in from other areas over time

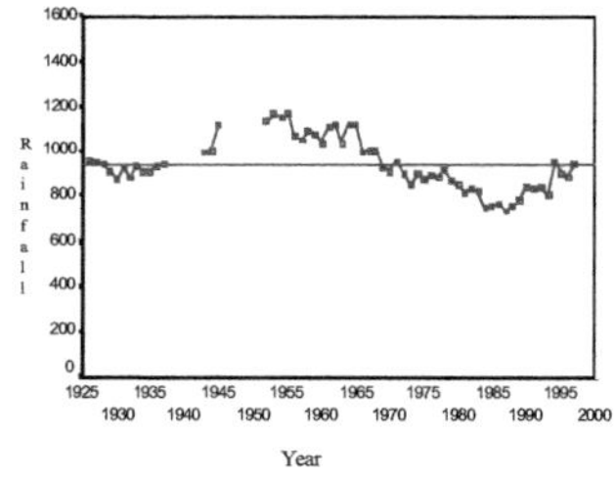

Figure 13.1a Annual rainfall in Koutiala 1926-1996 (Put & De Vos 1999)

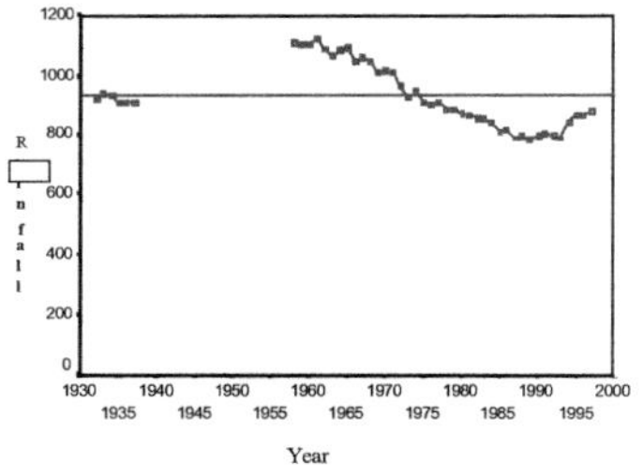

Figure 13.1b 11-year moving average of rainfall in Koutiala (Put & De Vos 1999)

The integration of cotton into the farming system does not seem to have affected food security for the large farming units. Most are self-sufficient with respect to basic food and are even capable of selling surplus production (Raymond *et al.* 1991). This is partly due to the introduction of new technology – mainly agricultural equipment and oxen traction. By now, most farming units possess agricultural equipment and the majority have at least one pair of oxen at their disposal. This has enabled farming units to expand their cultivated area and to combine cereal and cotton cultivation within the same unit. Livestock have become critical within the farming system. Given the high pressure on land and the almost permanent cultivation, the risks of soil erosion and the depletion of the soil's fertility are high. In this situation, organic manure is indispensable for the maintenance of soil fertility, the stability of soil aggregates and to improve the efficiency of chemical fertilisers.

Consequently, the importance of livestock has increased tremendously since the introduction of cotton and the increase in population density. The cotton-growing area, Mali-sud, has surpassed the region of Mopti as the area with the largest number of cattle. The main livestock species is cattle, followed by small ruminants. Donkeys and horses are also raised for transport purposes. It is estimated that the number of livestock in Koutiala District far surpasses its carrying capacity as determined by agro-ecologists (Bosma *et al.* 1996, Hijkoop & Van Der Poel 1989). However, there is no evidence that this has led to degradation of the vegetation cover in terms of productivity. On the contrary, according to Mosely (2001: 109-10), biomass has increased in most of the south of Mali and remained constant in other parts over the 1982-84 and 1996-98 periods due to some improvement in rainfall figures during this time.

Institutional context

Linkages with other areas were limited in the past. There was little political centralisation in Minyanka society, in contrast to neighbouring areas such as the Inland Delta of the Niger and the Senoufou Kingdoms to the south. No major trade routes ran through the area.

During the colonial period, major changes swept through the area. After failed attempts in the pre-war era, a parastatal agency, the *Compagnie Française de Développement des Fibres Textiles* (CFDT) was established in 1949 to promote the production of cotton. Its success was based on stable and acceptable producer prices and a monopoly on the purchase, grinding and export of cotton, a system that has been labelled 'coercion of the markets' (Roberts 1995). After independence, the CFDT gradually took on other tasks, such as rural infrastructure, the extension for cereal cultivation, credit provision, farmers' organisations and literacy courses. In 1974, the Malian government was given 60% of the shares (Degnbol 1999: 115; The mother company of CFDT/CMDT, the French CFDT, is controlled by the French government that originally held 64% of the shares. Private interests owned the rest).

During this period the area under cotton cultivation increased dramatically. Since 1960, the area has increased six-fold, and in the decade between 1984 and 1994, the area under cotton almost doubled (Degnbol 1999: 115). Cotton production increased from 3,900 tons in 1958 to 68,000 tons in 1972 and 561,000 tons in 1998 (Benjaminsen 2001: 262). The devaluation of the CFA franc in 1994 helped boost cotton production from 310,000 tons in 1992 to 561,000 tons in 1998. Productivity, that was around 200-250 $kg.ha^{-1}$ in the 1950s, reached its highest point in 1990 with 1,376 $kg.ha^{-1}$. Since then, levels of production have declined. In the main area around Koutiala, productivity decreased by 6.8% per annum during the 1993/94 to 1997/98 period. Total production was halved in 2000 following a farmers' strike when the CMDT lowered farm-gate prices to accommodate the collapse in world market prices (Tefft 2001: 218).

The CMDT success story until 1999 was based on an intensive model of vertical integration of the production chain. The CMDT controlled all the components of the production chain from the sale of inputs to the provision of credit and farming equipment, the processing and marketing of cotton and the execution of subsidiary programmes designed to combat soil erosion and deforestation. Since the 1980s the sale of inputs, credit provisioning and the purchasing of cotton have gradually been organised at village level by means of village associations (*associations villageoises*).

The reverse side of this success story is the small margin between profit and loss. Any prolonged drop in world market prices pushes the CMDT into the red and farmers' prices go down. The slump in world market prices of the last few years has forced the CMDT to lower farm gate prices substantially. A further weakness in the CMDT model is the amount of control it exerts over the rural economy. Corruption and abuse of power by local CMDT agents and members of the boards of village associations, right up to the highest levels of management, have led to a growing discontent among farmers. This, in turn, has led to the emergence of a countervailing power in the form of a farmers' union, the SYCOV (*Syndicat Cotonnier et Vivrier*) (Bingen 1994), though critics argue that they are more outwardly oriented than acting as power brokers for their members ("*L'une des faiblesses des syndicats paysans est leur faible capacité de communication. Ils n'ont pratiquement pas de référence (siège, numéro de téléphone, etc.). Ils sont plus connus à l'extérieur du Mali qu'à l'intérieur. Il y a une confusion entre les fonctions de la coordination des AV et celle des syndicats. Par ailleurs, on a assisté à une bureaucratisation précoce des responsables de ces syndicats. Ils passent plus de temps dans les forums, ateliers et colloques que sur le terrain à renforcer leur syndicat. Ce faisant, ils n'arrivent pas à mobiliser les ressources internes. Ces syndicats sont obligés de recourir à des financements extérieurs pour survivre, ce qui menace leur indépendance. SYCOV est intégré à un réseau international et c'est la raison pour laquelle ses responsables sont assez souvent en voyage à l'extérieur du Mali »,* De Bruijn *et al.* 2001: 51-52). This contributed to the outbreak of strikes amongst farmers and rioting in Koutiala, where an important element of the CMDT infrastructure is vested. Others have argued that CMDT policies have contributed to a growing differentiation in wealth and vulnerability among CMDT clientele although the evidence is contradictory (see Moseley 2001, Degnbol 1999, Nikiéma 1999, Jonckers 1987, 1995).

2.3 Douentza

Just as Koutiala District is at first glance the wealthiest district in Mali, the Douentza study area is one of the poorest areas. It is part of the Mopti region and is the poorest in Mali economically and in terms of human development. The study area (see map), which belongs to the districts of Douentza, Koro, and Bandiagara, is located between the 14° and 15° N and 2° and 4° W. The climate is Sahelian and rainfall amounts to 500 mm per annum in the south of the study area and 350 mm per annum in the north. It is close to the thermal equator, with an average annual temperature in Hombori just north of the area of 29.8°C. Total annual rainfall shows a decreasing trend from the 1950s and 1960s onwards, recovering somewhat in the 1990s (see Figures 13.1a & 13.1b).

Altitude varies between 250 and 800 metres above sea level and the area consists of three agro-ecological zones. The first is the Bandiagara Plateau, consisting mainly of rock that rises east of the Inland Delta of the Niger up to 700-800 metres high. The plateau ends in an escarpment at an elevation of 200-300 metres above the sandy plains of the Seeno and the Gondo. This escarpment with its watercourses at the bottom is the second sub-region. The third sub-region consists of the plains east of the escarpment. Near the escarpment, there are eolic dunes from the quaternary giving way to the east and the north to outcrops of the mountain formation. Here, clayey soils can be found. The relief levels are spectacular, with the Bandiagara Escarpment rising almost 300 metres above the Seeno and Gondo plains and the *Inselberge* that connect the plateau with Mount Hombori, the highest mountain in Mali.

The population density here is much lower than in Koutiala District. However the population distribution is extremely uneven. In Douentza District it is around 10 inhabitants per km whereas in the south of Koro District and along the escarpment in Bandiagara District population densities can be found that are higher than in Koutiala District. Annual population growth is around 1%, indicating that emigration is higher than immigration. In some sub-districts along the Bandiagara Escarpment the population is actually decreasing. Dogon farmers mainly inhabit the study area, with a large minority of FulBe, who are concentrated in the northern half of the area and on the sandy plains.

Cultivable land amounts to around 50% of the total area, as the rest of the land is susceptible to intra-seasonal drought because of its low water-retention capacity. Agricultural occupation further depends on the density of the population. In the south of the Seeno and Gondo plains and near the escarpment, 90-100% of the land is under permanent cultivation. To the north where population densities are lower, the level of agricultural occupation is lower, although it is on the increase. Large tracts of land are still available as pastureland for livestock rearing. On the plateau, only 10% of the land is cultivable, the rest being rock or soil which is too shallow to support a crop. All the land that can be used is being cultivated on a permanent basis.

The diversity in soils is huge, ranging from deep sandy soils in the dune areas to clayey or loamy soils of differing depths on the plateau and the outcrops of the mountains. Their agricultural properties vary accordingly, as do the chemical properties of the clayey soils. Some are more fertile than the sandy soils, but where there is a lot of gravel or laterite in the subsoil, water for plant growth is low and crops are susceptible to drought. Further crop damage may occur when these soils are poorly drained and the water stagnates after heavy rainfall. The sandy soils are much better drained but in general low in nutrients and, when the loam content is low, extremely susceptible to intra-seasonal drought. However, many farming units prefer the soils in the depressions as they contain some loam, are easier to work, fewer weeds grow and, when sufficiently fertilised, they can produce acceptable harvests. In most areas, millet is the main crop and is grown in monoculture in the north. Towards the south of the study area, millet may be found mixed with beans, while sorghum is sometimes grown on the heavier soils in depressions and valley bottoms. Market gardening takes place along watercourses and reservoirs created by dams on the plateau.

The economy of the region is based on the cultivation of cereals for subsistence and the rearing of livestock for meat and/or milk. For some on the plateau, market gardening is their main source of income. Remittances from migrants are an indispensable part of people's income because in most years parts of the study area do not produce sufficient grain and/or livestock to sustain the population. Even in urban centres, such as the district capitals, (8,000-20,000 inhabitants) most people have fields to produce millet. The urbanisation rate is low and does not exceed 10%.

Over the past decades, technological development has led to an accelerated growth in the amount of land under cultivation. Agricultural equipment in the form of ploughs drawn by any kind of draught animal – oxen, donkeys, camels or a combination of the three – is common throughout the whole study area. Though it seems that over a long period of time the productivity of the land declines under the impact of drought, total production has not declined because of the rapid increase in the acreage under cultivation.

Nevertheless, the northern part of the study area and the districts to the north are structurally deficient areas with respect to primary cereal production. To compensate for this deficit, livestock is the main cash earner in the area. In the southern part of the study area deficits occur only in drought years. In other years, it is a surplus area, provisioning the north of the study area and the areas further north towards Timbuktu (see Rutgers Van der Loeff 2001). Given the high degree of cultivation and the low level of rainfall, there is a lack of opportunities to compensate risks with other crops except for horticulture in vegetable gardens on the plateau. Therefore, the economy of the whole study area is extremely vulnerable to external shocks in the form of drought and market failure for primary products.

Livestock offers no reliable alternative. Biomass production in the area fluctuates to such an extent (with a standard deviation of 40%) that the periodic droughts that cause these fluctuations result in massive losses of cattle and other livestock and the collapse of the cereal and livestock markets (Hesse 1987, RIM 1987, De Bruijn & Van Dijk 1995). Over the past three decades this has happened twice, paralysing the whole rural economy for a number of years and causing a massive shift in the ownership of livestock from small family-based transhumant livestock keeping units to urban traders and civil servants (De Bruijn & Van Dijk 1995).

The social organisation of the various population groups inhabiting the area differs markedly. The main agricultural population, the Dogon, is organised on the basis of lineages, wherein seniority and gender are the main criteria for differentiation. In some cases, status (freemen versus non-free) also plays a significant role. Within the main group of semi-nomadic livestock keepers, descent and gender are central. Political hierarchies are important in defining one's status, social position and occupation. These attributes are crucial for an understanding of decision-making and the pathways developed by individual actors. They also place people in a structurally different position with respect to all kind of risks in the environment in which they have to function (see De Bruijn & Van Dijk 1994, 2001, at press).

In addition to these main population groups, there are members of numerous other ethnic groups present in the area. Either they have a weak position since they have broken away from their own societies and are part of an inferior status group and immigrants in the study area or they belong to specific occupational groups such as weavers, blacksmiths, bards, wood-workers or leatherworkers, all of whom have a rather ambiguous social status.

3. PATHWAYS TO MITIGATE CLIMATE VARIABILITY IN KOUTIALA AND DOUENTZA

3.1 Koutiala: 'Forced' Technological Innovation

Intensification, degradation and declining productivity

It is clear from the previous section that climate variability is much less of a problem in Koutiala than in Douentza. Consequently, the pathways of the local farmers and herdsmen are influenced much more by other factors such as markets and institutions than by rainfall levels and their distribution over time and space. Given its size, financial power and the degree of control the CMDT has over farmers, prices and institutions, it is the single most important actor shaping pathways. This does not mean that climate and the vulnerability of farmers is not a potential problem. Especially in the north of the area, rainfall figures are approaching the limits at which cotton cultivation is still possible (see chapter 10, and 16 as well). Secondly, cotton cultivation may increase the vulnerability of households and individuals. Thirdly, the area has attracted a large number of migrants over the past few decades. How they integrated their strategies into the CMDT format and how they fare economically is largely unknown. Fourthly, there has been some debate about the nature and extent of poverty in the area. According to the statistics of the CMDT itself, the poorest categories of farmers, i.e. those without oxen-drawn agricultural implements, are slowly disappearing. The question is whether these statistics cover all the farming units in the area and whether these poorest units are disappearing because they become wealthier or because they become even poorer. Their members may eventually move away or their unit may have become unviable and may have been taken over by other units, or may have died out. Finally, discussions about farmers' strategies in the area have been dominated by neo-Malthusian views pointing at the failure of farmers and intervening agencies such as the CMDT to achieve integrated soil management in a situation of population growth, declining rainfall and degrading natural resources. This failure is attributed to farmers' attitudes to labour and a short-term horizon due to their poverty (Koning *et al.* 1997). According to this line of thinking, the current wealth of the area is based on soil mining rather than sophisticated technological innovations (Van der Pol 1991). However, the empirical basis for such conclusions seems weak since they are based on weak databases. Moreover, the problem is caused by the enormous amount of land under cotton production that in turn has led to the shortening of fallow periods. The returns on labour from cotton cultivation are lower than for cereal cultivation (Raymond *et al.* 1991: 193) and therefore do not justify further investments in soil conservation and soil fertility at farm level given opportunity costs in other sectors of the economy and the transaction costs for obtaining credit.

Increasing vulnerability: Climate and markets

Questions about the vulnerability of farmers and the ecological environment have become more urgent. Farmers' protests against CMDT policies (and corruption) have increased following dramatic drops in world market prices and the CMDT's adaptation of farm-gate prices.

This led to a strike by farmers in 2000, a 50% drop in cotton production and dealt a massive blow to the financial position of the CMDT, potentially influencing Mali's national budget and its political stability. Efforts are now being made to restructure the cotton production sector.

The whole question of adaptation to climate variability and possible responses to climate change can only be answered with reference to the development of the cotton production and processing sector and the assessment of the vulnerability of various categories of farmers. Contrary to official statistics, some observers have pointed out that extensive poverty still exists in Koutiala District despite the developments in cotton production (Degnbol 1999). Jonckers (1987, 1995) noticed that, following the introduction of cotton production, fundamental transformations in Minyanka society in the role of women in social and economic life, religion, land tenure and informal social-security relations.

One of the main results has been rising economic inequality between farming units and especially between villages that are part of the CMDT extension system and those that have not been selected for the promotion of cotton production. The latter are not included in the official statistics. Consequently the main basis of social organisation, the lineage-based extended family is breaking up into small family units. In the absence of a village or regional-level organisation, this gap is being filled with institutions put in place by the CMDT such as the *associations villagoises*, which are known to be corrupt. Internal cleavages and political divisions within the villages associated with the administrative decentralisation currently underway may increasingly hamper their functioning (Nijenhuis 2002).

Minyanka farmers

There seems to be a difference between large and small farming units with respect to food security. According to Nikiéma (1999: 119), smaller units are more vulnerable to crop failure and less food secure than larger units. Large farming units profit more from traditional forms of social security as well as from modern institutions like government services, the *associations villageoises* and farmers' unions. In addition, these larger units dispose of more livestock and are able to accumulate more capital savings by means of cash crops (cotton and also maize) and secondary activities. Smaller units also undertake these activities but they seem to be aimed at closing the cereal gap rather than accumulating savings and the formation of capital.

The principal difference between these poor and rich units is the level of farming equipment. Apart from rainfall, numerous other constraints affect food security such as access to land and labour, and marketing conditions. Rainfall variability is one of these important constraints because the units' structural positions do not allow them to develop more sophisticated strategies to counter this problem. For rich households, access to land is their major concern after rainfall variability, since in the main area of the cotton basin there is no more land available to put into production. The productivity of the land has also reached its limits. Cotton and cereal production is declining because soils are exhausted and less fertile land is being used for production. Nikiéma (1999) concluded that there is a progressive impoverishment of the poorer categories and only a few farming units that are managing to improve living conditions.

To deal with these problems, rich and resource-poor farmers follow different pathways and strategies differ widely even between farming units of the same category. Agricultural intensification remains limited to the category of wealthy farmers because of the capital outlays required. For resource-poor farmers the options are more limited within the domain of agriculture. Extensification, i.e. bringing more less-productive land into cultivation, is their main pathway. Besides agriculture, the main strategies are labour migration, the diversification of income sources and attempts to create larger farming units.

Case studies

Based on the material available it is not yet possible to differentiate between distinct pathways for various categories of decision-making in units in the Koutiala area. The current wealth situation of a household cannot be taken as a result of a specific investment strategy or an indication that units have been either poor or rich in the past. However, a number of trends can be observed in the manner in which households deal with variability, adversity and the minimisation of risks.

An in-depth analysis of individual households shows that chance plays an important role in a group's fortunes despite the fact that many households have managed to build up considerable stocks of food and other assets to cope with hard times. This is due to the fact that not only rainfall is variable but also cotton prices and the labour markets in urban areas and abroad. For example, the death of a family's principal labourer may lead to a dwindling of assets. Drought, pests, disease affecting family members and other contingencies can have an enormous impact on a family's wealth and stock of assets. The influence of such events may last for years or even decades.

Another observation that can be made from the available research results so far is that despite the fact that cash income from cotton cultivation is considerable, most households do not manage to acquire more wealth in the form of livestock, or to consolidate their stock of assets. Instead, stocks of cereals have diminished over time in most of the households studied (Nikiéma 1999 also mentions in his case studies that the ownership of considerable herds of cattle also existed in the past and that young men even worked as cattle herdsmen in other places. The role of cattle in the management of soil fertility is, however, becoming more important, meaning that cattle and livestock in general have become the favourite form of investment). Jonckers (1995: 233) found the emergence of a category of wealthy farmers, but also observed '*une lente détérioration des qualités de vie et particulièrement de l'alimentation de certaines familles. Les greniers de réserves céréalières, indispensables pour pallier les sécheresses, et signes traditionnelles de richesse, se font rares, alors même que la production céréalière n'a pas globalement régressés dans la région'*. This phenomenon is not limited to the decision-making of poor units only and there has been a general increase in vulnerability of food deficits, especially for rich households. Their risks equalled those of poor households in 1999 (Mosely 2001: 126), mainly due to the rate of indebtedness of these families.

The ways in which Minyanka decision-making units try to cover adversity are extremely varied. In agriculture, some of the pathways pursued are technological innovations, extensification, intensification, and new crop varieties. Cash income is also playing an increasingly important role. It is being generated by means of agricultural diversification (market gardening, fruit trees), wage labour (livestock herding, labour migration), trade and credit. For men and women, having different sources of cash income is the most important. Women engage in different income-earning activities and have other sources of credit to men (Nikiéma 1999).

Every household has its own specific mix of activities based on preferences, possibilities in terms of labour constraints and available capital and knowledge. Nikiéma (1999: 59-67) has shown, for example, how different the pathways have been for two brothers born in the 1940s and raised in the same compound and who farmed together on the same farm until 1984. Until then, the brothers undertook labour migration in turn. When they went their separate ways the elder brother took most of the family assets and had ten members in his compound. The younger had a family of five. Assets in the form of livestock and cereal reserves had been depleted. The cereal reserves decreased from five years to one and the family herd had been halved in the course of the 1970s because of drought and bad harvests and the large number of young children.

Cotton cultivation meant that the family of the older brother was able to rebuild the family herd somewhat during the second half of the 1980s. By speculating with the cereal reserves on the cereal market in drought years, they were also able to enlarge their assets and the family was able to invest in education for some of the children. The eldest son, having finished his studies as a veterinary technician, is working in the village as a private vet. This increases the family's social esteem. His wife prepares millet beer and one of his cousins, who lives in the compound, is the only carpenter in the village.

The family is among the wealthiest in the village. Its cereal harvest amounts to 200% of consumption needs, not counting the revenues from cotton and all the cash income derived from non-agricultural activities. The main problems for this family are related to their cultivation activities. Rainfall variability is a permanent problem and it is no longer possible to expand their fields given the pressure on village land. Soil fertility is declining and so the family, unable to obtain more land, either has to incorporate fallow parts of the fields (which is not feasible) or intensify its soil management. The production and application of manure is a time-consuming and arduous task. To entice the young to remain on the farm during the dry season, the head of the family is 'investing' money in luxury goods such as bikes, radios and a motorcycle in an attempt to bribe them to stay.

The fate of the younger brother has been quite different. Having started in 1984 with many fewer assets than his older brother, things did not go so well. His family was plagued by illness and that demanded a lot of time and resources. Their journey to find the right treatment for their son brought them into contact with the Catholic mission who gave them food rations after their conversion to Catholicism. Some families in the village also befriended them and offered help.

The production deficit they have cannot be bridged by manual labour alone. However, they are far from being able to invest in a pair of oxen and agricultural equipment. The livestock they possessed – sheep and goats – have long since been sold to buy food. The only way out seems to be their children who will, in time, add to the family's labour force and this will allow them to expand production and undertake more remunerative off-farm activities. For the moment the playing of drums at ceremonies, festivals and Catholic rituals brings in most of their cash. The wife collects twigs in the bush to make 'toothbrushes' which she sells at the market in Koutiala. This activity brings in some money each week.

In another example, Nikiéma (1999: 68-71) shows how the splitting up of a family can turn out well for the people involved. However, this family too, has faced the problem of lack of land combined with variable rainfall and declining soil fertility. They are considering taking over land outside the boundaries of the village territory. The family has diversified with fruit trees (mangos) and market gardening, new varieties of maize, and labour migration to the town and to rural areas to guard livestock. The women in the family engage in all kinds of trade. Despite all these efforts, their cereal stocks have declined over the years. At the time of the research only one granary was left of the two that were held in reserve.

A major concern is the family's growing rate of indebtedness. Increasing demands for consumption have led farmers to take out large credits and the local banking system has not taken any measures to prevent indebtedness (Nikiéma 1999, Tefft 2001). This, in combination with bad and fraudulent management of village associations by the village elites, who have to guarantee these debts, has caused the technical bankruptcy of a large number of these institutions. Consequently, investments in agriculture have virtually come to a standstill and farmers have no option but to expand the areas under cotton without applying the necessary inputs (Nikiéma, 1999). With the declining world-market prices, farmers' discontent finally reached breaking point. They went on strike and cotton production decreased by 50% in the 2000-2001 season. However, they have no choice but to produce cotton and their desire to have cash income has forced them to expand cotton cultivation even further, leading to a record harvest of cotton in 2001-2002 of almost 600,000 tons (FEWS 2002).

Migrants

There is less information on the position of immigrants in the region. Compelled by drought and attracted by the prosperous economy of Koutiala District, large numbers of people from the northern Sahelian zone have moved to the south. These migrants are mainly Dogon and FulBe but there are also migrants from neighbouring areas who belong to the Minyanka, and also to the Bamana and Malinke ethnic groups. Nijenhuis (forthcoming) found 102 Dogon, 57 FulBe and 10 Bamana in one village out of a population of more than 1,200 and in another village 92 Dogon, 71 FulBe and 12 Bamana among a total population of 896. Among the Minyanka there are a number of people who can be regarded as migrants since they only arrived recently and are regarded as strangers (Nijenhuis forthcoming). These numbers are nothing compared to the massive inflow of Mossi from the Mossi Plateau in Burkina Faso into South-western Burkina Faso where the number of migrants sometimes exceeds that of the autochthonous population (see Howorth & O'Keefe 1999, Breusers, chapter 14). Very little is known about their socio-economic position because they are often not part of the *association villagoise* and are not covered by CMDT credit programmes. Frequently they are assigned marginal lands within the village but live outside (ibid).

Migrants' pathways are distinct from those of the sedentary farmers who belong to the autochthonous population. This seems to be due to the way in which they are only partly assimilated into society in Koutiala District. They are regarded as strangers by the autochthonous population and have only secondary rights of access to land and other resources, making their position much less secure from a political point of view.

Dogon farmers

Not much is known about Dogon immigrants in the area. Some authors mention them in passing (Jonckers 1995), others (Koenig *et al.*

1998, 1999) have analysed their position but do not give specific information about their pathways. Their numbers must be quite substantial because their presence is not limited to the villages around Koutiala (see Cissé 1993, Jonckers 1995, Koenig *et al.* 1998, 1999). The Dogon migrants are different from other groups of migrants in the sense that they often come from the same area of origin and cluster around several families in small hamlets, rather than living scattered over the territory of the host villages as other groups of migrants do. Often they are related by kinship and/or marriage. These relations are also the basis for their recruitment, with new arrivals first settling in a Minyanka village with people they are related to.

In many cases these families of migrants are only temporary inhabitants of the villages. Despite the fact that most families are recent arrivals, a number have already left again for places further south where there is less pressure on land or better economic opportunities such as in the towns of Koutiala and Sikasso. Some have even returned to their home areas in Koro (see 3.2 for an analysis of Dogon pathways in Koro) or Bankass Districts.

Migration is not a recent phenomenon among the Dogon. Since the beginning of the 20th century they have colonised the Seeno and Gondo plains east of the Bandiagara Plateau (see Gallais 1975, Petit 1998), a process that still continues today. Dogon are also known as labour migrants in urban areas in Mali (see Mazur 1984). Mobility is therefore part and parcel of this population group's recent pathways though migration to the cotton-growing area around Koutiala.

There are few indications that the group of immigrants in the south of Mali is different in socio-economic and/or political respects to those that remain behind. They are reluctant to talk about their motives for settling in the south (Nijenhuis forthcoming). In general they arrive without any assets, i.e. with no livestock or farming equipment. In this sense they are not atypical in Dogon society in their home area because the head of the household often controls the stock of assets. Furthermore, people who depart are not allowed to take their share of the family's possessions with them.

The only way to escape this situation is by cultivating and building up a viable farming unit (Nijenhuis forthcoming). In their farming strategies they differ from Minyanka. They cultivate marginal land though no less than the Minyanka households (around 0.7 ha per capita) and are therefore less inclined to take out loans or assume risks for cotton cultivation. Nijenhuis (forthcoming) found that in one of her research villages only five out of nine families cultivate cotton and when they do they only cultivate a maximum area of 0.5 to 2.0 ha. However, the same could be true of other immigrant farmers (Bamana, Minyanka) living in this village.

Due to the low fertility of their parcels of land and the lack of livestock to fertilise or add manure to the land, most immigrant families are not able to grow sufficient food and often have to buy additional cereals. This is in contrast to Minyanka farmers who have a mean surplus of almost 50%. Given their weak tenure position, they are regularly chased off their land if the original owner sees fit to do so. There are also cases where they are victims of a conflict at a higher level.

FulBe herdsmen

There are two categories of FulBe herdsmen in the Koutiala area. The first consists of long-term immigrants who have either been herdsmen for the rulers of the Kenedougou Kingdom of Sikasso or are transhumants who have lived in the south of Mali for a long time. The main occupation of this group of FulBe is still cattle keeping. They are culturally FulBe though linguistically they have adopted Bamana as their mother tongue. No research has been done among this category of FulBe and no information can be found in the available literature. The second category consists of recent migrants. This migration wave started before independence, not only in the north of Mali but also in the northwest of Burkina Faso. Around one third of this category came before 1965 although immigration peaked after the droughts of the 1970s and 1980s (see also below).

The ways in which these migrants have assimilated into Minyanka society differ enormously. Some have been successful and have managed to build up a safe existence with more or less secure rights of access to resources. Van Steenbrugge (2001: 1-9) detailed the example of a family from the Bandiagara Plateau (see below) that managed to establish a complete well-equipped farming enterprise with fields planted with millet and sorghum, and a field devoted to growing cotton. The land was obtained from the local Imam. Their livestock is herded in the bush and they have been so successful that they earned sufficient money to enable the whole household to go on a pilgrimage to Mecca. Family members have travelled a lot and many relatives use this family as their point of entry into the region.

The more successful migrants develop an existence by trading livestock. Nevertheless, this can be difficult because of the capital required, the knowledge and the network of social relations needed and the potential risks. It is not uncommon for trading enterprises to collapse and go bankrupt (Van Steenbrugge 2001: 63-4). Others make use of the religious reputation of the FulBe and carve out an existence as a *marabout*, an Islamic cleric. They subsist on the payments and gifts they receive for providing religious services such as healing, the making of charms and amulets, education, and the agricultural labour and begging activities of their pupils. Such an existence implies considerable mobility: clients for religious services are widespread and the *marabout's* own religious knowledge has to be updated from time to time (Van Steenbrugge 2001: 66-67).

Other families are definitely in a less favourable position. They survive on temporary labour such as the herding of village cattle or they work for other FulBe in the region, or are involved in petty trade. Often these less successful families and individuals are much more mobile. They exploit opportunities in various places and make use of an extensive network of kin relations that gives them the chance to at least reside in some area on a temporary basis (Van Steenbrugge 2001: 53-65). Young herdsmen are an example of this strategy: they are extremely mobile and hard to locate (De Bruijn 1998).

Only a few manage to subsist from the herding of livestock and most FulBe arrived in this region with no or few livestock. Pressure on land is quite high and there are only a few areas of refuge where cattle can graze in peace during the wet season. Moreover, villages in the area are trying to close their territories off to outside herds (like in 1999).

Another salient feature of this group of FulBe herding migrants is its association with other migrant groups from their own area, Dogon, *riimaybe* (former slaves) and artisans. These groups both participate in the local economy and perform services for the FulBe based on the division of labour in the home area (Van Steenbrugge 2001: 67-68).

3.2 Douentza: La condition sahélienne

Pathways to mitigate climate variability and other exogenous changes in the Douentza area are varied. They can be subdivided according to agro-ecological zone although they differ according to ethnic group and within these ethnic groups according to the position of the status group to which an individual belongs. A further subdivision can be made according to the degree of mobility of groups and individuals. Recent migrants from more northerly areas have developed distinct pathways. Given the differences between the various agro-ecological zones we will discuss pathways here according to zone.

The Bandiagara Plateau: Cereal cultivation, marginal livestock keeping and onion cultivation

The Bandiagara Plateau is an extremely harsh production environment. Officially one third of the area is bare rock. Another third consists of soils with a depth of less than 10 cm. In addition, most soils are susceptible to seasonal drought given their shallow depth or physical properties (Cissé & Gosseye 1991, GOM 1991). Consequently, around 15% of all the land is cultivable (GOM 1991). Population pressure is high with around 25 inhabitants per km^2 or 190,000 inhabitants in the whole of Bandiagara District in 1991. Bandiagara District corresponds roughly to the area of the Bandiagara Plateau, with the exception of the northern edge of the plateau that belongs to Douentza District. These people have to live off less than 48,000 ha of cultivable soil (6.3% of the total area) (GOM 1991: 12). In view of the soil fertility and potential production (between 300 and 800 kg/ha), cereal production is insufficient to feed the population. Between 1976 and 1987, 30,000 inhabitants left the area, which resulted in population growth decreasing from 3.5% based on natural fertility to 1.35%. In some areas along the Bandiagara escarpment where population densities may reach 100 inhabitants per ha^{-1}, the population is even decreasing (GOM 1991).

In 1991, 66% of the population was entirely dependent on dryland agriculture. Given the lack of land and the pressure of population, most land is cultivated on a permanent basis. The fertility of these soils is the most important constraint on cereal production as labour is available in abundance. Chemical fertiliser is hardly used for the production of cereals. Consequently, the repletion of soil nutrients is entirely dependent on organic manure in the form of animal dung, compost and household refuse. Most farming units have some small stock at the homestead to produce manure and make compost. Farmers also buy manure from FulBe herdsmen who have settled on the plateau, and herds of small ruminants and cattle are enclosed in fields needing fertilising. In 1991, it was estimated that more than 85,000 TLU were present on the plateau, which exceeded its estimated carrying capacity by 100% (GOM 1991). However, it is highly questionable whether all this livestock is present all year round since most herds leave the area during the rainy season to pasture near the border of Burkina Faso, and leave the plateau during the dry season to make use of the pastures in the Inland Delta of the Niger. Only in the post-harvest season when harvest residues are present are large numbers of cattle to be found in the area. In the second half of the 1990s, Dogon farmers started to remove residues from their fields to feed their own stock at their homesteads, making it even less attractive for livestock to remain on the plateau.

Pathways of Dogon farmers on the plateau emanate from a complex interaction between available assets in the form of land and plots for market gardening and livestock, markets for cereals, labour and vegetables (onions) and FulBe herdsmen. The most marginal Dogon, by far the majority, depend almost entirely on the cultivation of cereals and some monetary income from labour migration. These units have a yearly deficit and can only survive on cereals purchased in the market. Keita (1992: 8), in Van Beek & Peters (1999: 109)) estimates that in the late 1970s and 1980s, millet cultivation could not feed more than 40% of the population.

Farming units on the Bandiagara escarpment are better off because there is more land available at the base of the mountain and the soils in this area are better watered (Van Beek & Banga 1992). This option has its limitations because population densities are very high on the escarpment and the productivity of the sandy soils of the Seeno plain under permanent cultivation (see below) is much lower than the land under permanent cultivation on the plateau.

Those who have access to plots for onion cultivation are able to generate some cash income to help close the cereal gap. However, the water resources for this type of strategy have only been developed in the central area of the plateau around Bandiagara and along the main transport axes from Bandiagara to Sangha, Dourou and Bankass. Between 1974 and 1987, 66 dams were built (Diawara 1987: 612) bringing the total area under irrigation to 863 ha (Keita 1992 in Van Beek & Peters 1999) for 70,000 gardeners (Krings 1991: 223), which would mean that every gardener has access to no more than 125 m^2. This is not a viable enterprise either for coping with the cereal deficit nor for increasing long-term income. Nevertheless, these tiny plots have become a replacement for staple food production. In some cases, extreme specialisation has occurred. Van Beek & Peters (1999: 108), for example, mention that the women in a particular village specialised in the production of seedlings and have been quite successful since most male onion cultivators in the area now buy their seedlings from them. In the Sangha area there are families who rely more on onion cultivation as a subsistence strategy than on cereal cultivation.

Over the past years, this pathway has become less profitable because of the competition from other areas in Mali and the importation of European (Dutch) onions. Revenues from onion growing fluctuate and are declining in relation to cereal prices (Van Beek & Peters 1999: 104-5). Moreover, there is a trade-off between market gardening and cereal cultivation because of the amounts of manure needed for gardens. In some areas livestock and cereal cultivation are not able to produce sufficient manure and compost for the gardens, let alone for the millet fields (Van Beek & Peters 1999: 109).

Revenues from labour migration and market gardening are invested in the cereal deficit, in livestock to assist in soil fertility and as a source of capital. However, the possibilities for livestock keeping on the plateau are limited. For the herds residing there, the balance is precarious. The accessibility of feed for the animals is limited in the rainy season because of the cereal fields. During the dry season the quantity and quality of feed are low, the uncultivated soils produce very little biomass and the crop residues are harvested.

A limited number of FulBe livestock keepers (around 10% of the population) are, nevertheless, trying to create an existence from livestock keeping by interacting with the Dogon. Often they live on the land of the Dogon, herd animals of wealthier Dogon farming units and camp on the margins of village territories. Although their pathways centre on livestock, they cannot manage without cereal cultivation as well. They either own too few cattle to be able to live off livestock products, or they do not own the livestock they herd and are therefore unable to market the animals (see De Bruijn & Van Dijk 1999).

Fluctuations in rainfall and market price have a tremendous impact on the welfare and food situation of the inhabitants of this area. Though cereal fields may be quite productive on the plateau, the shallow soils make the crops susceptible to intra-seasonal drought. Drought years are marked by a massive outflow of labour from the plateau with people hoping to find employment in town. Most of this outflow is temporary though permanent migration is also an important phenomenon. The latter is a sensitive and painful issue as it is mostly the young who move away leaving the old who are opposed to their departure. Young families even leave Dogon villages in secret during the night. Among the FulBe, young herdsmen just depart under some pretext, leaving their parents, wives and young children behind ('he took up his herding staff and left') (De Bruijn 1998).

The Bandiagara Plateau is a risky production environment. Given the high population pressure, Dogon farmers respond by intensifying cereal and vegetable production. As an alternative, seasonal or long-term labour migration or even permanent migration are options for procuring income and food. Given the volatility of labour, commodity markets and prices, one kind of risk is exchanged for another. In between, FulBe herdsmen and their families occupy a separate niche on the margins of Dogon society created by the acquisition of livestock by the Dogon and the presence of a large area of marginal pastureland and the enormous demand for manure by the Dogon farming and vegetable-growing systems. A major question remains with respect to the farming and herding populations in the inaccessible rural areas where no hydraulic infrastructure has been constructed. These groups are exposed to ever-greater risks because of the declining rainfall over the past decades and can only compensate for this by migrating on a temporary or permanent basis.

Plains: Livestock keeping, cereal farming, oasis, aid, trade

The constraints on pathways on the Bandiagara Plateau and escarpment have existed for a long time and have become an integral part of the most basic strategy of the Dogon for dealing with the chronic shortage of resources – agricultural expansion. The plains east of the plateau have been a major area for agricultural expansion over the past century (Gallais 1965, 1975, Brandts 2002, Petit 1998). 43 villages were established on the plains from 1905 to 1913, 46 villages from 1914 to 1940 and 23 villages from 1940-1945 (Gallais 1975: 111). An east-west migration movement of Dogon to occupy the plains originated from the sub-district of Dinangourou (Gallais 1975) and a migration of Mossi from Yatennga in Burkina who then settled to the east of Koro District

(Martinelli 1995). This colonisation of the plains is still going on in the northern part (Petit 1998, Nijenhuis 2002a, Brandts 2002).

In the course of the last hundred years, Dogon cereal cultivators have increasingly inhabited these plains. This colonisation has a different measure of intensity from south to north. At present, around 90% of the land is used for cereal cultivation near Koro and Bankass, in the south of the study area. In the north, near Douentza, colonisation is much less advanced and towards the east of the study area there are still large areas that remain untouched by cereal cultivation. In 1990, two-thirds of all cultivable land was in production. The population of the study area totalled around 250,000 people in 1987.

There are two reasons for this difference. In the first place, conditions for cereal cultivation were, and still are, much better in the south of the study area. Risks are lower and rainfall is more abundant near Koro and Bankass. A second reason is the availability of water resources. In the north of the study area there are large tracts of land that are just too far from any source of water to allow human settlement or even grazing by animals after the wet season. On the Seeno-Mango in the north of the study area, 42% of the land is more than 15 kilometres from a water point, whereas this is not a problem near Bankass (Gottschalk & Krasovskaia 1987: 87). Given the physical properties of the soil in the north, around three-quarters of the area have a high drought risk compared with less than 50% in the southern part of the plains (GOM 1990: 9).

Specific cereal and livestock production systems have developed in these areas. In the south, the extensive cultivation of millet with the help of animal traction is the most dominant. In the north, farmers have only begun to invest in agricultural equipment since the last drought (1982-85). A plough with animal traction (camels, oxen, donkeys or any combination of these) allows a farmer to expand the cultivated area significantly. This remains the most important way of increasing agricultural production and the sandy soils dominant in this area can easily be worked with this type of equipment. The majority of the land is tilled in a mono-cropping system so as to avoid competition for water between crops.

The productivity of fields varies according to area (north-south), type of soil (clay-sand), year (dry-wet), sampling technique and chance (See De Bruijn & Van Dijk - 1995: 262-65 - for a discussion of the practical and methodological pitfalls). Even within homogenous samples with respect to soil and rainfall, huge variations occur with standard deviations of productivity varying between 48% and 100% (Hesse *et al.* 1985, De Bruijn & Van Dijk 1995: 273). Production estimates in the north of the area varied between 159 and 81 kg/ha for the FulBe, and between 87 and 54 kg/ha for the *riimaybe* in 1982 and 1983 (Hesse *et al.* 1985). The years 1982 and 1983 were drought years. In 1990-91, De Bruijn & Van Dijk (1995: 273) found production levels of around 350 kg/ha. For the southern plains, figures varied between 843 and 417 kg/ha (Harts-Broekhuis & De Jong 1993: 194) during the 1980s and around 340 kg/ha in the 1970s (Eskelinen 1979: 258-59). On the plateau, productivity amounted to 630 kg/ha (Eskelinen 1979: 258-59). Total cereal production in Koro District varied between 19,000 and 51,000 tons during the 1980s (Harts-Broekhuis & De Jong 1993: 194)

With respect to livestock production, a similar story can be told. Livestock production varies with all kinds of ecological factors. Biomass production fluctuates with a standard deviation of around 40% (De Bruijn & Van Dijk 1995: 284). Time and again disasters, such as the death of large numbers of animals, occur due to dehydration and starvation. This happened in the 1970s and again in the 1980s. The total number of cattle decreased by probably as much as 60% from 1970 to 1985. After 1985, numbers began to increase again although exact cattle figures are difficult to assess since the number present varies according to the season and to specific pasture conditions in the area within the seasons. It is often not clear where the home basis of the owners of many of the animals is and how much time they spend in the study area compared with other areas.

Case studies

The whole study area is one of variations in a climatic and an economic sense, both on the plateau and on the plains. This renders the population extremely vulnerable. Some examples of the diverse ways in which various population groups respond to these changes are considered below, although the data only partly reflects reality. There is much more variety than can be shown in a small number of short case studies. Nor is it possible to give an account of the people who did not manage to survive in the study area and have since left, or those who gave up or died under the impact of ecological stress. A number of families have been followed closely for 13 years by the authors and on a recent visit, in January 2002, this process became apparent. See also the case studies of the FulBe below.

Dogon farmers

On the Bandiagara Plateau, cereal cultivation has become so marginal that there is an increasing number of production units subsisting almost entirely on non-subsistence agriculture supplemented with income from labour migration (see Van Beek & Peters 1999). Investments are made in the expansion of their cattle herds, while outside agencies occupy themselves with the physical infrastructure for cash cropping. This general evolution towards commercialisation may well be a dead-end street. Given the tiny plots available for onion cultivation (average 125m^2), the limited scope and the increasing unit costs for the expansion of irrigated areas and the competition in the onion market, the likelihood of sustained income growth is low.

Prospects for the expansion of production are somewhat better on the plains. However, growth can only be sustained if land is still available for agricultural expansion. If this is the case, this strategy could be successful. Cereal production on the plains is quite high given the quantities (around 3,000 tons per annum) being sold to people from the north who are coming into the area specifically to purchase cereals (Rutgers Van Der Loeff 2001) The cereals transported to the markets in Bandiagara, Mopti, Douentza and Booni on the backs of donkeys, using donkey carts and by truck are not included in this figure. These quantities are also considerable.

As the examples show, even in marginal conditions in the north of the study area, a concentration on cereal cultivation can be quite successful if it is supplemented with other sources of income and/or labour. The Dogon believe that if they have a large family they will be more successful as the number of children under the control of the head of the household determines how much land can be cultivated. The only way to do this is to expand production and work very hard. Maas (2001) discusses a household with 49 members of which 15 are pupils of a family member who is an Islamic teacher. This household is located in a village on the plain that grew out of a temporary cultivation hamlet. This hamlet became a permanent village when the government livestock service established a well for the pastoral FulBe in the area. Incidentally, it promoted the settlement of Dogon cultivators.

This household compensates for bad years using the income of the Islamic teacher. In 1997, after a bad harvest, he supplied the family with 250,000 CFA francs and 2 tons of cereals. The other (young) men in the household went into the cereal trade and family members gathered *Boscia senegalensis* berries as supplementary food. In the end, no food had to be bought and capital reserves in cattle and agricultural equipment were not depleted. In a normal year, the food needs of this household are largely covered and cereals can be stored for bad years. In 1998, this household harvested 8,375 kg or 168 kg per capita (243 kg per capita without the Koranic pupils who would have been away during the dry season) (Maas 2001: 14).

Small farming units can also be successful, especially when adult labour is abundantly available. One family of five recently set up a separate family unit when its head came back from a prolonged period of labour migration to Bamako, the capital of Mali, and no longer wanted to work for the less productive members of his family. During the slack season, he and his brother work for a butcher in Mopti, the regional capital about 200 kilometres from the village. They produced 3,600 kg of millet in 1998 or 720 kg per capita (Maas 2001: 18).

Both families own a number of livestock (more than 20 TLU; TLU = Tropical Livestock Unit, 1 head of cattle is 0.7 TLU, 1 camel = 1 TLU, 1 goat or 1 sheep = 0.1 TLU). Livestock is used as a cash reserve and to apply organic fertiliser to the soils. Both families enclose their livestock on their fields and try to persuade FulBe herdsmen to settle temporarily on the fields with their animals.

Farming units are in a different position. Brandts (2002) discusses the case of a middle-aged farmer in a village close to the Bandiagara escarpment who is the descendant of a migrant. He has only small fields, which he borrows from a family of autochthones. The fields are far away from the village and he does not have the means to invest in agricultural equipment. During the dry season all his sons work in town to make ends meet and he himself works as an agricultural labourer during the growing season.

FulBe herdsmen

The droughts of the 1970s and 1980s have had a tremendous impact on the pathways of FulBe herding groups in the area (data on the FulBe were gathered from 1987-2002 over several fieldwork periods (1987, 1990-1992, 1997) and short visits (1995, 1998-2002) by the authors). Before the droughts theirs was a relatively easy form of subsistence based on a herd of cattle which yielded sufficient milk to cover either a significant part of the food needs of the family or, when bartered, brought in enough cereals to feed the family for part of the year. With the droughts, they were not only confronted with famine and a massive loss of livestock, but also had to face increasing competition for land from the Dogon who expanded cultivation and have to compete for pastures with incoming herds from the Inland Delta of the Niger. A lot of FulBe failed to cope with this situation and migrated to areas like Koutiala District (see above) or to major towns. Numerous families collapsed and became dependent on urban and sedentary livestock owners for employment and some (mostly very little) social protection. These families ended up on the margins of society. Young men, in the absence of livestock to herd and without prospects of employment, left in large numbers with sometimes devastating effects for the women, children and the old who remained behind (De Bruijn & Van Dijk forthcoming, De Bruijn 1998).

Those who stayed in the study area slowly started to rebuild their lives after the droughts. Their living conditions have improved slightly although their experiences have been mixed. For example, one family had the chance to herd a trader's cattle which enabled them to drink the herd's milk while economising on cereals and to build up a herd of goats to act as a buffer in times of food shortage in subsequent years. If necessary, the goats could also be converted to cash. Given their political position, they also had preferential access to a dry-season grazing area. These factors promoted the reconstitution of the cattle herd. Investments were made in a plough with camel traction to expand cereal cultivation and to facilitate transport. Some years later, a donkey cart was bought to transport water to temporary cattle camps in the bush, to enlarge the herding orbit and to open up access to faraway pastures for the cattle and goat-herd. Quite unusually, none of the sons left the family to try their luck in town or elsewhere. In 2002, 17 years after the drought, the family was prospering and had doubled in size.

The fate of their neighbours was quite different. This family, consisting of three brothers, their wives and children, also lost most of their livestock during the drought. One brother left after a quarrel, the second decided to move south to settle on the land of a Dogon farmer and to concentrate his efforts on small ruminants, and the eldest brother decided to bet on reinvesting in cattle in combination with cereal cultivation and to remain in his cattle camp. None of them was successful. The family of the brother who left suffered tremendously and only survived on a combination of herding labour for Dogon farmers away from home, seasonal labour migration, charity, the gathering of bush products (very humiliating for a noble pastoralist family) and cereal cultivation. In the end, the girls were decently married off, the father and the sons returned and they resettled in their home camp. The second brother remained in the village of his Dogon patron for 10 years, surviving on the most meagre means imaginable. Upon his return, he and his wife were so exhausted by the suffering they had experienced that they both died within a couple of years of each other.

The eldest brother also had a hard time. He decided to stick to cattle and to work hard at cereal cultivation with his sons to rebuild the herd. This proved an extremely difficult strategy. The gain in herd size during good years was cancelled out in bad years when he had to sell cattle to buy food. His sons ran away all the time to work as cattle herdsmen elsewhere and their contribution to the family was limited because their wages were low and they preferred to spend their earning on radios and cassette players. However, he was lucky in that one of his daughters, contrary to cultural conventions, married a well-to-do Dogon trader who helped the family during hard times. A second daughter, though unable to bear children, became the favourite wife of an important Muslim scholar in the area and was able to give some things to her parents as well. During the post-drought period, no investments were made to expand the family's productive base. Some gains were made in the number of livestock owned but the family fields were not fertilised and cereal yields started to decrease.

Riimaybe

Traditionally, the *riimaibe*, the former slaves of the FulBe, were at the bottom of the social hierarchy in the area. In the past, they were the first to suffer in situations of scarcity and the last to benefit from a period of prosperity. In a number of respects, they are still more vulnerable than other groups in FulBe society. At the same time, given their status and their attitude towards wage labour, they are able to respond much more flexibly to new opportunities than their FulBe overlords (De Bruijn & Van Dijk 1994). However, a distinction must be made between those who still belong to the network of clients of the noble FulBe and live near their former masters and those who have always lived in separate villages. Whereas the former still have to make contributions in kind and labour to the welfare and the household of their masters, the latter have managed to gain a certain measure of manoeuvring space both in an economic and a political sense.

Revenues from cereal cultivation are generally insufficient to cover the food needs of a family. All kinds of activities and seasonal labour migration are undertaken to close the cereal gap and to cover other expenses. For example, a family of nine people might cultivate three millet fields that yield between 30 (1997) and 80 (1996) bundles of millet (about 600 and 1,600 kg of millet respectively when the bundles are large). One of the sons settled permanently in town and does not contribute to the family. The second son is married and works temporarily in town each year as a weaver. This does not bring in a lot of money but at least he no longer makes demands on the household budget. His wife has a personal field, which yielded another 6 bundles of millet in 1998. She cultivates about 50 kg of rice in the pond next to the village and waters an onion garden every day. She and her mother-in-law also weave mats of palm leaves after the harvest, bringing in 125-150 CFA francs a day if she works very hard. In addition, she took out a loan with a women's credit association, which allowed her to buy goats. She is paying back the money with the revenues from her mats and by speculating on the livestock market with her goats. The revenues from this large spectre of activities are small and do not offer much scope for expansion. Misfortune in the form of a sick family member or a bad harvest may seriously jeopardise the family enterprise.

Among the *riimaybe*, the diversification of activities is enormous. Griep (2001: 92-93) mentions a woman who raises small stock at her homestead, cultivates rice and millet on her personal fields, engages in petty trade and participates in activities organised by development organisations, all in addition to her household chores. In other places where cereal production is better, the women will engage in fewer activities.

An important aspect of the development of new pathways is investing in dry-season water resources. Ponds and wells are being managed in new ways so that better use can be made of the water, and even food crops, such as beans and cash crops like garlic, perennial cotton, pumpkin, mangos etc, are produced in oases-like units. While developing these water resources, the *riimaybe* make creative use of the contributions by NGOs to development, to the extent that these have become an integral part of the income of *riimaybe* families (Griep 2001).

4. ANALYSIS AND CONCLUSION

In the concluding section of this chapter, two sets of questions are discussed. The first relates to the commonalties and differences between the two study areas. How can the structural differences between the areas be appreciated and how can these relate to the exposure to risk and the pathways people develop to mitigate these risks? The second set of questions is related to discussions concerning policy development and focuses on the specific role of climate variability within the study areas and the possible consequences of climate change. As a starting point one of the scenario baselines is used, as outlined by Van Den Born *et al.* (2000; also see chapter 5), predicting a decrease in rainfall in Koutiala of approximately 25% and in Douentza of approximately 50% and a substantial increase in variability and risk.

It is obvious from the discussion of the two study areas that the differences are vast and pervade all aspects of life. The differences are not only confined to climate and climate variability and the physical properties of the areas but can also be found in the institutional and economic context and the socio-cultural set-up of the areas, though the latter have developed due to interaction with climate and the environment.

A point common to both areas is the high level of risk in agricultural and economic activities in general. These risks are the result of the variability of the climate and of the economic context in the form of cereal and cotton markets. Moreover, in the local and even in the personal context, further risks can be identified pertaining to health, micro-climatic conditions and pests. Solutions are found in the diversification of activities in combination with mobility in the form of labour migration or permanent migration, and even in the set-up of multi-spatial livelihoods (cf. Foeken & Owuor 2001). In a number of cases, therefore, a much wider range of factors should be taken into consideration than the conditions in the study area alone. Individuals also make decisions with respect to investments in agriculture and their home area not only with a view on the conditions in place but also while bearing in mind the possibilities elsewhere. This will only become clear when considering the individual case studies as collected over the years in more detail than is possible in this chapter. The role of individual differences as an explanatory factor for the diversity of specific pathways of households and individuals and any other decision-making units cannot be stressed enough in this respect. Hence there is a need for a methodology that goes beyond this enormous variety in responses and focuses much more on the processes of decision-making under high-risk conditions (See Chapter 18).

Comparing the two study areas yields the same focal points as the most relevant for the understanding of pathways to mitigate climate variability and climate change. In both areas a number of factors seem to structure pathways: access to resources, ethnic and social identity, and life-history events. There seems to be no single development path related to resource endowments, ethnic identity and/or life-cycle events. It can be argued that specific pathways cluster in certain groups of people, such as the preference for livestock keeping and a high degree of mobility among FulBe pastoralists, as well as the preference for cereal cultivation and a more sedentary mode of existence among Dogon, *riimaybe* and Minyanka farming groups. However, this does not rule out the fact that large numbers of farmers are also mobile both within the study areas and beyond. In general, people in Douentza seem more mobile than those in Koutiala, given the higher degree of risk embedded in the ecological environment and the technology available to mitigate these risks.

However, the role of contextual factors differs substantially between the areas. Rainfall variability and, in its wake, soil quality (expressed in water-holding capacity because differences in soil fertility are not that important) play a much larger role in Douentza than in Koutiala because of the more irregular rainfall patterns. These differences are directly reflected in food security. In Douentza most farming and herding units have a food deficit in most years, whereas in Koutiala most farming and herding units are food secure in almost all years. Cereal productivity is much higher in Koutiala, especially that of maize that cannot even be cultivated in Douentza, and there are fewer fluctuations.

In contrast, governmental and non-governmental interventions seem to play a more important role in Koutiala than in Douentza in shaping individual pathways. In the Douentza area, outside interventions are not able to produce more than a marginal increase in income, whereas they have a direct impact down to the village level and on a number of aspects of the functioning of the farming unit in Koutiala. Because of the particular form of interventions in Koutiala, they seem to enhance the risk through the increasing indebtedness of farmers and their reliance on volatile markets for export commodities. The low labour productivity of cotton cultivation and constant pressure from the Cotton Company to produce more puts a premium on the depletion of soil fertility. One cannot blame farmers for being price responsive and for not investing in soil fertility if this lowers their marginal returns on labour below their opportunity costs.

An important feature of individual pathways is the mobility across regional, agro-ecological and social boundaries. This is most significant in the case of the Douentza area. This mobility occurs at all levels –

local, regional, sub-regional, national and international. There is no household without experience of mobility and there is no continuity without income from outside, except for some individual cases where consumption demands are still limited.

Though specific pathways cluster in certain areas with particular groups, the range of options is enormous. Within the FulBe, a distinction can be made between the nobles and the *riimaybe*, between the people living in their home area and those on the fringes of sedentary villages. Within the sedentary groups a distinction must be made between what can be labelled the autochthones and the immigrants, the latter in general being in a less favourable position with respect to access to resources. This does not mean, however, that those within the same cluster follow similar pathways. The examples of the three FulBe brothers in Douentza and the two Minyanka brothers in Koutiala show how different life can be even for people with the same starting position. Pathways also cluster on a regional basis and within specific regions. They develop in response to distinctive contextual factors, in this case rainfall, landscape and associated soils, and the rainfall gradient (see also Brons *et al.* chapter 16).

This grouping of pathways and the patterns that emerge from a detailed overview make it possible to look to the future and project existing trends of mitigation of climate variability in the future. We have chosen for a scenario of a 50% (Douentza) and 25% (Koutiala) decrease in precipitation. This is a worst-case scenario. However, the reverse increase in precipitation would create as many but different difficulties for these changes would upset the water balance of the whole region and lead to increased flooding and the need to adapt cropping patterns, which would require as much adaptation as would a decrease in rainfall.

Douentza

The consequences of the drought scenario are huge and would result in a major shift in pathways. With rainfall of around 200 mm in the Douentza study area, current and already low levels of cereal production could not be maintained and would be reduced to almost zero. Only in depressions and near water holes would there be any possibility for oasis-type farming systems and dry-season cultivation systems, as presently found on the Bandiagara Plateau and in some places on the plains. Those with no access to these water sources would have to shift from cereal cultivation in Douentza to extensive livestock keeping, in combination with the gathering of wild grains such as *fonio* (*Panicum laetum*) and *cram-cram* (*Cenchrus biflorus*).

The development of these oasis-type agricultural systems in the Douentza area would require large investments in hydraulic infrastructure and the development of marketing channels for the cash-crop-oriented production. Given the failure of local cereal production systems, these investments could only be made with the help of outside agencies or on the basis of the remittances of migrants from the area. Neither source of investment would be assured and, given the current levels of investment, it is highly unlikely that sufficient production capacity could be created in this way.

In addition it is to be expected that more ecological refugees from the north would arrive in the study area and clog up the towns (at least this was the case in the 1970s and 1980s) or they may start wandering through the area looking for food and water. This would put extra pressure upon production systems that are already under stress.

Given the current lack of non-agricultural sources of income in Douentza it is improbable that labour-intensive non-agricultural activities would develop here. The other options, i.e. gathering activities and extensive nomadic livestock keeping, would not be able to absorb the surplus labour from cereal cultivators who had been forced to abandon their activities. Because of the decreasing amount of rainfall and the increase in variability, the uncertainty about the productivity of the natural environment would augment and its absolute value decline. The capacity to produce sufficient food in the form of wild grains and livestock products would diminish. Food would either have to be imported from outside or people would have to move out of the area.

Koutiala

A different scenario can be drawn for the Koutiala area with no fewer consequences and with a severe impact on the national economy of Mali. A decline of 25% of mean annual precipitation would make the cultivation of cotton and maize impossible in large parts of the study area. Only in the south, where rainfall levels of around 800 mm are present, would cotton and maize continue to be important crops. This would deal a major blow to the economy in the area and in Mali as a whole, since cotton presently accounts for 50% of export earnings and it would also virtually wipe out most of the employment in the industrialised cotton sector. Urban areas would no longer be the motor for economic development in the area and would not be able to absorb surplus labour.

The most viable alternative for the farming and herding population would be to develop the intensive cultivation of cereals See, for example, Toulmin (1993) for a description of similar processes in more northern areas of Mali during the 1970s and 1980s). The main bottlenecks for this intensification process would be the variable amount of rainfall and the maintenance of soil fertility. It would not be profitable to use chemical fertilisers for food crop production and, given the high population density and the amount of land already under cultivation, there would not be sufficient space to produce organic manure from livestock.

Moreover, the Koutiala area would almost certainly become a major area of refuge for people from the north who would have to move when their home area becomes too dry. Already there are problems with absorbing the numbers of FulBe and Dogon migrants in Koutiala. In other parts of the south of Mali, there is still some space but there too production would decline as a result of deteriorating rainfall.

An enormous increase in population mobility would act as the main pathway for coping with change. Farmers and herdsmen who lose a substantial part of their cash income would try to compensate for this with seasonal labour migration to nearby coastal countries. Permanent migration of families and individuals would increase dramatically. Those who remain behind would become increasingly dependent on external sources of income and aid.

We will not venture here into the possibilities of increasing conflict due to more intense competition for resources because the relations between scarcity and conflict have not yet been established with any certainty. However, the chances of civil unrest are considerable given the near total disruption of rural and urban production systems, the adaptations that would have to be made at all levels of society and the emergence of large numbers of people wandering around the countryside looking for food. It is almost certain that such changes would have some impact on relations with neighbouring countries too, because there would be an enormous outflow of people from both the study areas and the whole zone would suffer from climate change. Providing the affected zone with new sources of livelihood and providing all migrants with a new basis of existence would be the main problems for policy makers in this part of Africa if the climate were to change to the extent that such major disruptions in these vulnerable economies occurred.

Chapter 14

RESPONSES TO CLIMATE VARIABILITY IN THE KAYA REGION, BURKINA FASO

Mark Breusers

Abstract: This chapter provides a detailed historical account of changes in the Kaya Region of northern Burkina Faso, an area that was selected as an example of the most difficult conditions in the West-African drylands: semi-arid, rural, high population density and relatively high levels of land degradation. Climate change and climate variability are put in a perspective along with many other factors such as the impact of colonialism, population growth and long-distance migration, mainly to the Cote d'Ivoire, the world market for coffee, and cocoa, and recent political developments in Burkina Faso itself. The droughts of the 1970s and 1980s resulted in a shift towards diversified livelihoods and multi-locational security networks and to changes in kinship relationships. Flexibility has been the key to survival and this is something that is often not very well understood by development planners involved in the area.

1. INTRODUCTION

The Kaya region is located in the Sudano-Sahel zone of Burkina Faso and covers the Bam province, most of the Sanmatenga province and smaller parts of the Yatenga and Soum provinces. The region is essentially rural, with only Kaya, the sixth town of Burkina Faso, being urban in character. Although diverse agro-ecological niches are encountered throughout the region, the overall variability in agro-ecological conditions is relatively limited. The tree savannah vegetation changes from north to south, with thorn bushes, which are gradually receding for dispersed trees and with a more dense tree vegetation occurring along the valleys of seasonal rivers and rivulets. Soils vary mainly in accordance to slopes and erosion and deposition by seasonal rivers. Sandy and gravely soils, with a low water retention capacity, dominate higher up the hills. They are often heavily degraded, have a

A.J. Dietz et al. (eds.),
The Impact of Climate Change on Drylands: With a Focus on West-Africa, 207–241.

low organic matter content and many of them are considered unsuited for crop cultivation. Nearer to the depressions, soils tend to become limy and further down, in the so-called *bas-fonds*, they are more fertile and clayey, but heavy to cultivate and they run the risk of inundation in years of abundant rainfall. In the north of the region, where ancient dunes have formed, there are deep sandy soils, with a limited fertility but allowing good millet and sorghum crops if rainfall and manure are sufficient (Brons *et al.* 2000).

The main activity of about ninety per cent of the households is crop cultivation or animal husbandry. Secondary activities are generally agricultural as well: animal husbandry or, in the zones where it is feasible, the cultivation of vegetables. Moose farmers constitute the large majority of the population. Their cropping systems are based on the cultivation of millet and sorghum and various other crops, while livestock rearing generally constitutes a crucial element of their production systems as well. FulBe, the largest minority group in the region, focus most on cattle herding but are also involved in the cultivation of cereals and some other crops. The importance of trade and off-farm activities within the region remains small in terms of monetary income, but are engaged in by many households , notably in the form of various artisan activities of which weaving is the most prominent. Other off-farm activities include cloth dying , bicycle repairs, forging and pottery, cart renting and trade, while there are also butchers, tanners, teachers, basket weavers, etc. In the second half of the 1980s, gold digging temporarily gained in importance. There is almost no industry. However, long-distance migration, particularly to the Côte d'Ivoire, is of crucial importance especially for Moose's livelihoods (Broekhuyse 1974: 94; Département du Centre-Nord 1978; DRPC 1990: 11, 90).

The chapter starts with a short history of the weather during the majority of the twentieth century, followed by an outline of the possible links between climate variability, on the one hand, and environmental degradation and water availability, on the other. Before turning to a summarised presentation of a number of possible pathways enacted by actors in response to climate variability and other environmental processes, we focus first on food security and yield development and secondly on changes in key aspects of the production systems in the study region. In the concluding section, we attempt to clarify how pathways are related to climate variability and to certain institutional changes. Finally, some major trends and possible policy implications are discussed.

2. CLIMATE CHANGE OR CLIMATE VARIABILITY? A SHORT HISTORY OF THE WEATHER

The Kaya region is situated in between the 600 and 700 mm isohyets. After a long dry season of about eight months, the rainy season generally starts in June and lasts, on average, for four months. Average yearly rainfall in Kaya in the period 1921-1997 was 684 mm, whereas the number of rainy days per year averaged 48 in 1959-91. The months July and August are the most humid, and account for 60 per cent of annual rainfall. Moreover, these are usually the only months during which rainfall exceeds potential evapo-transpiration. The force of evapo-transpiration is such that water quantities in the soils are low during most of the year (Marchal 1983: 45-7).

A crucial characteristic of the region's climate is the variability and irregularity of rainfall. Total rainfall varies considerably from year to year. Extremes for Kaya were recorded in 1943 (1,008 mm) and in 1985 (454 mm). The beginning and end of rainy seasons are irregular. However, the rains are unevenly distributed over the season, which is often marked by prolonged dry spells of several weeks. Showers are often heavy and highly localised, explaining the substantial rainfall differences from one village to another. Eighty per cent of the rains falls in high intensity showers of 10 mm and more, surpassing the infiltration capacity of the soil and entailing run-off and erosion, sometimes with devastating effects (De Graaff 1996: 151; Marchal 1983: 43).

Although not exempted from irregularity and variability, the 1940s and 1950s were marked by a high frequency of good rainfall years, meaning that rains were often well distributed in terms of space and time and/or were relatively abundant, resulting in frequent excellent harvests. In the first half of the 1960s, rains became increasingly irregular and this situation worsened in 1966-69. Nevertheless, rainfall insufficiency was not widespread throughout the region. This changed in the years 1970-72, which were exceptional in the sense that rainfall deficits occurred everywhere, although they were most pronounced in the east and the north, with both total annual rainfall and the number of rainy days remaining substantially below average figures at all stations. Following a temporary recovery in 1974-75, rainfall deteriorated again in the second half of the 1970s. The erratic distribution of rains in time and space had disastrous consequences locally, notably in the vicinity of Barsalogho (Breusers 2002; Marchal 1983: 46). Still, if the years 1970-72 were exceptional, this was even more so the case for the 1980s as a whole. Only in 1988, and at some places in 1981, were rainfall figures not far below long-term averages (see Breusers 2002; Reardon *et al.* 1988: 1066). Especially in the years 1983-85 and 1987, the scantiness of both total rainfall and number of rainy days throughout the region was without precedence in recorded history.

It did not take long for the drought of 1970-72 to prompt observations about the possibility of a change of climate taking place. Lahuec & Marchal (1979: 102) presented maps, comprising the Kaya region, that showed the movement of isohyets to the south from 1951 to 1972 and the acceleration of this movement in the period 1966-72. They stressed that the effects of decreasing rainfall were further exacerbated by its growing irregularity. No wonder, then, that the extremely low rainfall of the 1980s was often interpreted as a further indication of climate change. Van Zutphen (1991: 118), for instance, argued that statistics for the 1970s and 1980s confirmed the decline of annual rainfall and showed a shortening of the rainy season, both of which negatively affected the planting and growing season and the cattle carrying capacity in the Kaya region. Zanen (1996: 63) supported this view and maintained furthermore that not only annual rainfall levels had dropped, but that desiccation was also tending to become more severe from south to north.

Not everybody, however, was as convinced about the fact that climate change was indeed occurring. For instance, although agreeing that the Sahel zone had shifted southward since the 1950s, thus covering vast areas previously considered Sudanian and now classified as Sahel-Sudanian, Broekhuyse nevertheless maintained that if one examined a period of forty to sixty years, rainfall had remained the same or had decreased only slightly. He understood the problems of drought and desertification in the Kaya region first and foremost as the consequence of the destruction of the hydrological system, caused by unwarranted land use and entailing increased run-off and decreased infiltration (Broekhuyse 1985: 2-4; Broekhuyse & Allen 1988: 332).

If, then, there is little doubt that rainfall in the Kaya region during the 1970s and 1980s was substantially and almost consistently below the averages calculated on the basis of measurements during previous decades, as well as being very erratically distributed in terms of time and space, it is difficult to draw conclusions about whether or not the pattern observed during this period should be understood as a sustained trend of climate change or merely as cyclical climate variability. This is all the more so since data for the 1990s indicates at least a partial recovery of rainfall despite the fact that good years continued to alternate with years in which rainfall was relatively low and irregular (see also Brons *et al.* 2000). Moreover, historical information and, more specifically, data available for the twentieth century as a whole suggest that short-term cycles of dryer and wetter years have for years tended to succeed one another. For instance, the years 1910-1935 were relatively dry as well and were marked by several droughts (notably in 1913-14 and 1925-26) and harvest failures.

3. CLIMATE VARIABILITY, ENVIRONMENTAL DEGRADATION AND PROBLEMS OF WATER AVAILABILITY

The relationship between, on the one hand, drought and climate variability in general and, on the other hand, environmental degradation, is a complex one and mediated by factors such as population dynamics, land use practices and urbanisation. For instance, soil degradation, identified for decades as one of the region's major problems, is said to be closely related to the progressive disappearance of fallow, which occurred at such a pace that, by the 1970s, farmers were simply abandoning completely exhausted soils. The concomitant disappearance of the tree savannah eliminated the sponge effect of the hydrologic system. Large quantities of rainwater rapidly ran off, infiltration on slopes was reduced, superior soil layers eroded fast and lowlands became inundated (Broekhuyse 1983; Broekhuyse & Allen 1988: 333-4). If, then, soil degradation is firstly related to a continuous extension of cultivated areas – itself partly caused by population growth – and an absence of measures to intensify land use, erosion processes were aggravated as a consequence of the altered rainfall pattern of the 1970s and 1980s, when not only annual precipitation decreased but rainfall also became increasingly irregular. Fewer and more intense rains were even less well absorbed by the soil and had a greater erosive impact, especially at the start of rainy seasons (Marchal 1983: 44; Webb 1995: 9).

Run-off and erosion were further intensified because of vegetation degradation; trees and shrubs retracted as a consequence of the exploitation of wood reserves, the grazing by livestock and, especially, the extension of the cultivated area. Grass vegetation was affected even more spectacularly, and barren spots extended (Marchal 1983: 143-9, 226). As early as in the beginning of the 1970s, pastures were homogenised and the distinction between 'natural' and 'fallow' pastures had lost its meaning since, with the exception of zones unsuitable for agriculture or irremediably degraded soils, all land had been under cultivation during the preceding thirty years. Thirty years is considered the minimal period needed for the reconstitution of soil fertility when all conditions are favourable. Perennial species had disappeared and the majority of annual grasses were tenacious. The drought of 1971-72 would have played no major role in this regressive dynamic which, for that matter, did not tend to stabilise (Benoit 1982b: 75-6, 108). However, other observers pointed to the remarkable resilience of herbal vegetation in semi-arid regions in which perennial grasses at a certain moment during the droughts could disappear almost completely to reappear swiftly when rainfall circumstances became more favourable again (see De Bruijn & Van Dijk 1999).

Finally, droughts and climate variability have, of course, had a serious impact on the availability of water for human and animal consumption – a major problem in the region. The Kaya region is poorly endowed with water resources. It is crossed by merely one 'river' of importance, the Nakambé and to make matters worse this river is only flowing for part of the year. Several natural lakes exist scattered over the region, the *Lac de Bam* with a length of about fifteen kilometres being the largest. Besides, there are numerous small artificial lakes and temporary pools. Most of those dry up well before the end of the dry season and many of those that do not become muddy are no longer usable from January or February onwards. Furthermore, people and livestock depend for their water provision on wells and pumps and on a number of smaller and larger barrages. With regard to subterranean water resources, the region is one of the most deprived zones of Burkina Faso. Groundwater is situated in local basins, isolated from one another during the dry season but partially connected during the rainy season (DRPC 1990: 69; Ducellier 1956).

The droughts of the 1970s and 1980s caused a further serious decline in water availability. Groundwater reserves were fed only occasionally because of high evapo-transpiration, high run-off, and irregular, short and heavy rains, and in the dry season of 1975 groundwater was at its lowest level for some fifty years. This decrease could not be attributed solely to the reduced rainfall during the preceding decade. It was exacerbated by increased run-off following deforestation and by the increased demand for water (Benoit 1982b: 77-9; Marchal 1983: 46). Because of the persistence of both drought and soil degradation (reduced water infiltration), the groundwater level continued to decline by 0.5 metres per year from 1978 to a minimum in 1985, after which it stabilised and in 1990 even seemed to increase again slightly. The consequences for both human and animal populations, which were used to acquiring water from shallow wells, were dramatic with several water points running dry. By the end of the 1980s, the population had become more dependent than before on 'modern' water points (deep wells, bore holes) and surface water (barrages and pools) (DRPC 1990: 69-71; Lahuec & Marchal 1979: 132). Over the last ten years, government services and development projects made a considerable effort to improve the water situation through borehole and well construction programmes.

4. FOOD SECURITY AND PRODUCTION VARIABILITY

Since pre-colonial times, food shortage and even famine have been a recurrent phenomenon in the Kaya region. A notorious example is the Zogoré hunger of the 1830s, which lasted at least six years and killed tens of thousands of people in Yatenga alone. Another famine struck the region in 1887 (Izard 1982: 373; Marc 1909: 33). Several famines occurred in the colonial era (1897-1960) as well, notably in 1913-14, 1926-27 and 1933-34. Besides rainfall, other events were also important in explaining the harvest failures that entailed food shortages. Hence, the famine of 1926-27 can only be understood by taking into account the impact of colonial policies, which inhibited the building up of reserves (e.g. forced labour recruitment and forced cotton cultivation by which labour was drained from food crop production to other activities), while in 1933-34 insufficient and irregular rainfall and locust plagues combined to destroy harvests (see also Gervais 1990: 247). Although localised food shortages continued to occur (see, for instance, Dubourg 1957), the population's food situation was relatively favourable during the 1940s, 1950s and the first half of the 1960s. As the rainfall pattern deteriorated from the end of the 1960s onwards, so did the food situation. A series of years of insufficient and irregular rainfall and the concomitant exhaustion of cereal reserves resulted in the severe famines of 1970-72 and 1983-87.

During the crisis of the early 1970s, most people depended on the market or on food aid to provide for their subsistence, but it was nevertheless argued at that time that in normal years farmers were generally able to produce their own sorghum and millet subsistence. In the following years, this image changed, and it became more and more accepted that cereal production in the Kaya region was structurally deficient. For instance, Broekhuyse & Allen (1988: 333-4) calculated that an average, 'traditional' production unit was confronted with an average food shortage of 57 per cent, which was compensated partly by the purchase of food with money earned from livestock sales, marketed crops (groundnuts) and off-farm activities. However, even if farmers were to spend their complete monetary income on food, they would generally not have been able to meet the requirement of 2,000 cereal Calories per person per day. During the 1980s, the cereal balance in the Kaya region showed a deficit for all years, with the exception of 1988. Over the period 1975-87, the probability of a cereal shortage occurring in any one year was found to be almost 1, whereas the expected shortage amounted to 40 per cent of cereal demand. Problems of cereal deficits continued, for that matter, into the 1990s (ADRK 1995: 73; DRPC 1990: 139; Maatman *et al.* 1992: 161; Ministère de l'Agriculture et des Ressources Animales 1995b, 1996b, 1997, 1998).

In the 1970s and 1980s, food production results again appeared to be closely related to rainfall circumstances. The population's food security, however, evidently depended not only on the harvest results in any one year or on the presence of food reserves built up in previous years, but equally on the possibilities for food importation and aid, the availability and prices of food in markets, the existence of alternative sources of income and the population's purchasing power in general. In this regard it is important to note that during years of severe cereal shortages, the supply to the local markets was sometimes nil because traders were more interested in selling their reserves in urban centres where the purchasing power was higher and/or because the parastatal cereal-marketing organisation OFNACER (*Office National des Céréales*) could not supply cereals in time (Maatman *et al.* 1992: 166-9). OFNACER was established in 1971 to supply cereals to regions and cities in case of shortage on the private market and to stabilise consumer prices. It faced many problems (the organisation of the distribution of the cereals, the financing of the purchase from farmers, the financing of safety stocks, etc.) and was abolished during the structural adjustment programme concluded in 1991 between the Burkinabé government and the World Bank and the International Monetary Fund.

Yield development

Broekhuyse & Allen (1988: 332-3) maintain that 'original' (*sic*) sorghum and millet yields were about 700 to 900 kg per hectare, which yields progressively declined in line with population growth and the shortening of fallow periods to a level of 400 kg per hectare in the 1980s. A yield of 400 kg per hectare would be considered a good yield, with fields being abandoned only at a 200 kg per hectare production. However, a closer look at available productivity figures shows a different and much more blurred picture of cereal yield development. First, changes in rainfall amounts and distribution obviously played a role. The 1950s were followed by a period of some three decades during which average yearly rainfall decreased and the growing period of crops shortened. Moreover, as a consequence of the greater erosive impact of rains, water was less thoroughly absorbed by the soil and thus less available to nourish crops. Between 1975 and 1985, cereal yields decreased and, because of population growth and despite an extension of the cultivated area, per capita agricultural production declined even more sharply. However, in the first half of the 1990s, thanks to improving rainfall, millet and sorghum yields of well above 500 kg per hectare were again regularly recorded in the Kaya region (see Ministère de l'Agriculture et des Ressources Animales 1995a, 1995b, 1996b, 1997, 1998). Moreover, rainfall decline and variability constitutes only part of the story. In Yatenga, for instance, yields increased and the cultivated area decreased in 1975-85 (Maatman & Schweigman 1995: 26; Van Zutphen 1991: 118), suggesting processes of land use intensification taking place.

5. CHANGE OF PRODUCTION SYSTEMS

Although, in the early 1990s, the availability of land as well as the level of occupation of cultivable land varied greatly, farmers were said to suffer from land scarcity throughout the region. The gloomy perspective for land use was further compounded by high animal pressure on natural resources. For instance, for the three provinces Bam, Sanmatenga and Namentenga, the 'carrying capacity' was estimated to be only 65 per cent of the number of livestock actually present in the region in 1990 (ADRK 1995: 49; DRPC 1990: 142-4). Although during the droughts of the 1970s and 1980s, livestock losses were high, many herdsmen migrated away and many people were forced to sell their livestock in order to survive, animal pressure on natural resources in the end did not diminish. Herdsmen from the north moved into the region and small livestock herds in particular were rapidly reconstituted.

A progressive saturation of space was observed in many studies of land use carried out in the Kaya region, especially in the decades following the Second World War. Population growth and the necessity for the population to cover its subsistence needs, together with the introduction – and during part of the colonial period forced cultivation – of cash crops (notably groundnut and cotton) were held responsible for the extension of the cultivated area, which was further stimulated by the so-called 'atomisation' of production units. Although farmers adjusted to growing population pressure, not only by an extension of cultivated areas and by shortening fallow periods but also by applying manure on a larger part of the farm land, by gradually using more mineral fertilisers and by an extension of soil preparation and conservation activities, this tendency was hampered by constraints of labour and manure (Maatman *et al.* 1992: 172, citing Prudencio 1987). Repeatedly – and this at least since the end of the 1950s – the prevailing land use system was seen as having reached the limits of its possibilities, given its sole reliance on the progressively disappearing fallow to reconstitute soil fertility (see, for instance, Boutillier 1964; Broekhuyse 1974; Broekhuyse & Allen 1988). It is deemed necessary that, for production to at least keep pace with population growth, both crop cultivation and animal husbandry have to be intensified, implying the implementation of a technological package including manure production and application and animal traction. Land use improvement programmes of development projects in the region have been consistently based on these premises, especially since the early 1970s.

Crop cultivation

Formerly, agricultural production among the Moose occurred in large social groups, albeit hierarchically organised internally into smaller sub-groups. First, all members of a *yiiri* (kin group) worked on this kin group's collective field (*puugkasenga*). Of secondary importance was the work by the members of each compound on the compound's collective fields and thirdly there was the work on the *beolse*, the 'individual' fields. The *yiiri* elder controlled the production of the *puugkasenga* and the compound elder the production of the collective field of the compound. The individual to whom the plot was allocated controlled the *beolga* harvest (Marchal 1983: 474-5). In the course of the twentieth century, agricultural production units became progressively smaller and individual production tended to gain on collective production: the collective field came to be managed at a lower level of social organisation (the compound or a nuclear family within the compound instead of the *yiiri*) and the relative number of *beolse* multiplied. Monetisation of the local economy, during colonial rule partly enforced by the imposition of a head tax, and male-dominated labour migration contributed to this tendency of what has been referred to as the 'atomisation' of production units and the 'individualisation' of agricultural production (Lallemand 1977; Marchal 1983: 476-7). Migration also accentuated labour scarcity as the bottleneck of agricultural production.

Furthermore, reciprocal labour exchange declined, among other things because of its incompatibility with the cultivation of cash crops such as cotton, and the signification of labour parties changed from an expression of solidarity, a way to actualise and to reinforce social relations, into instruments for the promotion of individual prestige. Changes in labour organisation and increasing labour scarcity contributed to an extensification of land use practices (Kohler 1971: 108-16; Marchal 1987).

In the 1970s and 1980s, studies of land use change time and again emphasised that cultivated areas were continuously extended while there was hardly any intensification of land use practices. Marchal (1983: 395-418) found that although, on the whole, the principles of cultivation (crops, techniques and tools) had remained unaltered since the beginning of the twentieth century, important changes had nevertheless occurred regarding the priority given to different crops, the agricultural calendar and the quality of crop care. First, crop diversity had diminished, with millet and white sorghum having become more dominant to the detriment of notably red sorghum and maize, but also of rice, groundnut and sesame. Rainfall conditions, which had become unfavourable from 1964 onwards and catastrophic from 1969 to 1974, motivated farmers to concentrate on short-cycle millet and sorghum varieties, making the cultivation of maize less necessary to cover the period preceding the main harvest. However, the cultivation of these short-cycle varieties was also a response to the decline of available labour provoked by migration. The enormous predominance of sorghum and millet in the cropping system continued into the 1990s (see Ministère de l'Agriculture et des Ressources Animales 1995a, 1996a, 1996b, 1997, 1998). Only a relative shift occurred from sorghum to millet in 1970-87, probably related to the further deteriorating rainfall circumstances of that period (Maatman & Schweigman 1995: 26). Second, compared to the beginning of the twentieth century, the agricultural season started later and had shortened, and the quality of cropping techniques had deteriorated. The preparation of fields demanded less time mainly because of the disappearance of long-term fallow. Sowing was repeated as many times as made necessary by the irregularity of rainfall. Clear-cut weeding rounds existed only when labour was relatively abundant, but were no longer discernible after labour became scarce. In short, the tendency was to an extensification of cultivation practices mainly because of a declining labour availability per production unit, itself due to migration and processes of 'atomisation' and 'individualisation'.

Despite the image of technological stagnation or even regression, a certain progress of mechanised plough agriculture has been regularly reported over the last forty years, both by researchers such as Broekhuyse and by the regional agricultural extension service (see also Breusers 2002). Attempts at innovation can furthermore be observed in the domain of fruit trees (papaya, bananas, guava and mango), the cultivation of vegetables (onions, tomatoes and sweet potatoes) and rice cultivation in lowlands. As long as light and loose soils in the vicinity of compounds were sufficiently available, lowlands had played only a secondary role in the agrarian system. Land scarcity, however, pushed farmers to the lowlands, and as early as in the 1960s the use of lowlands for the cultivation of rice and vegetables had become generalised. Often, at the initiative of extension agents, mango and citrus orchards are planted in enclosures and gardens are established and irrigated with well water. Some cassava, and during the rainy season rice, are cultivated in these gardens as well.

The scale of these innovative attempts generally remains relatively small. The adoption of animal traction and themes vulgarised by the agricultural extension service remains low, even in villages heavily exposed to development interventions. Only the treatment of seed before sowing is a fairly well adopted theme (Barning & Dambré 1994). Categorisations of production units and farmers, most often along a 'traditional' – 'modern' scale reflect this image. For instance, on the basis of the results of a national survey carried out in 1993, no 'modern' farms (integrated in markets through cash crop production) could be discerned in the provinces of Bam, Sanmatenga and Yatenga. About 98 per cent of all farms were classified in the same category of production units that practised an extensive agriculture, economised little in livestock and were weakly integrated into markets. The remaining two per cent consisted of the production units specialised in cattle herding (Ministère de l'Agriculture et des Ressources Animales 1996a: 65).

Animal husbandry

Animal husbandry in the Kaya region in general, and cattle rearing in particular, has for a long time been regarded as being the specialisation of the FulBe population group. Besides being related to the needs of their cattle FulBe mobility in the nineteenth century was politically motivated and caused by tensions with other population groups, it became mainly ecologically driven when colonial rule was established. Herdsmen practised transhumance both to the north during the rainy season, going as far as Bandiagara and Dori, and to the south during the dry season (Benoit 1982b: 117; Marchal 1980: 36, 66). However, by the 1970s, the mobility of pastoralists and their herds had been considerably reduced. The 'pastoral' population had become largely sedentary or at least not very mobile. However, transhumance, albeit over smaller distances, continued to be practised. During the rainy season it was linked to the pressure of crop cultivation and directed towards pastures which, during the dry season, were recuperated by other herds moving in search of water. The result was that the animal pressure on natural resources did not vary much over a year (Benoit 1982b: 95-9). Certain FulBe resigned themselves to settling in dependent association with Moose communities, with whom they multiplied herding and manure contracts and on whom they became 'entirely dependent' for their access to pastures (Marchal 1983: 610-2).

As a consequence of the droughts of 1971-72, many FulBe left the region definitively, mostly to the western frontier of Yatenga where population densities were less elevated and to the Volta valleys in the south and west of the country (Benoit 1982b: 99-122).The droughts entailed a temporary increase of mobility of the herdsmen who stayed but, by the mid-1970s, most FulBe in the Kaya region were 'sedentarised' and engaged in agricultural activities during the rainy season. Farmers pushed them to higher locations. Livestock remained on the tops of hills during the rainy season and were allowed to enter the fields and graze the stubble only from October onwards when harvesting started. Afterwards, they concentrated in the lowlands and when water there dried up they drank at barrages or village wells. In the latter case they spent the night in fields near the wells, although the herdsmen were also confronted with farmers trying to prevent them from accessing water resources for their herds (Marchal 1983: 156-7). The FulBe's economic situation deteriorated seriously. Formerly, their economic autonomy was based, to an important extent, on the production of milk, which was the main subsistence source and which was exchanged for cereals. During the droughts, it was consumed and traded in only small quantities thereby compromising the social and economic role of FulBe women, which was founded on the commercialisation of milk. In addition, a diversification of animal species occurred to minimise risks. Nevertheless, although small ruminants (sheep, goats) came to play an essential role in the pastoralists' subsistence, either through direct consumption or through sales and barter, their proportion in the herds remained small when compared to cattle. Around their settlements, the pastoralists cultivated small fields but these only made a small-scale contribution to the households' consumption. Furthermore, the herdsmen became more dependent on cattle entrusted to them by Moose farmers, while the herding contracts became less favourable (Benoit 1982b: 128-33, 142-4; DRPC 1990: 141-2, 188-9).

Perspectives for pastoral land use in the Kaya region are generally viewed to be extremely grim. In particular, the extensive use of pastures is considered impossible to reconcile with the extension of cultivated areas. Because the Moose hold the political control of territories, own the majority of water resources and retain the best soils, and because Moose cattle owners opposing the transhumance of FulBe animals limit the mobility of FulBe, the space for FulBe to live in has become more and more restricted. It has resulted in their impoverishment. Although access to pastures continues to follow the simple rule that each and everyone can have his livestock graze where he wants, cultivated areas excepted, this liberty is being more and more frustrated because the right to farm always overrules the right to graze. Moreover, during the rainy season, herdsmen see their access to certain lowlands, with large grasses suited for livestock, obstructed because Moose reserve these grasses for the fabrication of roofs for their houses. Also, Moose increasingly limit the grazing of crop residues from harvested fields to reserve them for their own livestock. Finally, land scarcity would have engendered a certain intolerance of Moose towards the presence of FulBe and the few fields they cultivated. Emigration out of the region has often become ineluctable for a herdsman wanting to maintain the size of his herd (Marchal 1983: 609-10; Benoit 1982b: 118-23, 152).

Moose own and increasingly keep livestock themselves as well. In the colonial era their herds were decimated by epizootics and droughts. Colonial taxation policies and animal requisitioning hampered and discouraged herd reconstitution, which happened only gradually after the Second World War and especially following independence (see also Marchal 1983: 584-6). Until the 1950s, then, cattle were a rare feature in Moose compounds. Nevertheless, owning a cow constituted for Moose a sign of wealth in that time as well. A Moaga who had a small amount of capital bought an animal from a Pullo living nearby and left the animal in the Pullo's herd. Although the large majority of their cattle continued to be herded by FulBe, the number of cattle herded by Moose themselves gradually increased from the 1960s onwards. First, cattle were increasingly kept as draught oxen for ploughing. Some of these oxen were effectively stabled near the compounds. In relation to the diffusion of donkey cultivators and carts, the number of donkeys increased as well. Second, Moose tended to take care of some of the cattle they formerly entrusted to FulBe. This was a direct consequence of the drought of the early 1970s that caused many FulBe to leave the region. Finally, the engagement of Moose in cattle raising was, in a sense, facilitated by the droughts when the terms of trade for cattle with regard to cereals deteriorated and FulBe were forced to sell relatively more animals in order to meet their subsistence needs in grains. This livestock was purchased mainly by farmers with cash at their disposal and certain farmers thus enlarged their herds enormously (Benoit 1982b: 89; Kessler & Breman 1995; Marchal 1983: 534, 579-82).

Small livestock (goats, sheep, and fowl) are found in almost all production units and are used as gifts, sacrifices and for the production of manure. Most importantly, however, sheep and goats constitute a source of wealth on which owners call whenever they face a food shortage or certain monetary expenditures.

6. OPTIONS AND PATHWAYS

Historical background: coercive colonial administration in 1897-1946

Especially during the first half of the twentieth century, French colonial administration disproportionately shaped the options open to actors in the Kaya region. What is more, many of the decisions actors had to make most certainly did not result from a choice from a number of options: labour and military recruitment were forced upon them, as was the production of cotton, and cassava, albeit much less rigorously,. The colonial agricultural policy, which consistently advocated the extension of areas for both food and commercial crops, contributed to relative land scarcity and to an acceleration of new land clearings and, concomitantly, of the movement of people and their compounds. At the same time, the region was pacified and security was improved in bush land that had remained uninhabited in pre-colonial times. Fringes of no-man's land that separated certain kingdoms and chieftaincies from one another, however, remained relatively unsupervised.Such areas existed until the Second World War especially in the north of the Kaya region and hiding in them to escape coercive colonial policies came to constitute an attractive option for which many chose.

While the 'colonial condition' was a primary factor in shaping both the portfolio of options available to actors and the decisions that were ultimately made, other factors intervened as well. Population growth, which further stimulated the search for new land to clear was one, but rainfall circumstances stand out in particular. Whereas the last years of the nineteenth century had been relatively wet, the first decades of the twentieth century were relatively dry and marked by runs of successive years of insufficient or irregular rainfall. Both in the hiding areas and elsewhere, Moose farmers sought to clear formerly uncultivated lowlands in particular. In certain zones, a kind of race for the lowlands took place in the 1920s and 1930s, which resulted in some fierce conflicts over land (Breusers 1999b).

The hiding from colonial coercion and the search for new and more humid land were accompanied by fission of compounds, the break-up of villages and the seemingly random settlement of farmers in isolated compounds in refuge areas. The image that arose was one of disruption, of a society in social and economic disarray. Still, behind this disarray, certain creative and reconstructive forces were at play. Land tenure regimes were progressively established in newly occupied areas. While hiding from colonial repression and responding to recurrent dry years, people set out to inscribe the contours of land tenure arrangements that came to constitute the dynamic context in which actors developed their pathways in subsequent decades. Moreover, the breaking up of larger kin groups into smaller and 'atomised' entities must be put into perspective. Indeed, while movements may initially have taken place somewhat chaotically and without preconceived destination, they later became more geographically, socially and politically circumscribed and relations between the refugees and only rarely were those who remained behind completely disrupted.

Social security arrangements were nevertheless under pressure and it was quite difficult for people, some chiefs excepted, to economise - let alone to accumulate wealth. This was in the first place true of food stocks. Not only did overall labour availability decrease (because of recruitment, migration, flight), the colonial cash crop policy (notably the forced cotton cultivation programme) was responsible for a disproportionately large decline of labour input into food crops. The building up of food reserves was further frustrated by a number of rainy seasons with low and/or irregular rainfall and by locust plagues. Savings were hampered because the preferred object of investment, that is livestock and notably cattle, was subjected to requisitions (for the army, for the transport of cotton). Relative labour scarcity also entailed an extensification of cultivation techniques and, thus, stimulated the extension of the cultivated area even further. Access to land as such did not constitute a problem, although gaining access and establishing control rights to lowland areas was the source of strife between farmers, at least intermittently during sequences of relatively dry years. There seems to be no doubt that members of *tengbiise* (first arrived) kin groups occupied favourable positions in these struggles because of the ritual, judicial and religious authority of their elders in land matters,. Conversely, the presence of *tengbiise* with the ritual and religious authority to allocate land, or of people to whom such authority was delegated, was necessary for new land clearings to be legitimate, especially when competitors from other villages had to be confronted. Nevertheless, the overall result of the pathways enacted by actors during the first decades of the twentieth century was an impressive extension for many if not all kin groups of the area in which they could assert use and/or control rights.

Years of relatively abundant rainfall: 1947-1966

Compared to the previous period, two crucial environmental factors radically changed. First, rainfall was in many years relatively abundant and well distributed in space and time. Second, from 1946 onward, colonial policy was relaxed in a number of domains: all forms of forced labour recruitment were abolished, the free circulation of persons was established and a start was made to a colonial development policy. As a consequence a number of options, which were merely theoretical for most actors during the previous period, now became relevant options for many or obtained quite a different meaning.

This is most obvious with regard to long-distance migration, which became a pervasive phenomenon. Whereas, before the Second World War, young men who went to the Côte d'Ivoire were recruited by force and those who left for the Gold Coast either had fled or needed to earn the money to pay taxes, those who migrated afterwards did so much more voluntarily. For several reasons this migratory movement was at first directed mainly at the Gold Coast/Ghana. The Côte d'Ivoire was strongly associated with the bad memories of forced recruitment. Moreover, the cocoa and coffee plantation economy in the Côte d'Ivoire had lagged behind the one in the Gold Coast/Ghana. Only after the Second World War did the former start to grow rapidly. An important additional aspect, during the 1950s, was that producer prices for cocoa rose more significantly in the Côte d'Ivoire than in the Gold Coast/Ghana (Teal 1986: 274). So, despite the French colonial government's attempts to attract migrants from the Kaya region to the Côte d'Ivoire (e.g. subsidising free transport of labourers on the railway line to Abidjan, allowing migrants to be accompanied by their wives and children), it was not until the Côte d'Ivoire came to be perceived as more profitable that it became the Moose migrants' primary destination. This was further reinforced because, at the same time, the climate for Moose migrants deteriorated in the Gold Coast/Ghana, with some of their leaders being involved in the conflict between political parties in the first half of the 1950s.

Migration, had, therefore, become an option which was, in principle, open to all men and women. However, it was seen as incompatible with being married and not suitable for women. Not only did norms and values militate against women and married men migrating (they did so as well against single men leaving), employment opportunities were also much scarcer for women than for men, whereas for a man to have his wife accompany him on his migrations would have meant an economic burden. Hence, the majority of people migrating to the Gold Coast/Ghana and to the Côte d'Ivoire were young, single men. It was overwhelmingly a temporary and circular migration, with migrants alternating stays of one, sometimes two years abroad with years of living and farming in their home village. In the coastal economies they were employed mostly as wage labourers. After their first marriage, they generally settled down in the village, like before joining the work on their compound's fields, albeit now assisted by their wife.

If young men gave in more or less en-masse to the attractions (adventure) and benefits (money, clothes, bicycle) of the migration-option promised their elders, on the contrary, patently considered migration as an option to be discarded. They resented the fact that their youngsters were leaving, whether to the Gold Coast/Ghana or to the Côte d'Ivoire,. Whereas the fear elders had because of their own experiences with forced recruitment during the previous epoch can explain their resistance with regard to the Côte d'Ivoire, this is less obvious for migration to the Gold Coast/Ghana. It is quite possible that, at the time, elders just did not find it necessary for their sons to leave in search of a monetary income to complement farm income. Given the relatively favourable climatic conditions, crop cultivation was a relevant option during the larger part of the 1950s and 1960s as well. In most years, it was not only possible to satisfy one's compound's subsistence, but also to build up stocks and/or to sell part of one's harvests in local markets and to invest in [small] livestock and cattle, no longer subjected to administrative requisitions. For the elders, then, choosing the option of not having their youngsters migrate meant more labour for crop and livestock production and the prospect of larger reserves and savings, with all the social prestige this entails. However, in the case of youngsters , even if they spent some years in Gold Coast/Ghana and the Côte d'Ivoire, surpluses from crop cultivation gave them the opportunity to start acquiring some wealth in the form of livestock assets.

Young men's migrations were, for a part at least, motivated by a longing for adventure. Of course, they wanted to earn money abroad – if only to pay for the journey back home. The economic boom, especially in the Côte d'Ivoire in the 1950s, made it quite easy to do so. Part of this money was spent on 'luxury' items (especially clothes), but investments in bicycles and small livestock were also important. Moreover, those who were young at the time claim that they were ready to assist kinsmen and friends at home who were in need of help. Still, migration did not generally leave the migrant, and even less so the members of his farm or compound or other kinsmen, with anything substantial or long-lasting (with the possible exception of a bicycle) other than the experience of having travelled and worked abroad.

Other forms of mobility some actors came to be involved in had a more profound impact on their subsequent pathways. Indeed, like in the first decades of the twentieth century, to move one's compound over a smaller or larger distance was an option for compound elders to consider. However, contrary to the previous period when such movements were often related to administrative coercion or the will to escape it, or to the search for [more humid] farmland, in 1947-66 they were related mainly to socio-cultural factors and/or linked to different kinds of calamities.

They were generally framed in institutionalised relations within or between kin groups.

'Misfortune', manifesting itself as the illness and/or death of, most often, young children and often explained in magical-religious terms, urged some to seek shelter with their mother's brothers, evading the destructive forces of earth spirits or discontented ancestors tied to the place they were leaving. The mother's brother was – and still is today - a preferred person to turn to in times of distress. Not only is it deemed an insult to the ancestors for a mother's brother to refuse a sister's son shelter, but movements to a mother's brother were also facilitated by the use right that a sister's son holds in the land controlled by his mother's brother (see also Breusers 1999a). Nevertheless, this is a weak right and therefore a movement to a mother's brother generally implied a relative loss of control by the sister's son over the land he and his dependants farmed. It must be kept in mind, however, that use rights to land at the former place of residence continued to exist provided the members of the sister's son's kin group remained behind, as was often the case. The special relationship between the mother's brother and sister's son and its importance for actors' social security, was also expressed by the widespread practice of fosterage of the latter by the former. An important aspect of fosterage, besides the distribution of caring responsibilities, was labour redistribution among related compounds. Often, children were fostered when they reached the age of six or seven and were capable of fulfilling important tasks such as the herding of small livestock. Furthermore, the fosterage of younger children constituted an investment expanding the compound's future labour force - an investment possibly strengthening the relation between mother's brother and sister's son (future mutual support).

The rights and obligations institutionalised in the mother's brother – sister's son relationship therefore provided actors with options to which they could turn at times when they were confronted with smaller or larger calamities, caring/nursing problems or labour shortages. Such options continue to be of primary importance today as well. There is, however, nothing optional in fosterage for the child actually involved. A fosterage is arranged between adults without seeking the child's consent. Still, when a fostered boy (a girl is expected to leave anyway when she is married) reaches adolescent age, he is allowed to choose between staying with his foster parents and returning to his father's compound. If he chooses to stay, he will be allocated farm land by his mother's brother to which, by farming it, he will lay an initial claim for himself, his kinsmen and his descendants (see also Breusers 2001).

Recurring droughts, 1967-1987

In 1967, a prolonged decline of rainfall set in, entailing an upsurge in people's mobility. Nevertheless, certain actors did not leave the village they lived in. Among them, important differences regarding land use options could be distinguished between members of kin groups who had arrived earlier and 'strangers' who had arrived relatively late. All of them attempted to gain access to at least a parcel of lowland, whether or not they had to move their compound to another ward in order to do so. The dryer climatic conditions of the 1970s and 1980s evidently made lowland scarcer. Strangers' claims to use rights in the territory of the village where they lived were often limited to the land they had farmed since their arrival in the village, often distant from the village's lowland. They did not stand much chance of being allocated a parcel of the latter and could be forced to ask for land from kinsmen or affines beyond the village territory. In contrast, *tengbiise* held several options to farm lowland – although not each to the same extent (which depends, among others on their seniority rank; see Breusers 2001) – since, in principle, each of them had a use right to all land in their vast kin group territory. Many *tengbiise*, then, moved their fields several times during the 1970s and 1980s. Also, and again in contrast to 'strangers', they sowed several fields per farm dispersed over different soil types, of which, at the end of the rainy season, only a part was harvested. The series of successive dry years undoubtedly contributed to this practice of sowing 'security fields', which implied a decline of labour input per hectare, especially for soil preparation and sowing and hence an extensification of land use practices, itself further reinforced by the increased migration to the Côte d'Ivoire (see below). The movement of fields over the village territory was accompanied by fission and the splitting up of larger production units into several smaller, separate farms. Important relationships were, however, maintained between [newly created] farms, manifested in the frequency of fosterage and in the later recombination of compounds.

Concomitantly with the rush to the lowlands, which started in the early 1970s and reached a peak in the 1980s, sometimes violent conflict erupted - for instance, within villages between members of early arrived kin groups (*tengbiise*) and relative 'strangers', or between farmers from neighbouring villages. Hence, while access to relatively humid land, offering a better chance to obtain sufficient harvests in times of drought, constituted a primary motivation, the urge to move was further reinforced by the fact that farmers needed to reaffirm their claims on lowlands (for instance farmers from one village *vis-à-vis* farmers from neighbouring villages). Although such conflicts over land, especially along borders between chieftaincies, occurred throughout the twentieth century, the dry years of the 1970s and 1980s accentuated the ever-latent tensions, as had also been the case in the 1920s and 1930s (see Breusers 1999b). The Revolution of 1983 and the land reform proclaimed in 1984 only added to these tensions. The land reform made all land in Burkina Faso state property. At the same time, the regime led by Thomas Sankara condemned 'traditional' systems of land management as feudal and created a network of revolutionary defence committees (CDRs) in the villages, which attempted to assume responsibility for the allocation of land and thus undermined the *tengbiise's* authority regarding land issues.

The reappearance of drought in the second half of the 1970s, after only a few years of less disastrous harvests, caused farmers to despair. The situation became extreme and particularly during the larger part of the 1980s the population became uprooted. From the end of the 1970s, certain actors, destitute after repeated harvest failures, decided to migrate with their families to the south or west of Burkina Faso. They regarded their situation in the Kaya region as hopeless and it seems they were no longer able to provide for their familie's. Either they applied for a place in the Volta Valley resettlement schemes that were initiated by the government and donors after onchocersiasis in the valleys of the Volta rivers had been eradicated in the early 1970s, or they moved elsewhere asking for land controlled by other population groups. Moving to the resettlement schemes was an option for those who conformed to the conditions posed by the AVV (*Autorité des Aménagements des Vallées des Voltas*) regarding, for instance, age of the farm head and family composition. Migration of one or more young men abroad to compensate with remittances for the shortages at home, or to at least alleviate the food needs of the compound during the *soudure* (the hunger season) was generally not an option for these actors, since they had only young children. The aid they were promised in places within the resettlement scheme made this option more appealing.

By leaving for the south or west, the migrants did not forsake their rights to land in the villages they left. Often, their kin groups were firmly 'rooted' there and had acquired extensive control and use rights that the kinsmen who remained behind would hold in custody. Usually, those who left had been farm head before, so the migration did not directly affect the labour available either for those who remained behind or for the migrating farm. A major loss the migrants were to endure was the absence of kinsmen, friends and acquaintances. However, chain migration was soon established. For instance, in the case of migration to the AVV schemes, first-comers, settled by the authorities, came to constitute relays for kinsmen who followed later and who, since they often could not be accommodated within the scheme, established their farms as 'spontaneous migrants' on its fringes after having solicited use rights to land from local earth priests, thus creating a new possible destination for yet other kinsmen from their home village who might want to follow. For some migrants, the land they had been allocated by the AVV appeared, after some fifteen years, to be insufficient for accommodating for the growth in the number of their dependants and they opted to leave the scheme, migrating further to the south, handing over the farm in the scheme to kinsmen. In the south and west the migrants were generally able to subsist, and even produced a surplus, of which part was sent to the north to alleviate the food shortages of non-migrated kinsmen (Agricultural production in the south and west generally differed considerably from what migrants were used to in the Kaya region, e.g. oxen traction, cotton cultivation, more intensive use of fertilisers).

However, a lot of people lost out during these years. Some lost all the wealth (often manifested in livestock and especially cattle assets) they had acquired by means of crop cultivation or otherwise in the more favourable circumstances of the 1950s and 1960s. In most cases, young men's migrations to the Côte d'Ivoire could not counter this impoverishment. Often their leaving meant, at best, that remittances provided a compensation for food deficits at home or, at worst, it meant a few less mouths to feed during the difficult *soudures*. It was no exception that for the migrant himself migration amounted to a zero sum operation. To pay for the journey south, he would sell his bicycle (bought when he returned from his previous stay abroad) while at the end of his stay in the Côte d'Ivoire he would barely have earned enough money for the trip back home. In those years, the meaning and pattern of migration to the Côte d'Ivoire started to change substantially. First, seeking adventure was no longer a primary motivation, especially not among those who had left once or more before. Whatever the ultimate result of their stays abroad, migrants left with their main goal being to earn money. Second, men continued to migrate after their marriage and were sometimes accompanied by their wives. Migration also became more embedded in kinship networks as is shown, for instance, by the frequent appearance of mother's brothers in migration stories. Finally, and of crucial importance, migrants' occupations in the Côte d'Ivoire began to shift from wage and contract labour to agricultural entrepreneurship. Hence, several migrants worked as share croppers in cocoa plantations, others rented land for rice cultivation and a few bought land to establish a cocoa or coffee plantation of their own. The Ivorian administration encouraged the establishment of plantations by immigrants, defending the principle of 'the land belongs to him who exploits it'. Those who were among the first to come into pioneer zones acquired land from the moment they arrived, others had to pass through different forms of share-cropping (Lambert & Sindzingre 1995: 108; Schwartz 1979).

Establishing a plantation required sacrifices, notably a prolonged stay abroad, at least for the years necessary for the plantation to start producing but often longer if the immigrant was to earn the money to purchase land or to work for a number of years as a share cropper first. During the 1970s and 1980s, long periods of absence from home were required probably because of the dead-end situation at home. Substantial incomes and remittances for kinsmen at home could be derived from such plantations. However, many migrants from the Kaya region, who succeeded in acquiring land, did so only in the second half of the 1980s or later and thus were not able to assist their kinsmen at home substantially during the drought period.

After the droughts: 1988-1997, a period of stabilisation?

Compared to the previous two decades, rainfall recovered somewhat in 1988-97, but nevertheless remained erratic in terms of space and time. Good and bad seasons alternated one another, with the extremely wet year of 1994 causing problems especially to farmers who mostly had their fields in lowland areas. More than before, Moose farmers stopped relying solely on agriculture at home, or for that matter on income-generating activities restricted to their home region. Remittances in cash and/or kind from kinsmen, in the Côte d'Ivoire especially but also in the south and west of Burkina Faso, came to be welcome if not necessary additions to farm income and production.

It became fully accepted by elders that young men had to migrate as soon as they reached the age of about nineteen and a young man's leaving is now even considered part of his education. Still, a young man's first migration is often postponed if the situation at home requires his continued presence. In addition, later migrations became attuned to the activities and necessities of the village farms and male members of one farm generally took turns in leaving, so as to safeguard both sufficient labour at home and monetary income from the Côte d'Ivoire. Whereas youngsters' first migrations were looked upon rather benevolently and those who remained behind did not expect to benefit substantially, after several stays abroad migrants were expected, particularly after they are married, to support the village farm with remittances and also to make progress, to build up something, either in the Côte d'Ivoire (typically from labour migration to agricultural migration with the purchase of land as the ultimate crown), or in the village (typically through the purchase of cattle). During first stays abroad, migrants worked as wage or contract labourers or on the plantations of kinsmen, while later on more independent activities were sought for (share cropping, renting of land, purchase of land). Overall, then, wage work declined in importance to become often only a secondary activity combined with another main activity for which individual access to land in the Côte d'Ivoire was to be obtained. Concomitantly, it became more usual for a woman to accompany her husband to the Côte d'Ivoire. Instead of being a burden on the man who had to take care of her when he engaged in wage labour, she was now a valuable labour force participating in the work in the agricultural enterprise abroad. Pursuing such a 'career' in the Côte d'Ivoire was facilitated by the increasing kin-group embeddedness of migration. Both maternal and paternal kinsmen were involved in providing migrants with financial assistance to pay for their journey, or were migrants' travel companions, or acted as employees or mediators for employment in the Côte d'Ivoire.

Generally, migration periods of two years alternated with one or more years in the home village. Migrants were able to stay away longer. This happened in the event that land was purchased and the migrant had to wait for the plantation to start to produce before paying a visit home. It also happened in case of illness or other misfortune, sometimes even preventing the migrant from paying his fare home. And finally it happened in the event that the migrant left with a 'stolen' woman, thus making it more difficult on the one hand for him to economise while, on the other hand, increasing the chances for a marriage to be recognised by the spouses' kinsmen *post facto* because of a postponed return. Sometimes, migrants stayed away for a year or less: in most cases they returned home prematurely because kinsmen had arranged a marriage for them. In other cases these short migrations were prompted by harvest failure during the preceding rainy season and served to earn money to allow for the purchase of cereals during the next rainy season.

With regard to migration to the south and west of Burkina Faso, it became apparent in the early 1990s that migration was not necessarily definitive. Several actors, who had left the Kaya region at the end of the 1970s or in the early 1980s, returned. This underscores the importance of the maintenance of relations with the home region. The process by which one farm head succeeds another on an AVV farm, when the latter either returns to his village of origin or moves further south, suggests that instead of the migration of individual farm heads and their dependants the establishment of a compound was permanent, which was then inhabited by families who succeed one another at that place. The compound constitutes a kind of relay and is associated with a territory to which members of a kin group can gain access and which is, as it were, added to the kin group's pool of territories. Hence, even if many migrants eventually return home, migration to the south (and to the west) alleviated pressure on land and other resources in the Kaya region. Returned migrants were smoothly reintegrated into their home village. They arrived at the beginning of the year, and therefore had time to construct their compounds before the onset of the rains. They resumed farming on land controlled by their kin group of which they had now themselves become influential elders.

Those men and women who had money to spare after subsistence had been guaranteed and other expenditures made (e.g. funerals, health care) usually bought sheep, goats and/or cattle. Especially those who could rely on substantial remittances from kinsmen in the Côte d'Ivoire or a flourishing trade were able to invest in cattle. For them, but also for others who owned only sheep and goats, livestock often became the main source of monetary income. Van Zutphen (1991: 119-20) relates the increasing importance of cattle in Moose production systems to the growing uncertainty of crop farming in circumstances where farmers are forced more and more to exploit high-risk areas and to cultivate on already poor soils. As it becomes difficult to continue cropping, farmers start tending cattle to earn cash. Currently, the purchase of cattle often constitutes the crowning of a successful saving-process based on small livestock. Frequently, one or a few sheep or goats are kept at the compound to be fattened for commercial purposes. They are fed on crop residues and household refuse. If sufficient, the earnings from the sale of small livestock can be used to purchase cattle, which are then entrusted to FulBe herdsmen or kept at the Moaga's compound. The latter is the case not only for draught animals but also for oxen and bulls that are fattened for sale. The droughts of the 1970s and 1980s contributed to the increase of cattle ownership among Moose and the entrustment of these cattle to FulBe in yet another way. Many cattle died particularly in the 1980s. Several FulBe left the region and migrated to the south, east or west of Burkina Faso because of a lack of food for both people and animals. Those who stayed asked for more remuneration than before in return for herding the Moose's animals. Moreover, during the droughts, the terms of trade of livestock *versus* cereals deteriorated, making it relatively easy for Moose – despite them only being few in number - who had the means at the time to invest in cattle.

For cattle owners to pursue their investment-in-livestock-strategy successfully, they had to entrust the larger part of their assets to FulBe herdsmen. Therefore, it was essential to have or to establish friendship relations with FulBe. Moose sometimes put forward labour scarcity and FulBe's expert knowledge regarding cattle herding as reasons to entrust their animals. Still, these were mostly only secondary reasons. More important was that cattle owners found it necessary not to display their wealth. Cattle must be 'hidden' from other Moose villagers in order not to arouse other persons' envy and the best place to do so was with trustworthy FulBe, that is with FulBe friends. Indeed, for an asset so valuable as cattle to be entrusted to another person, that person must be trusted. Trust is one of the defining characteristics of friendship (*zoodo*), which is a rather exclusive and often durable relationship. A person seldom has more than three 'friends' (*zoadamba*, sing. *zoa*) during his or her lifetime. Cattle entrustment is often a major element of friendship relations between Moose and FulBe. It is characterised by certain more or less formal arrangements: the herdsman receives the entrusted cows' milk production and a calf each three years. For each animal sold by the Moaga owner, the herdsman receives a share of 5,000 FCFA. If the herdsman's herd damages crops, the Moaga cattle owner shares in the compensation and fines to be paid proportionally to the number of cattle he has in the herd. Still, the relationship between herdsman and cattle owner cannot be laid down in formal-contract-like terms. Informal mutual assistance is at least as important in the shape of FulBe stocking millet or sorghum with Moose friends, gifts of milk and cereals, borrowing of money, etc. The grazing by cattle of crop residues in

Moose's fields is often organised on an informal, *ad hoc* basis as well (see also Breusers *et al.* 1998).

With regard to actors' decisions relating to crop cultivation, notably as to the location of their fields, the village-level distribution of control and access rights to land continues to be highly important. Relative 'strangers' in a village occupy a vulnerable position, entitled as they are only to land not wanted for use by a *tengbiiga* or any other actor who is a member of a kin group that arrived earlier. While, because of their residence in the village, 'strangers' like any other actor can claim access to sufficient land to provide for their subsistence, they cannot easily claim access to specific plots. Hence, it is sometimes the case that a 'stranger' is forced to move his field by actors who have stronger use rights. In contrast to the restricted possibilities for 'strangers' to choose the location of their fields within the village territory freely, the options for *tengbiise* are manifold, especially in more or less favourable rainfall circumstances. *Tengbiise* – and members of kin groups that have arrived early – frequently move their farm to a completely different place in response to, for instance, soil depletion, changing rainfall circumstances, other people's claims on certain plots, or a combination of factors. In doing so, they revalidate their own and their kin group's claim to the 'newly' occupied places (see Breusers 1999c). Non-agronomic considerations may prevail when they have to choose from the many options open to them, such as envy with regard to strangers' success in crop cultivation or other enterprises.

7. CONCLUSIONS: SEARCHING FOR LIVELIHOOD SECURITY IN A VOLATILE ENVIRONMENT

The preceding sections reveal a number of pathways. In certain actors' pathways, notably those of members of kin groups that arrived first or early at a certain place, the extensive access rights to land within the village territory that these actors hold were of primary importance. During the droughts, they were among the first to be allowed to choose part of the village's lowland for the establishment of their fields. Having used these lowland fields during a number of years, their claims to that land were reinforced when the rains recovered and they decided to return with their fields to land at a higher location. Otherwise, they were also in a position to move their fields frequently within the territory of their village in response to specific soil fertility and climatic conditions or social circumstances. In contrast, much fewer options were open to actors who were relative 'strangers' in their villages. Because of their weak rights to land, they found it more difficult to gain access to lowland areas to counteract the impact of the droughts on their crop cultivation activities. They were even more or less forced to sow their fields outside the village territory. Because, in such a case, they did not use land in the village territory for a number of years, their access rights to land in this village territory were weakened when they brought their fields back after the drought. Nevertheless, they were sometimes able to embark on a successful pathway, notably by engaging in trade and then pursuing an investment-in-livestock strategy involving friendship relations with neighbouring FulBe herdsmen.Even despite their weak rights to land, some of these strangers also invested in crop cultivation (purchase of plough and cart, transporting manure to bush land fields).

On the other hand, strong access rights to land were an insufficient guarantee for a successful pathway. Farmers who had been able to accumulate some wealth from farming in the 1960s sometimes saw their assets wither away rapidly in the 1970s and 1980s. This was especially the case when they themselves - or (closely related) kinsmen - were affected, not only by drought but by other resource-draining misfortunes (notably illness). Many of these actors were uprooted during the 1970s and 1980s. They responded to the droughts by migrating to the Côte d'Ivoire, to the south or west of Burkina Faso or to nearby villages in the Kaya region and back again to their 'home' village often without ever recovering from earlier losses. Still, some of the actors who migrated to the south or west of Burkina Faso were nevertheless able to generate enough to live on for themselves or their dependants. Like migrants to the Côte d'Ivoire, albeit to a lesser extent, they were also even able to support kinsmen who had stayed behind. Moreover, several of these migrants, even if they ultimately returned to the Kaya region without having increased their assets substantially, contributed to the progressive extension of their kin group's pool of territories. They therefore extended the range of optional destinations for members of their kin group whenever they wanted, or were forced, to move.

There were also many who, during the droughts, migrated to the Côte d'Ivoire. Most of them only worked abroad for a limited period. Often they were able to send at least some remittances home and support their relatives through the droughts. Moreover, as was the case with regard to the migration to the south and west of Burkina Faso, such migrations ultimately did not affect the assets or incomes of most of the migrants or their relatives at home, but nevertheless contributed to enlarging or consolidating networks to which kinsmen who were to migrate afterwards could have access. The latter was especially true for migrants who acquired a plantation in the Côte d'Ivoire. Moreover, the acquisition of such a plantation appears to be, one of the possible ways for actors in the Kaya region to obtain savings and accumulate wealth in addition to trade. A plantation could indeed come to constitute a relatively stable source of income both for the plantation owners and their kinsmen in the home village. The investment of savings in cattle (itself an important possible source of income) entrusted with FulBe herdsmen and the pooling of labour between the farm in the Kaya region and the plantation in the Côte d'Ivoire then typically lad to the constitution of what can be labelled a multiple enterprise.

Pathways, changing options-portfolio and institutional change

The above is a summary of a number of *possible* pathways, but certainly not of *all* possible ones. Actors' pathways are in a way each unique and depend on certain individual-related aspects of personality and once-in-a-lifetime contingencies. However, they also depend on contextual factors related to the physical and institutional environment that circumscribe the options actors can choose from to secure their livelihoods. The range and nature of options changed over time and the relevant options differed from actor to actor. The options that were relevant to which actors depended especially on events and processes in the physical and institutional environment and on the assets and entitlements of actors and of the smaller social entities (kin group, compound, farm) to which they belonged. For instance, in the first half of the twentieth century, options were restricted by colonial policies of recruitment, taxation, cattle requisitions and forced cultivation, as well as by sequences of dry years in the 1910s, 1920s and 1930s. The relaxation of colonial policy in a number of domains meant that, following the Second World War, a number of options, which had been merely theoretical for most actors during the previous period, now became relevant for many. This was true, for instance, of voluntary migration to the Côte d'Ivoire and the investment of monetary savings in cattle. Moreover, the favourable rainfall conditions during the larger part of the 1950s and 1960s made crop cultivation more attractive as well, both in terms of subsistence production and production of a marketable surplus. Conversely, the adverse climatic conditions of the 1970s and 1980s shifted the balance to the detriment of crop cultivation especially when compared with migration to the Côte d'Ivoire. This shift was reinforced because migrants gained access to more independent employment and because the migration of married men and women became more socially accepted. Furthermore, the deteriorating terms of trade of livestock *versus* cereals and the growing demand among the FulBe for cattle to be entrusted to them increased the relevance of investment in cattle as an option for Moose.

It can therefore be deduced that a number of institutional and more structural changes took place in the environment with which all actors in the Kaya region were confronted. These were changes to which all pathways taken together also contributed. Before discussing land tenure arrangements and migration in this regard, we would first of all like to examine kinship institutions, which existed throughout the period studied and which were important factors in actors' pathways. First, fosterage of a child, whether by a mother's brother, mother's mother or a patrilineal kinsman, was and continues to be a widespread practice. Second, the relation between a sister's son and a mother's brother stands out as being of specific significance. Typically, the mother's brother is the person to turn to for help and shelter in times of distress. Also, a mother's brother can rightfully claim assistance from his sister's son, often realised in fosterage of the latter by the former. Rights, duties and affections between a sister's son and a mother's brother are extended to the domain of migration to the Côte d'Ivoire, as is obvious from the assistance given by mothers' brothers to sisters' sons in the latter's migration enterprises and by the employment of sisters' sons on mothers' brothers' cocoa or coffee plantations. Third, even though males were the main ones to migrate to the Côte d'Ivoire and even though many have continued to migrate after marriage, *de facto* female-headed farms are extremely rare, except on farms of elderly women exempted from the work on their compound's collective fields. Women who do not accompany their migrating husbands join the farm of a patrilineal kinsman (elder brother, father) of their husband. Finally, kin group offices, especially the office of kin group elder, are often likely to intervene in actors' pathways in a very direct way. Whatever the geographical dispersion of a kin group, adherence to one's kin group's ancestral shrines implies that at a certain age one can be called upon to occupy a kin group office. Therefore one has to move, although it must be noted that the location of a kin group's ancestral shrines is not immutable. If the kin group's presence and position at a certain place has increased and has become more advantageous in comparison to a site at the place 'of origin', a decision may be taken to move the ancestral shrines to the new kin group elder instead or *vice versa.*

With regard to land tenure arrangements, it can be assumed that, at the end of the nineteenth century, effective control over farm land by elders of early arrived kin groups was restricted to a relatively small orbit surrounding the village nucleus within which crop cultivation took place. Kin group elders allocated land in their respective kin group territories, while the village's earth priest retained ultimate authority regarding farmland. In the following decades, it was not so much the major 'principles' of land tenure arrangements (double seniority principle, specific rights to land previously used by one's father; see Breusers 2001) that changed. Instead, the space within which actors were able to claim control and/or access rights to farm land expanded substantially. This process was intimately intertwined with the mobility of people, fields and compounds. As explained above, in the first half of the twentieth century, many actors moved into less controlled 'refuge areas'. In the process, the actors who moved established control and access rights to the newly occupied land, not only for themselves, but for their descendants and also for kinsmen who were now geographically dispersed. Actors obtained rights in kin-group-based 'pools of territories', dispersed over various places. In the twenty years following the end of the Second World War the movement of farms and fields slowed down and, since the dynamics of land tenure arrangements are closely related to such movement, changes in the distribution of control and access rights were less marked than in the previous decades. Nevertheless, some people moved and established farms elsewhere, thus altering the distribution of control and access rights to land both at the place of departure and at their destination. The droughts of the 1970s and 1980s gave way to new and numerous movements of fields and compounds to lowland. Conflicts regarding the control of lowland arose and distributions of control and access rights to land were rearranged and/or partly reaffirmed. Of primary importance during that period was that migration to the south and west of Burkina Faso became a new option. While internal rural migration remained an option throughout the period studied here, albeit mainly confined to the Kaya region (thereby including the former refuge area to the north and being closely related to the access rights to land that actors held in dispersed pools of territories), this form of migration increased both in terms of numbers and distance from the end of the 1970s onward. Migration to the south and west progressively became an option for an increasing number of actors as chain migration was established – firmly embedded in kin group networks (departures in groups of related compounds, migrants who arrived first canalised later migrations of their kinsmen, migrants supported relatives in the home region with annual gifts of cereals or money, participation in [funeral] ceremonies at home, marriage arrangements mediated by non-migrated kin group elders). The government and donor initiated AVV resettlement schemes contributed, for that matter, to the importance of this option, which resulted in a further extension of the pools of territories to which the migrants and their kinsmen could claim control and access rights. Finally it should be mentioned that, especially since the 1980s, a number of migrants succeeded in purchasing land in the Côte d'Ivoire. Their plantations are preferred destinations for their kinsmen who assist with the work on the land without receiving any kind of fixed salary, thus forsaking the higher monetary income which can be earned by working as a wage labourer or share cropper. The plantation owner, however, is expected to assist the kinsmen who work with him in their various enterprises (marriage, establishment of their own farm). Furthermore, it was asserted that the owner's sons would not simply inherit a plantation but that other kinsmen who worked on the plantation would be entitled to a share as well. This might lead to a transformation of individual private property into a kin group related corporate property.

In the course of the twentieth century, migration to the coastal economies of the Gold Coast and the Côte d'Ivoire not only increased considerably, but was also accompanied by important institutional changes. Until 1946, this migratory movement was largely conditioned by the imperatives of colonial rule. Afterwards, several options became available to actors. Following the abolishment of forced labour, migration to the Côte d'Ivoire rapidly became an institutionalised phenomenon. During the 1950s, mainly young, single men were involved in temporary and circular migration. In the Côte d'Ivoire they were employed mostly as wage or contract labourers. In the 1970s and 1980s, instead of seeking adventure, earning of money became the primary goal especially among those who had left once before or more often. The recurring droughts made subsistence, let alone the generation of a marketable surplus, from the farm at home more than precarious. Probably because of the desperate situation at home in many years, migrants tended to prolong their stay in the Côte d'Ivoire, which in turn allowed some of them to acquire a cocoa or coffee plantation. Again related to the difficulties in the home villages, but also to the widening range of options in the Côte d'Ivoire, it became easier and more accepted for women to accompany their husbands who did not stop migrating after being married. Migration was embedded in kinship networks, with patrilineal kinsmen or mother's brothers acting as travel companions or mediators for employment, or helping to finance migrants' journeys. Finally, migrants' occupations abroad shifted from wage and contract labour to agricultural entrepreneurship. By the mid 1990s, elders had come to accept young men's migrations and made decisions regarding who was to leave at which point in time. Migrants started pursuing 'careers' in the Côte d'Ivoire (typically from wage or contract labourer to agricultural entrepreneur) and care was taken to safeguard both the interests of the farm at home (sufficient labour for agricultural production) and the monetary income to be derived from activities in the Côte d'Ivoire. At the same time that migrants' agricultural entrepreneurship became more important, female migration became more frequent and accepted, since women now could assist with the work on share cropped, rented or owned land.

The development of multi-local livelihoods in hazardous circumstances and its limitations

As far as deciding what the driving forces behind the change were, one can only say that there were several. Undoubtedly, climate variability was an important force in the shaping of many, if not all, actors' pathways. Each and everyone had to devise responses to the recurrent sequences of dry years, and especially to the extreme drought of the 1970s and 1980s. Also, when rains recovered after a dry period, the actors do not resume the combinations of livelihood practices they had pursued before the drought. To what extent such changes can be attributed to climate variability or, for that matter, climate change is extremely difficult to establish. In each instance, several physical, socio-cultural, economic and political forces, which were sometimes mutually reinforcing, were at play: e.g. colonial rule, population growth, economic and political developments in the Côte d'Ivoire, offer and demand of cocoa and coffee on the world market, devaluation of the CFA-franc, land reform, etc. All these forces continuously combined to shape and alter the portfolio of options from which actors could choose.

A crucially important factor affecting actors' pursuit of food and livelihood security are their social resources, which are situated mainly in kinship networks and friendship relations. These relations allow the mediation of access to land, labour, marriage partners and migration opportunities and the bestowing of gifts and counter-gifts of cereals, and other items. With regard to cattle ownership, friendship relations with FulBe are of particular importance to Moose. These relationships also involve mutual gift-giving and help in the event that one of the two partners is in need. Second, livelihoods have tended to become increasingly multi-local: farms in the north, migration enterprises in the Côte d'Ivoire, farms in the south or west of Burkina Faso and livestock in the herds of neighbouring FulBe combine to form what can be rightfully be referred to as 'confederations of households' (Smith 1984), between which labour, livestock, rights to land, cereals and other things circulate. This leads us to the observation that an important, if not the most striking, result of actors' pathways enacted throughout the twentieth century is the mobility and dispersion of the population and the related extension of pools of territories in which kin groups have access rights to land. Here again one can only conclude that several environmental forces – physical as well as socio-cultural, economic and political – were involved: colonial rule, droughts, environmental degradation, population growth, economic opportunities abroad, large-scale development interventions, dynamics of land tenure arrangements, marriage and fosterage practices and rules of succession to kin group offices all contributed to stimulate, to facilitate and to perpetuate processes of geographic mobility.

Concomitantly with population dispersion and increased mobility, there has been an overall tendency for agricultural production in the Kaya region to take place in smaller entities (in terms of number of members). The labour force per production unit declined, a process reinforced especially by long-distance migration. With less labour available for crop cultivation, farming practices tended to extensify. Extensification, which, in the first half of the twentieth century had been 'encouraged' by colonial agricultural policy, was also stimulated by the sequences of dry years that incited farmers to apply certain risk-avoiding practices such as the sowing of 'security fields'. Still, although the vicious-circle-thesis according to which a downward spiralling process is taking place where extensification and environmental degradation are mutually reinforcing is widely adhered to, it has not yet been firmly established empirically (see Mazzucato & Niemeyer 2000). In any case, it would be too simple to conclude that the tendency is unidirectionally towards extensification. At least some farmers intensify their land use, in particular by applying manure and/or compost on bush land fields. Also, and related to the issue of applying manure, carts and ploughs, whether pulled by oxen or donkeys, are no longer rare items in the villages. It must be said, however, that carts are frequently used for water transport in the first place, whereas the main advantage of ploughing is considered to be the fact that it allows for timely soil preparation and sowing and for a flexible response to erratic rainfall, rather than its utility for land use intensification. Furthermore, it can be argued that the increase in cattle ownership among Moose, even when the larger part of these cattle is entrusted to FulBe, has entailed an increased integration of crop cultivation and animal husbandry - albeit not on the individual farm-level – since entrustment arrangements often also include agreements about the grazing of crop residues on harvested fields.

Whatever the extent and the diverse tendencies of extensification or intensification, after the droughts, no farm would any longer aim to secure its livelihood by crop cultivation alone. A clear and general tendency towards diversification of activities occurred. Besides engagement in migration, this led to animal husbandry and trade being most prominent. In the case of farms whose members own a plantation in the Côte d'Ivoire, one may in certain cases doubt whether the centre of gravity is still the farm at home or whether priority is given to, for instance, the labour demands of the plantation. Nevertheless, migrants only rarely fail, in the end, to return to their home village. In a way it should not be expected that Moose will start specialising in animal husbandry in the near future. There are clear conceptions about a division of tasks in this regard between Moose and FulBe, as well as social sanctions against Moose who, by herding their cattle themselves, would display their wealth too ostentatiously.

Finally, the pathways show a tendency towards a deepening of socio-economic differentiation expressed in unequal access to migration opportunities and remittances and the unequal distribution of livestock assets. While cattle ownership among the Moose population as a whole has increased substantially since the 1950s, not all Moose have become cattle owners. Wealth differences also account partly for differences in the relations maintained with FulBe herdsmen and for the particular shaping of conflicts based on crop damage and water resources. Simultaneously to the increase in cattle ownership by Moose, the cattle property relations changed to the detriment of FulBe herdsmen who, as compared to the period before the droughts of the 1970s and 1980s, often saw a larger fraction of their herds being owned by Moose. Government and development programmes and interventions risk marginalizing FulBe land users further if they continue to be incapable of acknowledging the importance of mobile modes of natural resource management and the importance of inter-village natural resources. The failure to do the latter in many applications of the *Gestion des Terroirs* approach entails the closing off of village territories for FulBe and the under- or non-representation of FulBe in decentralised bodies (see also Hesseling 1996; Marty 1993).

If it can be inferred from the pathways presented above that the diversification of activities dispersed over several localities - especially engagement in trade, migration and/or animal husbandry - can be livelihood- securing options, it must also be mentioned that the long-term sustainability of these options is far from certain. A first factor to be mentioned, and one which is of crucial importance, concerns recent events in the Côte d'Ivoire which have demonstrated the precariousness of even a long-established livelihood activity such as long-distance migration. Not only have thousands of migrants of Burkinabé origin been expelled following violent conflicts over land with 'autochthonous' population groups in the southwest of the Côte d'Ivoire, the new land legislation, voted in at the end of 1998 by the Ivorian parliament, excludes non-Ivorians from land ownership and puts an end to the relatively protected access to land from which strangers benefited during the Houphouët-Boigny era (Chauveau 2000). Second, the issue of control and access rights to land is far from resolved in Burkina Faso either. The growing number of land disputes in the south and west of the country point to the growing resistance among autochthonous population groups to further Moose migrations (see Laurent & Mathieu 1994; Mathieu 1993; Traoré 1996), while government plans to implement private titling programmes in the resettlement schemes made the option of migrating to the *terres neuves* probably less attractive as well. These developments highlight the fact that the expansion of kin groups' pools of territories may at certain times be halted or even reversed. In more general terms, the outcomes of the ongoing land reform process, in which large donor organisations such as the World Bank press, if not for individualised land titling, then at least for individualised security of land tenure, risk compromising livelihood-securing activities based on actors' mobility in otherwise unchanged circumstances (see Breusers 2001).

The increasing tensions between population groups, whether between Moose and FulBe or between migrated Moose and 'autochthonous' population groups in destination areas, are often suggested as being the consequence of the sheer pressure on natural resources. However, in many cases, this provides only a partial explanation. Processes of democratisation, decentralisation and privatisation seem to play a part as well. Instead of having established local fora where the use and control of natural resources can be negotiated, decentralisation and democratisation have caused a lot of confusion. The legal status of decentralised bodies remains unclear and there are no means available for these bodies to take up the responsibilities formally devolved to them. Moreover, decentralisation in Burkina Faso was implemented in such a way that it is difficult to take account of the interests of actors whose production systems imply a land use beyond single village territories. Also, as argued by Geschiere (2001), the process of Africa political liberalisation, coupled with an emphasis by the 'development establishment' on by-passing the state, has reinforced people's preoccupation with rural roots, as political patrons seek to establish a clientele in rural areas. This is also exactly what happened in the Côte d'Ivoire in the 1990s, where multi-partyism, new liberal legislation on land rights, changed conditions for obtaining the Ivorian nationality and an economic crisis all came together to undermine the legitimacy of strangers' land claims and to reinforce xenophobic tendencies among the Ivorian population (Chauveau 2000). The livelihoods of not only millions of migrants from Burkina Faso, Mali and other countries are thus threatened, but also those of millions of their kinsmen at home as well, since they depend at least partly on the remittances sent to them.

In both the recent and the more distant past, land use practices elaborated by West African population groups have regularly been condemned to extinction. They were judged as being too extensive and space consuming and totally unsuited to cope with population growth and the exigencies of modernity in general. People overcame crisis upon crisis, responding to their volatile environment with the elaboration of livelihoods that involved an impressive mobility of people, livestock and fields. Nevertheless, the price paid was also high, especially in terms of human lives. The extreme vulnerability of the livelihoods to climatic, economic and social crises stands out. What policy response is to be advocated and , more particularly, what can be learned from policy failures in the past? After all, it is not without reason that people in the West African Sudano-Sahel quite consistently have opted out from the kind of production systems and livelihoods – most often 'intensive' and sedentary – that have been proposed to them – or sometimes even enforced upon them, as during the colonial epoch – by administrative and development authorities.

What is needed are policy recommendations that refrain from the conventional directives for improved land use that emphasise sedentary and unimaginative intensified production systems, but that nevertheless also go beyond the simple celebration of livelihood practices as they prevail today. In one way or another, these livelihoods must be curbed and made more secure. Still, the policy initiatives should not assume, as they mostly do today, what they nevertheless may aspire to achieve. This point is most notably of importance with regard to the issue of mobility. Quite often, policy recommendations and interventions in the field presuppose the sedentary way of life they intend to achieve, and this is not only the case with regard to the actual land use practices (in which, for instance, certain recommendations with regard to soil improvement or land reform assume an immobility of actors' fields), but also with regard to certain administrative and political arrangements. In order for local bodies to be representative of the diverse stakeholders, a number of mobility processes should be taken into account in preliminary analyses and then the implementation of measures of decentralisation and democratisation. Likewise, land reform should be based on the flexibility of existing arrangements, which allow for mobility of land users and do not assume a fixity that might undermine the little livelihood security land users presently have.

In short, policy should allow for, and stimulate, the robustness of livelihood modes that hitherto escaped development and government planning exercises – or were merely negatively affected by them. This means modes of livelihood that often are unexpected and devised in complex ways by actors in interaction with their changing environment. The problems West African and development policy makers face are such that they call for regional solutions. The economies of several countries are closely intertwined, even if this intertwinement first and foremost concerns the countries' 'informal' economies. Likewise, the livelihoods of actors of different nationalities are closely dependent on one another and at times conflict violently with each other. Unilateral decisions, such as the one excluding retroactively Burkinabé migrants from ownership rights to land in the Côte d'Ivoire, are to be avoided. The multi-spatiality of actors' livelihoods must be a starting point as well - a multi-spatiality that implies the intertwinement of livelihoods of people of different population groups and nationalities.

Chapter 15

CHANGE AND ADAPTATION IN A LIVESTOCK-PRODUCING REGION: NORTHEASTERN BURKINA FASO

Fred Zaal and Asaïta Diallo

Abstract: Case study of the north-eastern part of Burkina Faso, the provinces Seno, Soum and Oudalan, representing a semi-arid zone with low population densities and relatively low levels of land degradation. This area has long been a zone of transhumant livestock keepers, mainly FulBe. During and after the drought period of the 1970s and 1980s, many of these FulBe and other people (Tamachek and Mossi), many of them immigrants, settled in the area and adapted to the new circumstances, with the result being an increase in the cultivation of millet and higher yields. Better rainfall levels in the 1990s and a quite adequate adaptation resulted in a higher population with higher food security than in the decades before. The situation was further improved via extra income through migration, gold digging and the irrigated cultivation of vegetables aided by an improving transport infrastructure. In addition, growing urbanisation in the region itself triggered a growing market economy. For most households, livestock is no longer the dominant occupation, although in terms of total income its importance is still high.

1. INTRODUCTION

The northernmost Sahel region of Burkina Faso is, in the true sense of the word, a ′coastal′ area, a zone where the cultivators meet the livestock producers and where any decrease or increase in precipitation is directly followed by the ebb and flow of cultivation and pastoral production and thus a shift in livelihood strategy. Whenever the annual rains decline, cultivation declines in importance too and livestock becomes a more important source of sustenance and income. Whenever the rains are adequate, cultivation contributes considerably to sustaining the population. Of course, this applies to those who practice both forms

A.J. Dietz et al. (eds.),
The Impact of Climate Change on Drylands: With a Focus on West-Africa, 243–265.

of agriculture. A number of groups focus specifically on cultivation rather than livestock production and they have additional strategies for avoiding a complete disaster whenever the rains fail. Other groups have focused almost exclusively on livestock and they form the first representatives of a system of production that extends north into the true desert of the Sahara.

In this chapter, data is presented on this ebb and flow of cultivation and livestock production, on population dynamics as a consequence and on other major trends following the changes in climate of the past decades. This data is primarily derived from an earlier report by Ms Asaïta Diallo, for the ICCD project, to which was added (statistical) information from government, meteorological and secondary scientific publications and databases. This data is presented in order to allow an assessment of the more detailed data in this region, as opposed to the higher-level analysis in earlier chapters (e.g. Chapter 7). Not surprisingly, this data had to be pieced together from a wide range of sources, some more unreliable than others. The areas all form part of the Sahel region (we take Seno Province as one, before the subdivision in two separate provinces, as most of the data are not yet specific for the new provinces). The representation of government and other intervening organisations is limited and so is the data-collecting infrastructure. However, the data that is available refers to past changes and fundamental adaptations in the past. Such changes indicate that climate change, as presented in the scenarios in chapter 4, may result in fundamental and permanent changes in livelihood strategies. The ′coast′ will be to the south and this area will no longer be part of that dynamic zone where cultivation and livestock production meet and interact. Major trends include rainfall patterns and population dynamics. We will discuss those first and link them to tendencies related to soils and land use, hydraulics and infrastructure, government and non-government interventions and finally to livelihood change. Map 15.1 provides an overview of the research area.

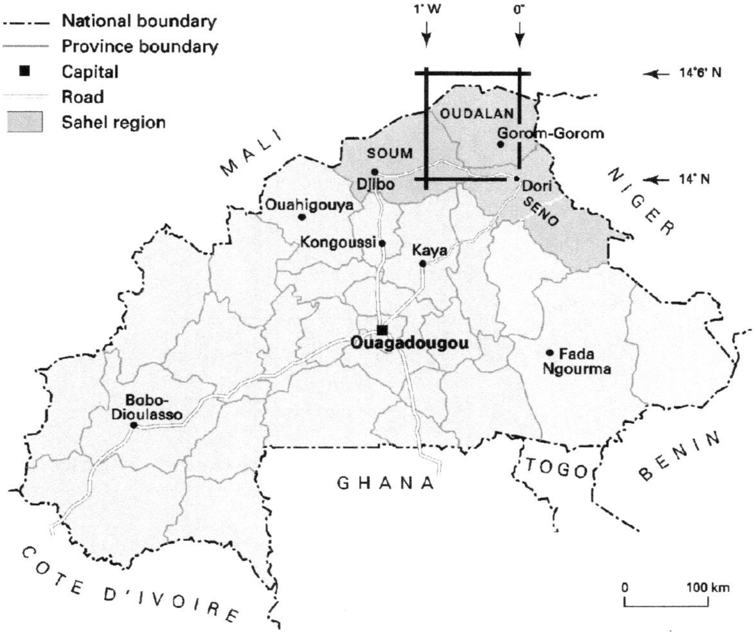

Map 15.1. The study area in Burkina Faso *(See colour section, p. 463).*

2. MAJOR TRENDS IN THE BURKINABÉ SAHEL: RAINFALL PATTERNS

The first trend, the one of climate, is certainly in keeping with the general patterns presented in earlier chapters. The following figure presents the precipitation for a number of stations in the Sahel, including the data for Dori, Gorom-Gorom, and the average for four stations (Dori, Gorom-Gorom, Gorgadji and Aribinda). As was found for other areas, the rainfall figures reveal high levels in the 1960s, reduced levels in the 1970s and 1980s and increasing levels in the 1990s (see figure 15.1).

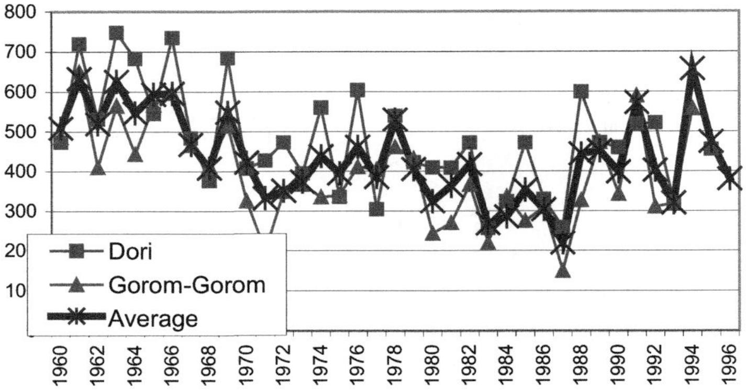

Figure 15.1 Rainfall in mm per year, Dori, Gorom-Gorom and average for four Sahel stations, period 1960-1996 *(See colour section, p. 464).*

Reliable but low rainfall, producing little but highly nutritive grazing in the northern arid areas has, for a long time, been a lure to transhumant livestock producers. Figure 15.2 shows that variability, expressed by the average rainfall for the Sahel stations divided by the SD, is generally between 2.7 and 5, with higher extremes in periods of higher rainfall, and lower extremes when rainfall is generally lower.

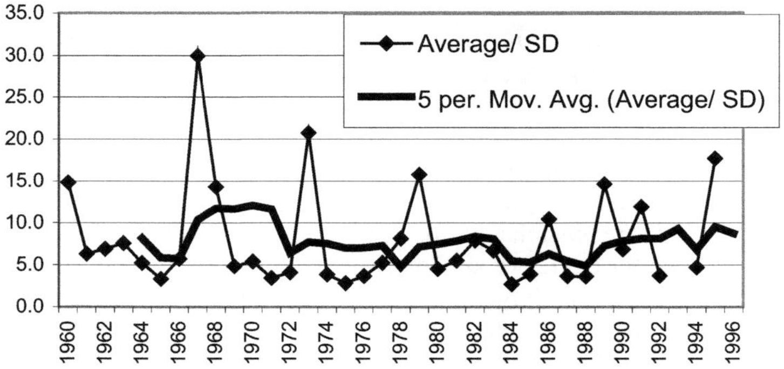

Figure 15.2. Average annual rainfall divided by SD and five year moving average for four Sahel stations during the period 1960-1996 *(See colour section, p. 464).*

Undoubtedly, the temperatures over the years will have been influenced by these trends in rainfall, but we do not have such detailed temperature data. Table 15.1 presents the average temperatures for Dori station for the period 1955-1990.

Table 15.1. Average minimum, maximum and average daily temperatures (in °C), for Dori station, period 1955-1990

	J	F	M	A	M	J	J	A	S	O	N	D	Av.
T max	37,5	41,3	44,7	47,2	46,2	43,8	39,8	39	39,4	44,2	41,3	37,2	41,8
T min	15,9	18,6	23,3	28,2	30,2	29,5	26,7	26,3	26,8	25,9	19,8	16,4	23,9
T Av.	26,7	29,9	34	37,7	38,2	36,6	33,2	32,6	33,1	35,0	30,5	26,8	32,8

These temperatures are very high, especially in the last part of the dry season, when they reach well above 40°C. During these months of March, April and May, food reserves are generally already low, while grazing is limited and water increasingly unavailable. Livestock has to be trekked over increasing distances, or kept away from the home in areas where grazing and water are still available. The following tables provide some information on evaporation and evapotranspiration.

Table 15.2. Daily evaporation, average per month (in mm.), at Dori station, average for the period 1961-1990

	J	F	M	A	M	J	J	A	S	O	N	D	Av.
Piche	7.21	7.65	9.53	9.98	8.6	7.06	4.89	3.47	4.12	6.65	7.17	6.8	6.36
Bac A	7.31	8.01	10.4	10.7	10.9	9.53	7.62	6.36	6.4	7.86	7.41	7.04	8.3

Irrespective of the method adopted (using the 'Evapometer of Piche' or the *'Bac classe A'*, both methods used by the *Direction Nationale de la Météorologie*), evaporation is highest during the hot, dry and windy months of March, April and May, while they are lowest during the rainy and cloudy months of August and September. Evaporation in the dry months is important for the availability of water from open sources (the *'mares'* and *'retenues d'eau'*) for livestock, while evaporation during the rainy months is important for the availability of moisture for plant growth. More relevant however is the evapotranspiration rate, indicated for the rainy season in table 15.3.

Table 15.3. Daily evapotranspiration, average per rainy season month (in mm.), at Dori station, average for the period 1981-1992.

	April	May	June	July	August	September	October
Average per day	6.10	6.20	5.74	5.33	4.83	4,93	5.02

If we assume that the length of the rainy season in each year is two times the number of rainy days in that year, as indicated in figure 15.3, we arrive at a comparison of evapo-transpiration and precipitation as presented in the following figure.

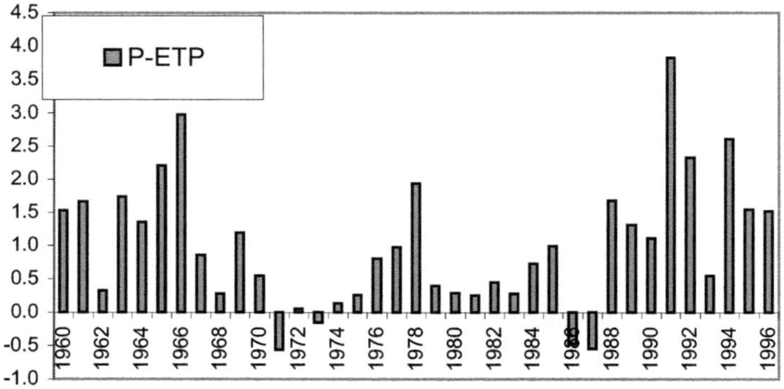

Figure 15.3. Estimated balance between precipitation (P) and evapo-transpiration (ETP) in mm per day during the rainy season, period 1960-1996

Surprisingly, there does not seem to be a difference in general balance levels between the 1960s and the 1990s. Both periods are characterised by a daily surplus of moisture of between 0.25 and 3.5 mm though over a smaller number of rainy days, as we have seen. The early 1970s were clearly a very bad period, but the very worst period was 1986-1987 when there were two years in a row with a negative water balance during the rainy season. The early 1980s had been slightly less adverse, but there was no large excess of rainfall, certainly not enough to provide a buffer of moisture to save a crop during short dry spells in the rainy season. It is interesting to note that in many peoples' views both the early 1970s and particularly the early 1980s were disastrous. Although, in terms of rainfall deficit, the late 1980s were worse they are considered less adverse compared to those earlier periods. By the late 1980s people had adapted their livelihood strategies.

Apart from these figures, which only give a global view of the climatic situation, the differences between the provinces are quite marked. Soum province is Sahelian in character, though relatively arid. Rainfall generally falls in short but heavy showers and is heterogeneously distributed over space. Oudalan has a similar climate and both provinces experience marked changes in wind direction. During the rainy season, the wind and rain come from the South-west, while in the dry season the Harmattan comes from the North-east. Seno province is slightly less arid, with a slightly longer rainy season and less erratic rainfall spatially. August has the highest rainfall, while the rainy season generally starts in July, or sometimes June, and ends in September. In the province of Oudalan, the rainy seasons are a few days shorter. It is the difference between being able to cultivate sorghum on the better soils (in Seno Province) and surviving on millet (in the other two provinces).

3. UNEXPECTED POPULATION DYNAMICS

The general demographic trends in Africa were discussed in chapter 8. They showed a steady migration away from the most marginal areas bordering true deserts and away from areas of conflict towards the coastal areas and major cities, both coastal and inland. In the Burkinabé Sahel, details on population dynamics show that local conditions do make a difference and point to adaptation behaviour to cope with the climatic trends described above. First of all, population growth is higher in some of the most northerly provinces than in the country as a whole: e.g. Seno during the whole of 1975-1996, and Soum in 1985-1996. This is contrary to what was expected. This means that either the natural provincial population growth was high, or that migration (families moving permanently into the area) into the area was considerable, particularly during the period of the droughts in the early 1970s and 1980s, or both. In the second period, the annual growth figures are lower than in the first period but higher than the national average, with the exception of Oudalan Province. However, compared to Burkina Faso as a whole and to many other rural dryland regions in Africa, figures show that population densities are still low.

Table 15.4 Population in Seno, Soum and Oudalan Provinces, and annual growth rates, 1975, 1985 and 1996.

	Area (km2)	1975		Annual growth 1975-85	1985		Annual growth 1985-96	1996	
		Pop.	Dens.		Pop.	Dens.		Pop.	Dens.
Seno	13570	157759	12	3,8	228905	17	2,7	320164	24
Soum	12665	143805	11	2,8	186812	15	2,8	253773	20
Oudalan	9931	80841	8	3,2	106194	11	2,3	136521	14
Sahel	36166	382405	11	3,5	521911	14	2,8	710458	20
Burkina Faso	274200	5638203	21	3,5	7964705	29	2,5	10465823	38

Source: STRAT Sahel 1998, INSD/ RGP 1985

In the first period, net migration is generally out of the northern provinces, in particular of Soum Province. For Seno and Oudalan province this is less clear and depends more on local agro-climatic conditions. Table 15.5 shows that in-migration has gone down, except for Oudalan between 1985 and 1991, and that out-migration has decreased as well. However, the net effect has been an increase in the migration ratio, making Seno and Oudalan provinces in-migration areas from the late 1980s onwards. The ratio is the relationship between the net effect and the total number of migrants. Migrants are those people who were born in a different province from where they lived during the census.

Table 15.5. Balance of permanent in-migration and out-migration for the Sahelian provinces and selected provinces with highest ratios, 1985, 1991

	In-migration (% of total population)		Out-migration (% of total population)		Migration ratio	
	1985	1991	1985	1991	1985	1991
Oudalan	3,1	4,7	9,7	2,9	-0,55	0,3
Seno	4,7	3,7	11,7	2,8	-0,46	0,2
Soum	7,8	3,6	14,1	6,8	-0,32	-0,3
Kadiogo	42,9	33,3	22,8	13	0,44	0,5
Sanmatenga	4,7	2,5	20,7	14,4	-0,68	-0,7
Sissili	38,5	24,3	7,4	5,8	0,77	0,7

Source: INSD, RGP (1985); Enquête démographique (1991) - STRAT Sahel 1998

Between departments, population densities vary considerably (see table 15.6) and indicate an increase in urbanisation going north. The highest densities are found around the larger provincial headquarters, the department of Djibo in Soum province having the highest density of 57 people per km^2. Relatively high densities can be found around Bani, Dori and Seytenga in Seno Province and Aribinda and Pobe Mengao in Soum province (between 30 and 35 people per km^2). In Oudalan, the highest densities are less striking, at around 20 people per km^2 in the most densely populated area around Gorom-Gorom. However, in Oudalan the differences in density between these departments and the others are more marked, followed by Soum and Seno province. Generally, the urbanisation rate is between 10 and 20 percent, with the exception of Falagountou (64.3 percent) in Seno, Djibo (46.6 percent) and Tongo Mayel (34.9 percent) in Soum, and Deou (52.5 percent) and Markoy (24.6 percent) in Oudalan. The reasons for high rates of urbanisation are mostly that the departments contain a regional administrative centre, a market centre or a gold mining site. The latter, in particular, attracts large numbers of people to towns like Essakan and Gassel in Oudalan, Falagountou and Goulougountou in Seno, and Fétékolé and Bouriel in Soum province.

Ultimately, net in-migration is of limited importance for explaining the high population growth figures. During certain periods of drought both in-migration and out-migration increased but had a minor effect on the balance in the region as a whole. If we compare the two periods we can conclude that a sharp decline is taking place in natural growth figures in view of the much lower annual growth in the period after 1985.

Table 15.6. Population figures: size, density and number of villages per province and département, 1985 and 1996.

Provinces	Départements	1985			1996		
		Pop.	Density	No Villages	Pop.	Density	No Villages
Seno	Bani	35139	22.54	64	49493	31.74	69
	Dori	67836	26.79	59	82020	32.39	85
	Falagountou	6644	11.02	11	13212	21.91	14
	Gorgadji	16426	18.05	15	22273	24.48	20
	Sampelga	9873	14.59	8	11742	17.35	9
	Seytenga	14861	21.29	16	24139	34.59	29
Soum	Aribinda	51048	20.03	36	68635	27.29	49
	Baraboulé	16892	17.74	18	22497	12.91	21
	Diguel	3809	6.95	6	7078	23.63	7
	Djibo	29287	38.28	14	43560	56.94	16
	Kelbo	-	-	-	-	-	-
	Koutougou	8914	5.87	11	13512	8.90	16
	Nassoumbou	13369	6.30	11	15546	7.32	14
	Pobe Mengao	13358	23.65	12	17431	30.86	14
	Tongomayel	50115	13.63	10	65514	17.81	47
Oudalan	Déou	12901	7.56	10	16797	9.84	16
	Gorom	15152	16.42	60	68905	20.51	87
	Markoye	18660	14.15	17	22096	16.75	26
	Oursi	9918	9.183	8	11998	11.11	18
	Tin Akoff	9563	3.88	8	16725	6.78	15

Source: INSD / Compilation STRAT Sahel 1998

Ethnicity in the three provinces and their (assumed) specialization is important, as data on the actual livelihood pathways these various groups are taking might shed light on recent changes in land use and cereal production (see below). In this area, the main groups vary per province as shown by Table 15.7.

Table 15.7. Resident population by language, per province.

	Fulfuldé (%)	Tamachek (%)	Mooré (%)	Others (%)
Seno	78,52	3,41	3,61	14,46
Soum	42,20	3,38	23,94	30,48
Oudalan	22,84	41,96	2,54	32,66

Source: INSD 1985

In Oudalan province, the Tamachek-speaking population dominates while in the Seno, with its large and well-established population of Liptako FulBe, Fulfuldé is widely spoken. In Soum, a steady migration of Mossi from the Central Plateau has increased the proportion of Mooré-speaking people to almost a quarter. The Tamachek and FulBe DjelgoBe and GaoBe are generally considered to specialize in livestock production. The FulBe Liptako and Mossi are considered more sedentary and cultivators.

4. OTHER RESOURCES: HYDRAULICS, MINERALS AND INFRASTRUCTURE

Since the droughts of the late 1960s and early 1970s, a large number of village and pastoral water projects have started and many have since been completed. Village drinking water projects contribute to the welfare of the whole population of, men, women and children and the welfare of all classes. It is usually among the first priorities next to cereal production and water projects have therefore attracted continuous attention. These sources of water are sometimes also used for the production of vegetables and other (cash) crops. Pastoral water projects are also multi-functional, providing drinking water for people and animals, but sometimes also for some small-scale cultivation. The first sign of agricultural diversification and the exploitation of market linkages in the Sahel is the appearance of irrigated market gardening around these pastoral water points. Water resources are crucial for many similar processes of intensification and some relevant data is presented below. Table 15.8 shows the number of sources and their character.

Table 15.8. Principal hydraulic infrastructure in the Burkinabé Sahel.

	Seno	Soum	Oudalan	Total
Modern (lined) wells	245	117	157	579
Pumps	688	248	196	1132
Enlarged natural depressions	6	43	34	83
Larger dams	13	14	11	38
Permanents water points	367	217	154	738
River basin and natural lakes	44	52	49	145

Source: DEP/Ministry of Water Resources (1998) and STRAT (1994)

Every type of water source has its own specific uses and limitations, but it is clear that Seno Province is best placed with regard to water for human, animal and agricultural use. Not all the pumps and wells are fully functional however. On the other hand, many traditional wells that exist are not included in this inventory and the situation may, on the whole, be better than is suggested here. It is around the permanent water points and larger dams that we find market gardening activities and some small-scale production of vegetables and rice for own use is found around the wells and pumps. The naturally occurring wild grass-type crop of fonio is found around all the river basins and natural lakes, where it offers the first opportunity to obtain food. The fonio is collected well before the millet and sorghum crops are ready and is the first new source of food after the hunger season, the *'soudure'*. Generally, people are ashamed to admit that they collect this grass, as it indicates that the harvest last year was insufficient. A harvest sufficient for the whole family for the whole year is a matter of pride and status. However, there is no other option in the event of severe shortages and few other alternatives.

There are a number of areas where mineral resources have become a source of income for the population and the country. Manganese is found in a zone between Tin Edia in Bouba until Oursi, near Gaigou and Tambao. In Tambao, the reserves are estimated at 50 million tons, including 50 percent of iron in the North-east. These are quantities initially deemed adequate for the railway to be extended from Ouagadougou to Tambao. After it had been built to Kaya, and the track had been cut until Dori, the project was abandoned after the World Bank withdrew its support. Calcite is found at two sites near the Beli River. This also is a mineral that is mostly exploited by bigger businesses or the government. However, the most important mineral in this area is gold which is mined by the population during severe droughts and other problematic periods. In Seno province the sites are near Tambiri, Solna, Mossiga, Helga, Tiena, Kankanfogol, Tangangari, Takatami, Aoura, Diouga, Gangaol, Bayildiaga, Goulougountou (800 diggers) and Goroul Kadol. In Soum province the sites are near Inata, Kéréboulé, Tongomayel, Tingalé, Aribinda, Fétékolé (100-200), Domsa, Kékéinéré, Sénékiaké, Tounté, N'Darga and Bouriel (1000). In Oudalan gold can be found near Assakan (2600), Batiala, Batiala, Toudikana, Kirohari (480), Markoye, Falagountou (300), Gosel and Gassel (1500) (STRAT Sahel 1998; the number of workers at the most important sites is indicated between brackets). The use of gold digging as a measure of last resort is also indicated by the starting dates of these gold-digging exercises. Much of the mining started in the first half of the 1980s. The sheer number of sites not only indicates the amounts found but mostly that a lot of possible sites are exploited based on the need of the population when drought strikes. Many hundreds of young men in Assakan work in pits often forty to fifty metres deep, extracting gold in very dusty and warm conditions. A French corporation (CEMOB) is also engaged in mining activities in Assakan and Gossy in a more mechanised manner. In other places in Seno and Soum industrial gold mining has ceased.

Of crucial importance for the development of this region, indeed any region, and the full exploitation of its resources is the network of roads that connect it with the bigger markets. It is clear from observations in areas slightly closer to Dori, Kaya and Ouagadougou that whenever the combined resources are there (labour, soils, water, knowledge, experience, social organisation), the connection of these resources with markets very quickly leads to rapid adoption of new technology increasing the efficiency of the use of these resources and the profitability for individual big and medium-sized farmers, and of groups of poor small farmers. Infrastructure in the region of Dori – Djibo – Gorom-Gorom is very limited in extent and quality however. The following tables show that both the length and the quality of the road network are limited.

Table 15.9. Route infrastructure in the provinces of Seno, Soum and Oudalan (in km)

	Seno	Soum	Oudalan
National route, permanent	50	-	-
National route, intermittent	45	-	-
Secondary route, permanent	158	196	-
Secondary route, intermittent	37	69	313
Rural ′piste′ (sandy path), intermittent	277	478	214
Total	567	743	527

Source: STRAT 1998/ DRPC Dori 1998

Table 15.10 Connections between capitals of provinces and departments.

Province	Connection	Distance (km)	Type	State of the routes
- Seno	Dori-Sebba	92	P.A.B	bad
	Dori-Bani	35	R.T.O.	very good
	Dori-Gorgadji	53	R.T.O.	good
	Dori-Seytenga	45	P.O	very bad
	Dori-Falagoutou	72	P.O.	very good
	Dori-Sampelga	45	P.A.B.	mediocre
- Soum	Djibo-Baraboulé	29	P.A.A.	mediocre
	Djibo-	40	P.O.	very bad
	Koutougou	45	P.O.	very bad
	Djibo-Nassoumb	18	R.T.O.	good
	Djibo-	27	R.T.O.	good
	Tongomay	70	P.O.	bad
	Djibo-Pobe Man	90	R.T.O.	good
	Djibo-Diguel			
	Djibo-Aribinda			
- Oudalan	Gorom-Markoye	40	P.O.	mediocre
	Gorom-Déou	72	P.O.	bad
	Gorom-Oursi	45	P.O.	bad
	Gorom-Tin akof	74	P.O.	very bad

Source: various project statistics at PSB and the CRPA in Oudalan

P.A.B. = ′Improved Piste′ type B; P.A.A. = ′Improved Piste′ type A; P.O. = Ordinary ′Piste′; R.T.O. = Non-surfaced Route (sandy surface).

This shows the limited extent and usefulness of the road network. Even the permanent roads can sometimes be blocked for weeks due to calamities and excessive rainfall. A lot of people are familiar with the unusual way of entering Gorom-Gorom during such periods: you have to phone the person you want to meet in Gorom-Gorom, who will go by car to the river valley between Dori and Gorom-Gorom and when you have arrived there you have to go on foot to the other side of the valley to be picked up and be transported to Gorom-Gorom.

This indicates the value of the other important network: the telephone. Though there are only telephone stations and small networks in Dori, Djibo, Gorom-Gorom, and Markoy, the latter two in particular are extremely important. They are both a means of arranging road transport and of finding out about market conditions elsewhere. Livestock traders use these networks to find out about prices in Ouagadougou and Abidjan and can then estimate the prices on the local market that will still give them a profit. However, the only official route for cattle to be transported between Markoye, Gorom-Gorom, Dori, Kaya and Ouagadougou is no longer operational or protected and taking the animals on foot to Ouagadougou is still a costly and risky endeavour. Especially in the rainy season, the encroachment of cultivation on the cattle routes makes it difficult to get to Ouagadougou. The railway to Kaya has not been operational for these purposes since 2001.

5. ADAPTING LIVELIHOOD STRATEGIES; CHANGE OF LAND USE AND NON-FARM INCOME

Though cultivation of dryland crops is probably still the mainstay of the three provinces, livestock production is in an important second place. The following tables show the occupational patterns for the year 1998. There are limitations to this data. Most people, and certainly most households (however defined) have more than one or two main occupations. For example, the tables does not include migration as a source of income through remittances or own earnings, nor pastoral production. This is strange since we saw that in-migration and out-migration affect a large number of people, especially during droughts. Moreover, the provinces of Soum and Oudalan are among the provinces with the largest herds of the country. Those people stating ′no income′ may well belong to those two occupational groups. The source of the data, the annual agricultural survey of the Ministry of Agriculture, focuses on cultivation and sedentary sources of income in particular, as the goal was to develop an annual grain production and consumption balance sheet, for food aid and trade planning purposes.

Table 15.11 Primary (P) and secondary (S) activity of the population in the rainy season

	Agriculture		Livestock		Maraîchage		Commerce		Artisan		Other		No income	
	P	S	P	S	P	S	P	S	P	S	P	S	P	S
Oudalan	61	1	4	17	0	2	0	0	4	2	5	7	26	70
Séno /Yagha	65	1	3	12	0	1	0	1	2	3	1	5	29	77
Soum	76	4	5	18	0	0	1	1	2	2	2	3	14	72
Sahel	68	2	4	15	0	1	0	1	3	3	2	5	23	73
National	86	4	5	13	0	1	1	3	0	2	0	10	5	66

Source : ENSA – STRAT Sahel 1998

Maraîchage is the production of irrigated crops, usually vegetables, usually in the dry season, partly for sale. Séno/Yagha are the two new provinces which were previously Seno province and have been combined in this table.

Oudalan Province the province with the least agriculture. About sixty percent of the household population states that their primary activity is cultivation, and in view of the above it may in fact be even less. Livestock production comes second and there is a whole range of other activities as well. The whole of the Sahel Region of Burkina Faso is much less engaged in cultivation than the country as a whole.

Table 15.12 shows the difference between men and women as regards their main source of income in the dry season. This could be a better indicator of the livestock orientation in this area. It shows the extent of livestock ownership among men to be similar to the percentage of people interviewed who do not engage in cultivation. It is safe to suppose that about 40 percent of the population in Oudalan Province is engaged in livestock production and could be called (at least part-time) pastoralists. What we also see is that livestock is much more important as a source of income in the dry season in the Sahel than in the country as a whole, but only for men. Women focus on artisanal production much more than elsewhere in the country.

Table 15.12. Primary activity of the population in the dry season, by sex.

	Cultivation		Livestock		Maraîchage		Trade		Artisan		Other		No income	
	M	F	M	F	M	F	M	F	M	F	M	F	M	F
Oudalan	0	0	40	1	4	2	1	1	2	25	21	25	33	46
Séno /Yagha	0	0	22	1	4	1	4	3	4	11	29	11	38	73
Soum	0	0	36	9	2	0	7	7	4	25	21	13	30	46
BF Sahel	0	0	31	4	4	1	4	4	4	19	24	15	34	57
National	1	1	25	3	7	3	6	12	6	10	18	25	38	47

Source: ENSA - STRAT Sahel 1998

We now have a general idea of the options chosen by the population to develop their present livelihood. The question is to what degree we can ascertain that there has been a *change* in livelihood in the last few years, or at least since the droughts of the 1980s. In particular, we would like to know whether there has been a change away from cultivation and towards livestock production, or away from both these options and towards maraîchage or artisanal production, migration or other occupations.

To start with cultivation, we have developed the following table to relate cultivated area, yields and production with population figures to control for population growth as a driving force for land use change. The main crops in this area are millet and sorghum, but sorghum is of such minor importance that we neglect it here. There does not seem to be a trend in area or yields for sorghum production either. This also applies to other grain crops and cash crops. However, in the last few years, groundnuts have become a cash crop. The following graph (figure 15.5) shows the development of the millet area in Oudalan Province, with a linear trend added.

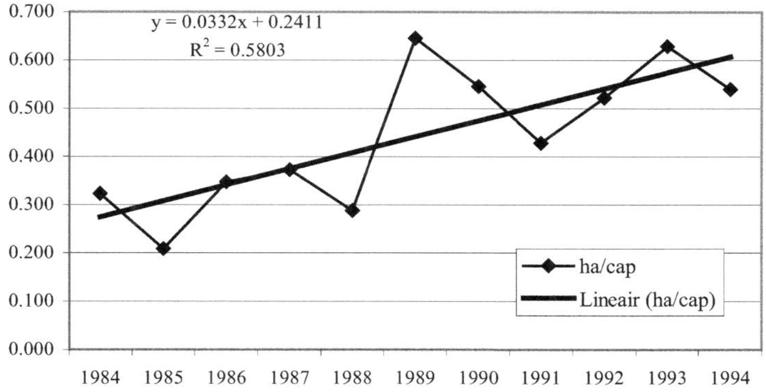

Figure 15.5. Change in land use in Oudalan province, millet area per capita and linear trend, 1984-1994.

Earlier studies have established that the average surface of cultivated fields per inhabitant in the northern part of Oudalan province (the area between Oursi and Gorom-Gorom) is around 0.73 ha per person (Claude, Grouzis & Milleville 1991), a figure close to an earlier estimate of 0.80 ha for the same area (Peretti 1977, quoted in Claude, Grouzis and Milleville 1991). This average hides a wide range of cultivated areas per person, which may vary from 1.98 to 0.15 ha, depending on the type of soil and the yields, the density of population and the extent of cultivable soils. Data from the National Survey on Agricultural Statistics (ENSA) by the Ministry of Agriculture and Livestock Resources (MARA/DSAP 1996) suggest that the household fields (of which some 81 per cent were considered to be 'communal' or under the control of the male representative of the household) were almost exclusively used for millet production in Oudalan and for combined millet and sorghum production in Seno Province. There is clearly an overall increase in land per capita used for millet production, while other crops are either insignificant (such as groundnuts) or do not show any trend (such as sorghum). The sorghum area per capita was around 0.03 ha/cap for the whole of the period, except for two peaks in 1985 and 1992 when the area doubled.

A doubling of the area under the main millet crop in such a short period is quite spectacular. Above all, it is contrary to what we would expect, though the period is perhaps too short to be certain about the relationship between rainfall trends and cultivation. Even though the number of ploughs in the province is relatively low (the average is 0.01

draught oxen and 0.03 donkeys per household, compared to 0.38 and 0.26 respectively for Burkina Faso as a whole), it may just be that technological change has caused this change in cultivated area, so that growth of the area compensates for declining yields. Therefore, we have also included the following figure, showing the yields per hectare and per capita (figure 15.6).

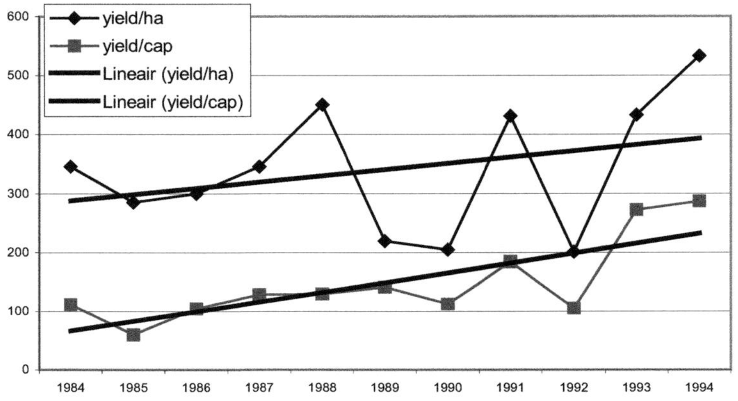

Figure 15.6. Yield per hectare and per capita and trends in yields in Oudalan province, 1984-1994.

The data reveals something quite remarkable which has also been noticed by others (Niemeyer & Mazzucato 2000, Zaal 1998, 80) namely a growth in yields per capita which would be compatible with increases in the area in which animal traction is used (of which there is little) and the use per hectare, which is not compatible. Inputs such as artificial fertilizer and herbicides and pesticides are practically non-existent in this province and this trend can therefore be ascribed to increasing levels of sophistication in tillage, crop choice, planting techniques and manure application. The latter is an interesting possibility in view of the availability of animals and animal manure. This is all the more interesting, as the population in this province is generally not considered to be the best in cultivation. We discovered that they belong to ethnic groups that are considered to be good livestock producers but not cultivators. Even if they were not, they have certainly managed to close the gap. The average yields per capita and per hectare in Burkina Faso are around 300 kg and 500 kg respectively. It may be that an increasing part of the population is engaging in cultivation. That would partly account for the growth in yields, as more labour is being put into the production of crops.

Another interesting figure is the yield per capita. This finally negates many years of laments that *'les bilans céréalières donnent la province chroniquement déficitaire, ce qui entraîne chaque année le recours a l'aide alimentaire'* (Diallo 2000). From a 50 percent coverage of per capita food needs from cultivation of millet alone, the production system has now improved to provide more than 100 percent of cereal needs per capita, though of course not on an annual basis and not for all groups. The rain deficit years will always push the self-sufficiency ratio down to levels where alternative sources of food need to be identified by the local population. This poses a challenge for the present study because it completely negates the initial ideas developed in the early phase of the research project of a deteriorating trend, following negative climate change. We will come back to this in the concluding part of this chapter.

Let us turn to the other mainstay of the province, namely livestock production, and examine whether this generates any more surprises. One point of interest is, at least, that the movement of cattle, transhumance, seems to have reduced. The causes are that people are increasingly engaging in cultivation and need to be with their crops for at least some of the time during the rainy season. More recently, they have had to be near the irrigated gardens at the small dams during the dry season. Labour shortage is hampering an optimal movement of animals as young people migrate south for part of the year. Migration (seasonal labour migration) as a potential source of money (or the reduction of consumption in the household) is least frequent amongst FulBe DjelgoBe in the northern part of the province, who need a regular input of labour in the herds and generally have a cultural aversion to migration (Hampshire & Randall, 1994, 2000, Zaal 1998). This is a more important influence than the size of the household, which is large (with an average size of 12 members per household). However, FulBe GaoBe living around Gorom-Gorom, who have only slightly smaller households (an average of 10.6 members), have more seasonal labour needs in view of the importance of cultivation in their production system and are less averse to migration culturally. To the south of Oudalan, the FulBe Liptako cultivate extensively. This causes a slack season in labour needs and migration is higher as a result. Migration is generally higher as well amongst the RimaiBe for example, the erstwhile slaves of the FulBe (GaoBe and Liptako, not DjelgoBe), who are more likely to cultivate in a given environment than the FulBe. From these variables (size of the household, ethnic background and production system), ethnic background seemed to have the most important effect, followed by size of the household and production system (Hampshire & Randall 1994, 2000). With the increasing importance of cultivation, particularly among the people in the southern part of the province, livestock movements have reduced. However, with increasing levels of rainfall in the 1990s, we noticed in the field that a lot of livestock keepers had returned to the area their parents had lived in earlier. They again occupied the central zone of their transhumance area, where their parents had migrated to the southern part of those areas to exploit those regions during the dry decades of the 1970s and 1980s.

There are varied views on livestock wealth in the Sahel and this is partly because of the different sources used. The annual agricultural statistics, based on surveys among cultivators, mostly reveal lower livestock ownership figures than sources based on surveys among pastoralists. Two tables show this. Table 15.13 is derived from the annual ENSA survey of the Ministry of Agriculture. Table 15.14 comes from a study of FulBe pastoralists in Oudalan Province.

Table 15.14. Average herd size (ownership) of sedentary households per province in the Sahel, 1998.

Provinces	Cattle	TLU	Sheep	TLU	Goats	TLU	Total livestock in TLU
Oudalan	4.2	3.0	6.2	0.62	9.3	0.93	4.5
Seno/Yagha	4.4	3.1	3.2	0.32	7.2	0.72	4.1
Soum	6.1	4.3	6.1	0.61	9.2	0.92	5.8
Sahel	4.9	3.5	4.9	0.49	8.4	0.84	4.8
National	2.8	2	3.6	0.36	5.2	0.52	2.9

Source: MARA/ ENSA - STRAT 1998

1 TLU equals 1.42 head of cattle, and 10 small stock.

These figures show that according to the rule of thumb that pastoral life is possible if about 4 TLU per capita are available, the people surveyed cannot be pastoralists, or more accurately, cannot live the lives of pastoralists. However, the following table shows that if households are selected for their pastoral way of life from FulBe DjelgoBe, GaoBe and Liptako backgrounds, livestock ownership is considerably higher.

Table 15.15. Average herd size (ownership) of pastoral households in the Sahel, Oudalan and northern Seno Provinces, per wealth category, 1996.

Wealth category	Female cattle in TLU	As percentage of total cattle in TLU	Total cattle in TLU	As percentage of total livestock in TLU	Total livestock in TLU
Poor	4.9	70	7	73	9.6
Middle	12.6	64	19.6	85	23.1
Wealthy	53.9	68	78.4	89	88.2

Source: Zaal 1998

With clear divisions between wealth categories (up to 1.5 TLU per capita, between 1 and 4 TLU/cap and above 4 TLU/cap), table 15.13 clearly pertains either to poor pastoralists or to cultivators with animals. These cultivators can still be from FulBe background, such as in the Liptako area, or they may be Songai, Gourmanche, and Mossi. Earlier figures show a remarkable similarity in the number of animals per household or per person. The most readily comparable figures are those presented for the early 1980s (1980-1982) by Claude, Grouzis and Milleville (1991). They found that the ratio of livestock to persons among the ethnic groups studied varied between 0.23 and 2.4 UBT/capita. The *Unité de Bétail Tropicale* (UBT) is mostly used in French literature. One UBT equals 1.17 bovine and 11.7 small stock. The figures given by Claude, Grouzis and Milleville (1991) above would translate into slightly lower numbers in TLU (about 10 per cent lower, or between 0.2 and 2.2 TLU/capita). The precise figures were 0.23 UBT per capita for RimaiBe (the erstwhile dependents or slaves of the FulBe), 0.4 for MalleBe (FulBe), 0.55 for Iklan (Tamachek), 1.9 for GaoBe and 2.4 for DjelgoBe (both FulBe groups). Figures by Barral (1977) were higher (as they were indicative of livestock holdings before the drought of the early 1970s). For the same ethnic groups, he found UBT figures for the same groups of between 0.8 and 6.2 (the figures by Barral were 0.8 UBT per capita, 0.8, 2.1, 4.3 and 6.2 for the same ethnic groups respectively; we cannot recalculate these figures into TLUs). Claude, Grouzis and Milleville (1991) also quote Lhoste (1977), who gave an average of 41 head of cattle as the average number (28 TLU per household) for the same area after the droughts of the early 1970s.

Another characteristic of a typical pastoral (milk- and subsistence-based) system is the high percentage of female animals in the cattle herd. We found that these percentages were 62, 69 and 67 for DjelgoBe, GaoBe and Liptako herds respectively in the mid 1990s (Zaal 1998). Lhoste 1977 (who presented figures from 1976 for the area between Oursi and Gorom-Gorom and the whole of the Oudalan respectively) estimated that 69 and 70 per cent of the herds consisted of female animals respectively, while the remainder were male animals of varying ages. Claude, Grouzis and Milleville (1991) found percentages of 68, 71 and 81 for female animals in three herds studied in Saba Kolangal, Totiri and Timbososo respectively, all in Ouadalan Province.

One interesting fact in this region relates to the position of women as owners of livestock, or as independent operators in the economy in general. Depending on the sub-ethnic group, women have considerable independence in some groups. Although livestock ownership is not equally divided, the following table (15.15) shows that women do own some.

Again, this is not the pastoral population, but here too women possessed considerable numbers of livestock as a source of milk. The situation is slightly better again for small stock, while women own more than a quarter of chicken and other birds, contrary to what is found in the rest of the country.

Table 15.16. Distribution of ownership of livestock over sedentary men and women, 1998, in percentages

	Cattle		Sheep		Goats		Birds	
	men	women	men	women	men	women	men	women
Oudalan	83.5	16.5	78.9	21.1	71.9	28.1	59.1	40.9
Seno/Yagha	88	12	74.8	25.2	70.3	29.7	71.7	28.3
Soum	90.3	9.7	76.8	23.2	74.4	25.6	74.9	25.1
Sahel	87.8	12.2	76.5	23.5	72.1	27.9	70	30
National	93.7	6.3	85.9	14.1	79.6	20.4	86.5	13.5

Source: ENSA - STRAT Sahel 1998

Incomes, a very firm indicator of the choice for certain options, are still very much based on livestock ownership. The following tables contain details from a recent study on pastoral livelihood strategies from which some of the above evidence was also derived. They show clearly that livestock is still a very important source of revenue, a situation not much changed from earlier periods. However, in recent times, cash sales have become more important and exchanges have declined and almost disappeared.

Table 15.17. Monthly gross incomes of pastoral men (in cash and kind), for each of the wealth categories, in FCFA, Burkina Faso, 2/95-1/96.

Wealth	Cash incomes				In-kind incomes			Total
Category	Cattle	Small stock	Other	Sub-total	Livestock	Grains	Sub-total	
poor	11675	4523	3092	19290	1334	4665	5999	25289
middle	14181	5313	562	20056	1401	6914	8315	28371
wealthy	49023	3184	1363	53570	4806	12069	16875	70445

Source: Zaal 1998.

'Cash transactions Other' consisted of the sale of other (livestock) products and income form gifts and (own) labour migration.

Table 15.18. Monthly gross incomes of women (in cash and kind), for each of the wealth categories, in FCFA, Burkina Faso, 2/ 1995 - 1/1996.

Wealth	Cash incomes				In kind		Total
Category	Livestock	Milk	Other	Subtotal	Milk	Sub-total	
poor	252	1473	193	1918	4794	4794	6712
middle	517	1435	480	2432	8324	8324	10756
wealthy	101	1213	652	1966	18103	18103	20069

Source: Zaal 1998

For pastoral men and women, most of their cash and in-kind incomes are still derived from sales of livestock and livestock products. As far as men are concerned, the sale of cattle, as well as small stock, provides most of their income, which are used to a large extent to fill the grain store. Milk sales are important to women and they spend much of the money earned on family care and food items.

Table 15.19 Development of area and production of market gardening crops, Oudalan Province, 1993-1998.

	1993	1994	1995	1996	1997	1998
Tomato						
- Area	-	1	3	4	0.46	0.15
- Production	-	12	54	72	6.9	2.17
- Yield	-	12	18	18	15	18
Onion						
- Area	2.02	2	3	3	2.2	1.07
- Production	40	30	99	45	41	16
- Yield	19.8	15	20	15	18	15
Cabbage						
- Area	0.15	1.5	4	3	0.27	0.35
- Production	1.5	15	58	42	5.2	14.35
- Yield	10.7	10	15	15	19	14
Potato						
- Area	5	1	3	2	0.99	0.48
- Production	13	15	48	30	4.9	7.2
- Yield	0.4	15	12	15	15	15
Aubergine						
- Area	0.02	1	3	1	0.02	0.02
- Production	0.5	10	29	10	3.2	0.24
- Yield	25	10	10	10	12	12
Sweet potato						
- Area	-	2	6	3	0.54	0.25
- Production	-	26	146	60	10.8	5
- Yield	-	13	24	20	20	20

Source: Direction Provinciale de l'Agriculture de l'Oudalan, various years

There are other activities that add to household income. although still insignificant, irrigated agriculture is increasing in importance in the northern part of Burkina Faso , in particular the production of vegetables for the market and own consumption. There has been an increase in area and investments, not only in the north but in many parts in Burkina Faso where water, good soils, available dry-season labour and marketing infrastructure to nearby or further away markets are found all together. In Oudalan, the irrigated area covers only 6390 hectares, with 6334 of it being used for the production of forage for animals, some 50 for rice production and only 6 hectares for vegetables. The latter area is increasing, however, although certain crops suffer regular setbacks. The following table shows details on this very fragile beginning of a new option.

The importance of this activity is not so much the earning at this moment but the fact that it exists and that it may grow should conditions improve, in particular road and market conditions. A similar situation was seen in the provinces to the south where, after a slow start, irrigated cash cropping exploded once the roads had been improved and traders from Ghana and Togo provided a ready outlet for the produce.

6. CONCLUSIONS

As in many other cases in this study, the trends at general level may be clear, but this does not mean that these correspond closely to developments on the ground at lower levels of scale. The findings in the area of Gorom-Gorom and Dori prove that local conditions and characteristics can influence general developments and it is precisely these exceptions, these adaptations of people to locally favourable conditions that are of interest. They form the possible points of departure for local dynamic change and economic growth and improvement in living conditions.

First of all, the climatic conditions did not seem to be very different from the general trend in the West-African Sahel at large, though it is interesting to see that a reduction in rainy days with increases in daily rainfall do generate similar agricultural conditions in the 1990s as in the 1960s, albeit for shorter periods. Farmers clearly adapted to the new situation. Apparently, the disasters of the 1970s and 1980s were serious enough, but similarly low levels of P-EPT for example in the late 1980s did not have the same serious and regional impacts as those earlier droughts.

This is reflected in migration trends. Even though population growth has declined and was lower by a full percent in the period 1985-1996 than in the period 1975-1985 (with the exception of Soum Province where it was similar), it is clear that both in-migration and out-migration declined between 1985 and 1991. Migration rates have gone from being negative in the northerly provinces (and positive in the urban and developed agricultural areas in the centre and south of the country) to being positive (again with the exception of Soum). This is remarkable, as precisely during that time, the drought in terms of negative P-ETP was probably worse than in the early 1970s and 1980s.

It must be said that the migratory flows were probably targeted on the 'urban' areas in the north where 'urbanisation' is higher than in the southern part of the Sahel. The northern rural areas are largely used as grazing for semi-nomadic and transhumant pastoral production and densities and general agricultural options are limited. In the southern area, these possibilities are better and densities are higher in the rural countryside. This follows ethnic lines, as the Liptako FulBe living in Seno with other local ethnic groups are generally considered to be cultivators, while in the northern Soum and Oudalan provinces, FulBe GaoBe and DjelgoBe and Tamachek are generally livestock producers, though the GaoBe also cultivate some fields.

Not only are other groups of people entering the area and introducing cultivation where earlier it was not, formerly pastoral people are also increasingly engaging in cultivation. Though the Sahel Region in Burkina Faso is less engaged in cultivation than the country as a whole, almost 70 percent still states that cultivation is the prime source of income, with livestock second. Total herd size seems not to have changed much in time and herds are generally not large enough to sustain the pastoral population. The surprising thing is that without a clear increase in the application of animal traction or mechanised traction, the area of millet per capita has steadily increased. Not only that, the yields per hectare have increased as well, while the yields per capita have increased even faster. Yields per capita are now adequate to sustain the population, as they have increased from about 100 kg per capita in the 1980s to above 200 kg per capita in the first half of the 1990s. They did not even show a significant decline in the period of extreme drought in the late 1980s, when P-ETP rates were even below those of the 1970s and 1980s. Two reasons were suggested, namely the use of manure, indicating a closer integration of cultivation with livestock production (or at least the use of manure previously left to cultivators or in the grazing areas for cultivation by pastoral producers) and the selection of land with better soils and water retention capacities, made possible by the lower rainfall. No longer does the runoff caused by the rains cover the whole of the river basin with water and parts of the river areas are now cultivated using adapted strains of millet and sorghum. As a result, people are now facing high risks during wet years when abundant rainfall causes damage not by drought but by flooding. The cumulative effect is an increase in cultivation and yields, leaving the population better off despite the possible drying trends in these areas.

Chapter 16

LIVELIHOOD STRATEGIES AND DEVELOPMENT PATHWAYS AT HOUSEHOLD AND VILLAGE LEVEL

Johan Brons, Fred Zaal, Lianne Kersbergen and Ruerd Ruben

Abstract: The analysis in this chapter focuses on the effects of climate variability on village development in southern Mali and on household livelihood in Burkina Faso. The case of Mali showed that villages with relatively high and stable rainfall have developed commercially oriented crop production systems that benefit from market policies. Under more variable climate conditions, villages with relatively good land availability have pursued intensive cereal production or livestock husbandry sometimes supplemented with non-agricultural activities. In villages with a highly variable climate and limited land availability, food security depends on institutional and social exchange networks. The Burkina Faso study illustrates important similarities and changes in household livelihood strategies during two periods (1981-85 and 1993-1998). Crop production remained an important factor in household resilience and has been intensified by new crop management technologies. Livestock continued to be important for stabilising income and became additionally important for managing organic matter balances. Off-farm income relatively increased but remained an important factor in income inequality.

1. INTRODUCTION

Understanding local strategies regarding food production and land use requires the identification of critical factors that explain differences in resource allocation under variable and changing agricultural production conditions faced by farmers and their communities. Therefore, attention should be given to the decision-making on resource use and portfolio choice, the different reactions to rainfall variability at farm-household and village level and the derived implications for food security and sustainable land use.

A.J. Dietz et al. (eds.),
The Impact of Climate Change on Drylands: With a Focus on West-Africa, 267–291.

Making use of farm-household and village-level data, specific socio-economic and biophysical factors can be identified that influence decisions regarding choice of livelihood options, labour allocation and input use. The assessment of the impact of climatic variability on resource use and livelihood strategies enables the identification of different pathways of agrarian production in West African drylands. For this purpose, two analytical concepts were used: (i) portfolio choice defined as the composition of activities and options that contribute to food security and sustainable land use at farm household or village-level (Dietz & Verhagen 1998; see also Chapter 1) and (ii) development pathways, that include the selection of particular livelihood strategies that explain income and productivity differences amongst farming systems over time (de Bruijn & Van Dijk, 1998; see also Chapter 13).

The regional analyses for the ICCD research programme were conducted for different regions in Southern Mali and Northern Burkina Faso and at two different levels:

1. spatial analysis of different local strategies for portfolio diversification and risk-coping behaviour, paying attention to the determinants of inter-village diversity in selected development pathways in a case study of the sub-humid region of Southern Mali (Brons *et al.*, 2000a; Section 16.2)
2. household level adjustment of livelihood strategies under conditions of changing rainfall illustrated using a case study in the semi-arid region of Northern Burkina Faso (Brons *et al.*, 2000b; Section 16.3).

This chapter presents an analysis of differences in farmers' behaviour and portfolio choice and identifies arguments for farmers' selection of development pathways. We discuss the structural and behavioural factors that are the basis for different farm household strategies aimed at coping with climate variability and risks and the possible adjustments in the areas of crop production, income generation (e.g. activity diversification), investment behaviour and consumption patterns (Figure 16.1). This analysis enables a further discussion of farmer strategies for coping with climate variability and suitable policy instruments for reinforcing farm household development pathways towards sustainable land use and food security (see Chapter 21).

Different procedures of analyses were performed, making use of statistical data obtained from national agencies (IER and CMDT in Mali; ENSA and INERA in Burkina Faso), local research centres (CEDRES in Burkina Faso) and international organisations (FAO, World Bank, ICRISAT-SH). Analytical procedures are based on factor analysis and principal component analysis to identify development pathways; portfolio analysis to review the composition of risk-coping strategies and profit estimates to detect determining factors of input use and erosion control.

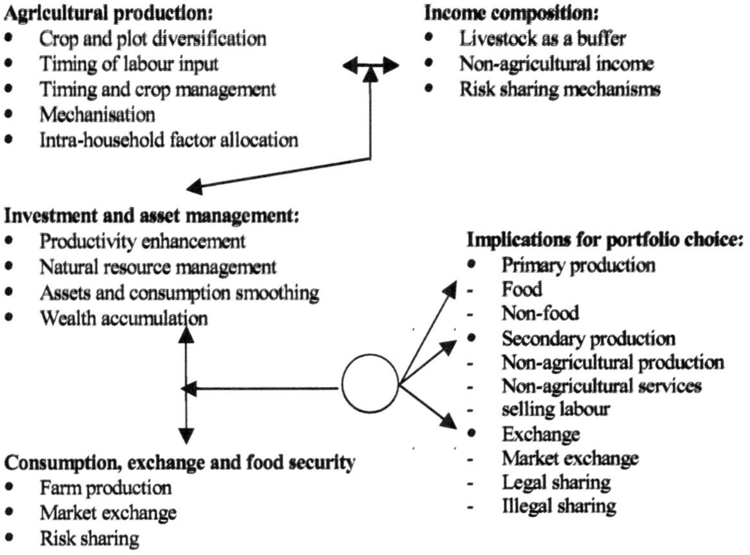

Figure 16.1. Farm household response to climate variability.

Local consultants provided assistance with the data collection, but additional data was compiled through the incorporation into the project of a Dutch research assistant based in Mali and Burkina Faso. Data on rainfall was derived from local weather stations and converted into a drought index (see Chapter 3). Soil characteristics were included using data on soil degradation and topo-sequence.

2. DEVELOPMENT PATHWAYS IN SOUTHERN MALI

The Koutiala and Sikasso region in Southern Mali (see Map) can be divided into five different zones according to bio-climatic (climate and rainfall) and agro-ecological (land quality and land use) conditions:

1. The northern range land zone (north of latitude 12.80°), where cotton cultivation has disappeared or is disappearing (10-20 percent of the area);
2. The zone within 35 km around Koutiala (Koutiala zone), where almost all arable land is cultivated with few possibilities for diversification;

3. The central zone (between 11.80° and 12.80° latitude), with very little room to expand the cultivated area and a cotton area that has reached its rotation maximum (40% of the cultivated area);
4. The southern zone (south of 11.80° latitude) where there is still room to expand the cultivated area, although the cotton area is approaching its maximum (25 – 35% of the cultivated area) but agro-ecological conditions allow for a diverse cropping pattern (vegetables, rice, sylviculture);
5. The zone within 35 km around Sikasso (Sikasso zone) with considerable potential for growing vegetables, potatoes, rice and fruit trees and an important nearby urban market. Farmers also grow cotton, but this crop is less important compared to regions 2, 3 and 4.

The field sample consists of 43 villages equally divided across the five zones. Tables 16.1 and 16.2 provide basic information on the socio-economic and agro-ecological conditions in each of the five zones. The zoning principles are similar to those used by local administrative organisations (especially the CMDT Cotton Board). The aggregate information at zone level provides information on the available potential within each region, while principle component analysis is used to identify specific development pathways that are selected for coping with local climate, demographic and market features.

Households in the villages in the proximity of Sikasso cultivate a significantly smaller acreage with labour intensive crops (rice and vegetables). In the southern zone, and in the Sikasso zone, maize is an important crop providing an early harvest in the food deficit season (1- 2 month prior to the cotton harvest) and generating cash income through the sale of fresh corncobs. In the northern zone, sorghum and millet provide the subsistence basis. The major cash crop, cotton, occupies an equal share in all zones except for the most northern zone. This indicates that the expansion of the cotton crop in the southern zones is reaching its limit of rotation requirements. Crop yields are highest in the Sikasso zone, which may indicate a positive relationship between cotton production and intensive rice cultivation in this area. Cotton yields are remarkably low in the Koutiala zone indicating that agronomic (soil nutrient content and texture) and perhaps economic conditions (preference for market gardening) are becoming less favourable to cotton cultivation. The Sikasso zone outperforms the other zones with respect to gross revenues per ha and production per household member while the northern zone stays behind in this respect.

The Koutiala zone is characterised by high levels of equipment and high stocking rates. Expenses per hectare (variable costs) are also highest in this zone. In fact, the stocking rates almost allow self-sufficiency based on livestock production provided all the milk and animals are sold at the market for grain purchases. The smaller number of small ruminants in the Koutiala zone indicate increasing scarcity of fodder availability (natural resources) and perhaps a decreasing need to maintain small ruminants for precautionary saving motives. There is a significantly high negative correlation between labour availability (number of persons per ha of cultivated land) and the farm equipment level (-0.456***) and livestock ownership (-0.695 ***).

Table 16.1. Agro-ecological conditions and land use by zone.

	North	Koutiala	Centre	South	Sikasso	All
Number of villages	*11*	*6*	*7*	*7*	*12*	*43*
Rainfall 30-year average (mm)	735	857	810	1,062	948	877
Drought index 1997	2.8	1.5	1.0	1.7	1.8	1.9
Standard deviation 30 year rainfall (% of mean)	18.6	17.4	16.0	12.6	17.3	16.6
Ratio arable land / cultivated area	2.36	1.16	2.65	2.75	1.70	2.12
Ratio non-arable land / cultivated area	1.55	0.15	0.31	1.55	0.44	0.87
Ratio total area / cultivated area	4.25	1.71	3.16	3.87	3.45	3.52
Livestock pressure (heads per 100 ha cultivated area)	69	99	33	44	45	56
Land degradation index 1998	23.3	24.3	11.9	13.1	18.2	18.6
Land degradation index 1988	13.3	6.8	13.5	8.1	11.4	11.0
Area cultivated (ha person^{-1})	0.67	0.80	0.76	0.76	0.52	0.68
% millet and sorghum	62.3	44.1	40.7	26.5	20.2	38.7
% maize	4.5	8.4	15.5	29.9	25.0	16.7
% rice	2.1	0.1	3.5	5.5	10.4	4.9
% cotton	16.2	36.5	34.5	31.4	36.1	30.1
% groundnut	6.5	4.1	3.6	4.7	1.6	4.0
Number of ploughs per 100 persons	14.5	16.8	15.6	9.6	10.4	13.1
Number of seed drills per 100 persons	3.0	4.6	4.0	1.7	3.7	3.4
Number of cattle per 100 persons	43.5	85.7	59.1	62.0	57.9	59.1
Number of small ruminants per 100	60.2	63.9	64.9	56.1	33.9	52.6
persons	7.7	12.5	12.4	11.8	13.2	11.4
Cropping expenses (FCFA 1000 ha^{-1})						
Cereal yield (kg ha^{-1})	1,001	1055	1095	980	1781	1,239
Cotton yield (kg ha^{-1})	943	908	1024	1139	1210	1,066
Gross margin (FCFA 1000 ha^{-1})	79	91	104	97	148	106

Source: Brons et al., 2000a

Typical characteristics of village economies concern the occurrence of non-agricultural activities, migration patterns and incidence of poverty and food insecurity (Table 16.2). Poverty and welfare indicators are used to approximate the need to develop mechanisms to cope with insecure livelihood situations. Development organisations monitor the number of months during which food shortages are experienced and the percentage of households experiencing food shortages. In addition we assess the total cereal production per inhabitant by using the yield and area estimates from the survey.

Table 16.2. Village characteristics by zone.

	North	Koutiala	Centre	South	Sikasso	All
Gross margin (FCFA 1000 person^{-1})	54	73	75	70	79	69
Cereal production (kg person^{-1})	472	439	484	398	504	467
Number of food shortage months	2.6	2.0	2.3	2.4	2.3	2.3
Food deficit households (% of	47.0	25.7	23.1	19.0	29.7	30.6
population)	21.4	18.9	17.6	46.5	28.0	24.9
School attendance rate (% of children)						
Distance to education	2.5	1.2	3.9	3.5	3.3	2.9
Distance to health services	6.6	6.3	8.5	7.5	6.8	7.0
Distance to nearest bank	16.3	6.8	5.8	19.0	13.2	12.5
Distance to market	8.8	6.7	6.6	3.8	7.8	7.2
Non-agricultural activities (# per 1000 inhabitants)	25.4	20.6	22.1	10.3	9.0	17.9
Migration patterns (% of population)	7.3	1.0	3.5	1.5	1.6	4.7

Source: Brons et al., 2000a

The higher level of non-agricultural activities per inhabitant in the northern and the Koutiala zone suggests that their importance increases when agricultural resources become scarce in terms of quantity (Koutiala) or quality (the northern zone). Table 16.2 suggests that out-migration is particularly important in the northern zones. This is confirmed by negative and significant partial correlation coefficients with rainfall (-0.48**), gross revenues per ha (-0.52**), cotton yield (-0.65***), percentage cotton in cultivated area (-0.65***) and expenses per ha (-0.67**). Out-migration and distance to cities are positively and significantly correlated (0.57***). Food shortages appear in all zones but particularly in the northern zones. School enrolment is relatively low with a positive exception for the southern zone. Income levels per capita are comparable across the zones, except for the northern zone where the average income per capita is about two third of the other zones. In summary, there seems to be a relationship between the north-south gradient and indicators of poverty: the more south one comes, the lower the average poverty level at village level.

Principal component analysis was used to identify specific factors that determine portfolio choice in each of the zones. Results are presented in Table 16.3 where 27 structural village characteristics are reduced into seven components that explain almost 80% of the variation. Only significant variables with a score higher than 0.4 are included in the main components. The seven components reveal different, not necessarily exclusive, pathways of village-level coping strategies with risk and climatic variability. Each component describes a specific common portfolio strategy at village level, while individual households may still exhibit particular combinations of strategies.

Table 16.3. Development Pathways in Southern Mali (Results of PCA).

Farming results	Gross revenue ha^{-1} 97	0.945	0.796	0.337		0.391			
	Gross margin ha^{-1} 97	0.945	0.817			0.414			
	Gross margin per $person^{-1}$	0.946	0.311	0.393	0.370		0.680		
	Quantity cereals per $person^{-1}$	0.943	0.451				0.777		
	Families with food shortages %	0.472				-.624			
	Months with food shortages	0.848							0.849
	Degradation index	0.421						-.394	0.399
Crop yields 1997	Millet / sorghum	0.749	0.609			-.452	0.312		
	Maize	0.794	0.853						
	Cereals	0.937	0.910						
	Groundnut	0.686	0.672						
	Cotton	0.789				0.665			0.377
Climate	Rainfall May to October 1997	0.799		0.711		0.467			
	Rainfall 30 year average*	0.699		0.440		0.630			
	Drought index 1997	0.710		-.822					
Farm characteristics 1997	Cattle per 100 persons	0.758			0.792				
	Oxen per 100 persons	0.892			0.843		0.414		
	Farm equipment level	0.570			0.486				-.403
	Area % cotton	0.893		0.814	0.401				
	Area % rice	0.833	0.473		-.340				
Input use	Oxen per ha	0.818			0.857				
	Expenses per ha all crops	0.770		0.776	0.345				
	Persons per ha cultivated	0.944	0.446		-.384		-.725		
Village characteristics	Non agricultural activities	0.417		-.347			0.424		
	Distance to city	0.646	-.707	-.310					
	Distance health services	0.716						0.814	
	Distance to education services	0.730						0.810	
PCA results	*Own values*		*5.3*	*3.4*	*3.2*	*2.7*	*2.3*	*1.7*	*1.6*
	% of variance explained		*19.6*	*12.9*	*12.1*	*10.3*	*8.7*	*6.5*	*6.1*
	Cum % of variance explained		*19.6*	*32.6*	*44.7*	*55.0*	*63.7*	*70.2*	*76.4*
Farming results	Gross revenue ha^{-1} 97	0.945	0.796	0.337		0.391			
	Gross margin ha^{-1} 97	0.945	0.817			0.414			
	Gross margin $person^{-1}$	0.946	0.311	0.393	0.370		0.680		
	Quantity cereals $person^{-1}$	0.943	0.451				0.777		
	Families with food shortages %	0.472				-.624			
	Months with food shortages	0.848							0.849
	Degradation index	0.421						-.394	0.399

*) Period May to October

The first factor, labelled as *'commercial cereal production'* is characterised by a substantial village area of diversified cereal production for commercial purposes (rice but also maize, millet and sorghum) that are able to realise high gross margins and revenues. Moreover, the joint high yield levels of groundnut points to some diversification taking place into the production of leguminous crops and positive effects of generally high input levels. The strategy of intensification of food production based on high factor productivity levels and rather controlled applications of external inputs can be used to satisfy food self-sufficiency at household level and also generates surpluses for market exchange when the market place is relatively nearby. It should be noted that this strategy can be followed only when rainfall risks (e.g. low rainfall but also high variability) are limited or controllable and do not require high levels of management expenses. It is thus accessible to farmers who possess sufficient labour resources to permit reliance on intensification practices (mulching, composting, crop residue management, improved fallow, etc.) applied to land that is scarcely available in relative terms. This often results in a further reduction of the vulnerability to climatically adverse conditions.

The second factor includes variables related to *commercial cotton production*. Farmers practising this cropping system are located throughout the zones below the 800 mm isohyete, which clearly indicates that the former cotton basin has extended southwards (as discussed in Chapter 10). This factor indicates high investments for input purchases in commercial cotton production realising a surplus that can be used to satisfy consumption expenditures. Specialisation in cotton production is mainly feasible due to favourable agro-climatic conditions (high rainfall and low rainfall variability) that enable substantial investments for input use. While cereal production per person is relatively low, food security is guaranteed through the market purchase of cereals. Since the factor scores on annual rainfall and drought are higher than average climatic conditions, this indicates that there are possibilities for adaptation of input use according to real climatic conditions. The high score for climate indicates that this strategy of specialisation in cotton production can only be followed under relatively favourable climatic conditions. In addition, favourable policies for input delivery (fertiliser supply through CMDT), credit and marketing and high levels of price certainty enable further specialisation into cotton production. These favourable conditions are partly a consequence from the relative proximity of the cities Koutiala and Sikasso. This would also explain the rather isolated occurrence of a high incidence of non-agricultural activities as part of this component.

The third category includes variables related to *livestock development*. Villages where farmers possess relatively high amounts of cattle and oxen can be found firstly in the Northern region (where cotton cultivation is disappearing) and secondly in the ancient cotton areas characterised where important crop revenues allowed farmers to purchase cattle. Investments for cattle purchases are usually financed out of (earlier) cotton revenues. Inputs for livestock production are limited and returns per hectare are low. Part of the livestock production is located close to villages and specialises on dairy products. Only a small number of families suffer from food shortages. Strangely enough, these villages are not significantly affected by soil degradation. This can be understood from the recent theoretical understanding that pastoral and livestock production as a system may be better adapted to local ecological restrictions and a high market involvement than was previously understood. In view of the low gross revenues, this component is considered to be mostly related to low-input low-output livestock keeping and not to the commercial ranching type of production.

Commercial production of *rice and other cereals* scores high on the fourth factor. Villages that diversify into rice cultivation and specialise in it as far as grain production is concerned are strongly dependent on stable rainfall patterns. Rice was introduced originally as a major diversification crop and initial investments were financed out of cotton revenues, which are relatively high per area unit. In the medium term, the farming systems become more dependent on rice as a major crop, thus leading to a new type of specialisation at village level. Commercial rice production is mainly selected by farmers with limited land resources who are able to exploit a labour-intensive cropping system. While land productivity and labour intensity are high, there is no significant effect on labour productivity. Consequently, this segment of villages is in a favourable position with respect to food security, in particular because of the increased production of rice and maize. This again highlights the favourable relationship between market involvement and food security, which depends on a specific set of variables, such as distance to markets.

The fifth factor is labelled as '*Extensive cereal production*'. Villages that have access to a large area per person can still rely on extensive cereal production for food self-sufficiency. A low population density and the availability of oxen facilitate the required land preparation and weeding activities for an extension of the cultivated area. Reliance on extensive cereals production mainly occurs in villages with low, but stable rainfall patterns. This system guarantees high net revenues (low external input costs) and adequate food security levels. No systematic relationships between productivity levels and the degree of mechanisation are registered. Apparently, the high-income levels encourage activities in the non-agricultural sector (as a relationship vice versa could not be established). This is the component with the highest score on the number of non-agricultural activities.

The sixth component can be labelled as '*sustainable subsistence farming*'. This component does not appear with any significant factor score in any of the agricultural indicators or rather, it appears as the only component which scores high on indicators of what it may not be: it is not degraded and not close to services. The joint remoteness of health and education services points to less favourable socio-economic conditions. With little agricultural development (scores for agricultural input use and equipment are low), low land use intensity and a favourable state of the natural resource conditions compared to the other components (the negative coefficient points to low levels of soil degradation), it is a component which seems characteristic of an early stage of agricultural development in the region, now only found in remote areas.

Finally, the seventh component is labelled as '*marginal subsistence farming'*. In the less developed villages with relatively few farms equipped with oxen traction and little non-agricultural activity, farmers maintain diversified farming systems without particular implications on yield levels. These farmers have a small number of livestock and often only incomplete equipment. The considerable distances to the cities limit off-farm employment options. Low levels of education and health also inhibit participation in the labour market. The combined effect of the features in this component is a regular occurrence of food shortage.

The observed pattern of portfolio choice indicates that diversity in land use and activity choice closely follows rainfall variability. Commercially oriented cotton and rice production systems are concentrated in villages with higher and more stable rainfall and their development can probably be further reinforced through market-based policies (*i.e.* infrastructure, price information, *etc.*). Under more variable climatic conditions combined with relatively favourable resource endowments, either extensive cereal production or livestock farming appear to be interesting options, the former combined with some engagement of non-agricultural employment. Improved risk management for these systems can be perceived through strategies of resource intensification (*i.e.* crop choice, soil moisture management and livestock feeding regimes or alternatively better marketing conditions for output, inputs and food). When highly variable climatic conditions coexist with an increasing pressure on the resource base, food security strongly depends on institutional support for human development (*i.e.* education, health care) and reliance on social exchange networks.

Different pathways can be classified according to major agro-ecological factors (drought risk and land availability) and socio-economic factors (market development and accessibility). Table 16.4 provides a general framework for the analysis of the feasibility of farming systems under various conditions. Depending on the locality-specific combinations of these dimensions, an appropriate strategy can be devised that builds on the actual choices for certain options made by individual people and households.

Table 16.4. Feasible farming systems.

Market Development	Population Density	Drought Risk	
		Low	High
High	High	Cotton / Rice	Livestock (migratory)
High	Low	Commercial Cereals	Livestock (dairy)
Low	Low	Extensive Cereals	Sustainable Subsistence
Low	High	-	Marginal Subsistence

The upper left quadrant indicates that when drought risks are low, conditions for commercial production prevail. Such a development will follow the direction of cotton / rice systems or of cereal systems depending on population density. The upper right quadrant suggests that economic development in risk-prone environments will be oriented towards the livestock sector. When population density is low and markets are less developed, farmers resort to extensive cereal cultivation. This kind of system is likely to move upward when population density increases. In a risk-prone environment, households may obtain a subsistence-level income provided that population density is low (lower left quadrant). However market development is less likely to take-off. Finally, the lower right quadrant highlights problem situations where past economic opportunities may explain the high population density. Drought risk is a relative concept so we may find these villages in zones where production opportunities meet transition problems. Some cells remain empty because of simultaneity reasons. First, a highly risky environment is not conducive to dense populations and actors do not invest in market development. Second, when risks are low and population density is high (attracted by low risks), markets are likely to develop, hence this cell remains empty (and production systems evolve towards the upper left quadrant).

These feasible farming systems are closely related to agro-ecological zones and predominant types of farm households within each of these zones and also reflect specific portfolios of options in a rural and peri-urban semi-arid area, as previously described. Table 16.5 presents the average village scores on the 7 principal components for the 5 zones. The pathway descriptions for the zones as presented earlier correspond remarkably well with what one would expect to be major characteristics of different agro-ecological zones. Although the components enable us to describe each zone according its specific traits, the relatively high standard deviation on each factor (within the zones) reveals important oversights when the zoning would be limited to geographically clustered villages.

The high positive score for commercial cereal production (component 1) in de Sikasso zone confirms the comparative advantage for production of rice and other cereals. Favourable climate conditions guarantee relatively good cereal yields. Commercial cotton production (component 2) is particularly relevant in the rural central and southern zones and in the proximity of Koutiala town. Clearly, the northern zone distinguishes itself on the negative side of this component, indicating the declining comparative advantage of this region as regards cotton cultivation.

Livestock farming (component 3) dominates in the regions near to Koutiala town, while to the south villages score negatively on this component. The robustness of the fourth component (commercial rice production) is confirmed by the high scores in the two southern zones and the negative scores in the two northern zones. Rice is successfully grown only in the southern villages. The fifth component (extensive cereal production) shows that villages in the southern zone usually do not opt for this pathway. Favourable agro-ecological conditions as well as the scarcity of arable land permit more intensive cropping systems in these regions.

Table 16.5. Villages component scores by zone.

	Component*						
Zone (Number of villages)	1 Commercial cereal production	2 Commercial cotton production	3 Livestock farmers	4 Commercial Rice production	5 Extensive cereal production	6 Sustainable subsistence farming	7 Marginal subsistence farming
North	-0.243	-1.165	0.029	-0.670	0.075	-0.131	0.267
(11)	(0.634)	(1.106)	(0.641)	(0.453)	(1.094)	(0.730)	(1.458)
Koutiala	-0.023	0.395	0.558	-0.583	0.229	-0.465	-0.520
(6)	(0.427)	(0.570)	(0.823)	(0.381)	(0.502)	(0.826)	(0.668)
Central	-0.225	0.434	-0.169	-0.140	0.419	0.360	-0.323
(7)	(0.501)	(0.440)	(0.495)	(0.956)	(1.069)	(1.324)	(0.687)
South	-1.107	0.542	-0.410	1.050	-0.002	0.099	0.318
(9)	(0.911)	(0.323)	(0.532)	(1.647)	(1.053)	(0.949)	(1.004)
Sikasso	1.011	0.302	-0.032	0.375	-0.427	0.085	-0.018
(10)	(0.891)	(0.740)	(1.597)	(1.157)	(1.021)	(1.160)	(0.725)

*) Average score and standard deviation between parenthesis
Source: Brons et al., 2000a

The central zone, particularly villages east of Koutiala, is distinguished by component six (sustainable subsistence farming), characterised by relatively low productivity levels combined with sustainable cropping systems. Otherwise, the zone in the proximity of Koutiala town scores negatively on the sustainable subsistence-farming component. Clearly, farming systems in this zone are neither subsistence oriented, nor sustainable with respect to the natural resource base. Similarly, the Koutiala zone does not have a marginal nor a subsistence character (component 7). Villages in the southern zone score higher on the marginal subsistence component: farm technology is less widely adopted and income levels are subsequently lower than in the villages of the central zones.

Since these pathway descriptions are by definition independent, agro-ecological zoning definition can be improved when information regarding village level characteristics is used. Different zones do not necessarily comprise a geographical joint cluster of villages but rather a cluster of villages with the same specific niches for potential development or necessary policy interventions. Interestingly, the lower level of degradation related to livestock production is reflected in a considerable body of literature pointing towards flexible and pragmatic land management strategies that are often at the foundation of the production system (Scoones 1989, 1995, Sandford 1995). In addition, the positive effects of an established relationship with markets that provide adequate outlets for buying and selling livestock and crop products is confirmed (Zaal 1998, Bates and Lees 1977, Little 1992, Ensminger 1998).

Box 16.1 Regional development in a bird's eye view

The evolution in agrarian development in the Koutiala-Sikasso region can be divided into four different phases (Fok, 1994). Before 1973/74, rural areas were hardly integrated into world markets and agricultural research did not systematically cover rural production systems. Some efforts to systematise agricultural research for cotton and maize emerged and production systems were relatively traditional in character. Three quarters of the cattle population were held north of the 600 mm-isohyete, while south of the 1000 mm isohyete hardly any livestock activities were undertaken. The period 1950 - 1971 was characterised by improvements in the veterinary sector (control of sleeping sickness), the creation of various water sources (wells) and relatively abundant vegetation due to favourable rainfall. The result was that, shortly before the droughts of 1973/74, some livestock activities were already taking place in southern Mali. However, there was no integration of livestock and crop production at any level. Finally, after 1960, population growth grew much more than before.

The period 1974 – 1986 is marked by the creation of the CMDT, which led to a take-off of cotton and maize production. The increase of cotton production was made possible by i) the technical progress in cotton production (varieties, fertilisers and insecticides), ii) the introduction of animal traction iii) and the extensive CMDT service network (credit, extension, education, veterinary care, etc.). After the dry years of 1973/74, considerable numbers of livestock shifted to the region with between 600 - 1000 mm rainfall. This shift accelerated the dissemination of the use of animal traction. Livestock and crop production started to be integrated , predominantly at market level.

The 1986 – 1995 period is marked by a rapid expansion of the area cultivated (both rain-fed and irrigated), made possible by increased dissemination of animal traction. The installation of irrigated rice fields also started in this period. At national level, the cereal markets were liberalised. Increased crop production was mainly achieved through area expansion (extensification). This period is characterised by the emergence of various natural resource management programmes (gestion de terroir, lutte anti-érosif) and increased efforts to integrate livestock and crop production systems, this time at household level (fodder production, crop-residue recycling, animal traction). Changes in livestock systems took place as well. Cattle became more important on sedentary farms, as did the sedentarisation of transhumant livestock systems and the appearance of new actors (owners) in the cattle sector (traders and government employees from urban areas and the Minyanka farmers themselves). These factors contributed to increased scarcity of common resources (pastures, water). The societal organisation changed dramatically for the pastoralists as well as the sedentary farm households. For pastoralists, competition for resources meant reduced opportunities for herd management. The shift of ownership exacerbated this situation. The impact this had on the local livestock keepers is difficult to describe, due to the limited availability of data on these FulBe groups. The situation also changed for sedentary households. Individualism, which had been strong among the Minyanka farmers anyway, increased. A rural exodus started, mainly due to increased wealth differentiation in society. Conflicts took place between young and older generations. Changes at community level: traditional organisations started loosing power (groups of elder, hunters, earth chief, etc.) while at the same time new organisations developed (CMDT village associations, work-groups, etc.).

Differentiation processes prevalent in previous periods accelerated after 1995, particularly in densely populated areas which, in relative terms, were not very diversified. Since almost all arable land is cultivated and all pastures are exploited, the rural population actively seeks to i) intensify crop management (higher input levels, more labour input); ii) invest in land conservation measures (anti-erosion stone bunds, improved fallow, etc.); and iii) integrate livestock and crop husbandry systems. These household level strategies are accompanied by community strategies for natural resource management. While in former periods these strategies strongly depended on external support, during the last period a more endogenously driven adoption of intensification occurred. Local work groups seem to be very important in mobilizing labour.

3. RISK-COPING STRATEGIES IN NORTHERN BURKINA FASO

This section takes a closer look at the dynamics of household behaviour during two distinct periods, one of critical rainfall and one of favourable rainfall. The comparison of the detailed micro-level panel data on Northern Burkina Faso with information from distinct periods – though not necessarily covering the same villages and farmers – offers the opportunity to identify trends in farmers' behaviour and portfolio choice and to develop an argument on farmers' selection of development pathways. In the following sections we present a comparative analysis of different farm household strategies for coping with climate variability and risks. We first illustrate the climate contrast between both periods and then present the results concerning agricultural production, income composition, investment behaviour and consumption patterns respectively.

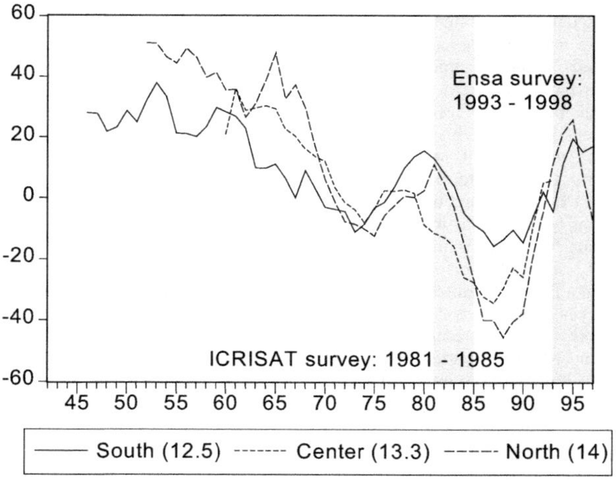

Figure 16.2. Northern Burkina Faso : Five-year moving average rainfall deviation (% of long term average). (Between parentheses the latitude of the weather station is indicated).

The analysis concerns two sets of panel data from the Kaya region in Burkina Faso for two markedly different periods (see Figure 16.2). This panel-data provides information on resource use by rural households, made available by ICRISAT (1981 – 1985) and by ENSA (1993 – 1998). In the first period (1981 – 1985), rainfall was just below the long-term average and shows a clear downward trend that continued until 1988. During the second period (1993 - 1998), rainfall was just above the long-term average and presents an upward trend with extreme high rainfall in 1995, though after that year rainfall seems to have declined again.

Four major dimensions of rural household livelihood strategies are taken into consideration, namely:

- Production, labour allocation and land use: decisions regarding land use, crop and technology choice, input and labour intensity of production and their implications for yields and incomes.
- Income composition and portfolio management: decisions regarding the choice of production activities and the allocation of labour to farm and off-farm or non-farm activities.
- Investment behaviour and asset management: strategies regarding the maintenance of livestock and other reserves and criteria for decision-making on (long-term) investments in soil conservation.
- Consumption, exchange and stocks: decisions regarding stocking or (non)-market exchange of food and non-food commodities and their implication for household food security and nutrition.

Figure 16.3 illustrates cereal yield and rainfall patterns from 1981 to 1998. In a least square multiple regression equation with cereal yield as the dependent variable, the coefficient for rainfall in the southern part is 2.51 (**) while the trend variable coefficient for $t = 1$ to 18 is 16.2 (***). The equation explains 55 % of total yield variability. In other words, the increase in cereal yields can be partly explained by rainfall conditions although a more significant share of the progress originates in technology development.

Despite the devaluation of the FCFA in 1994 by 50 percent, nominal prices have hardly changed across the two periods (Figure 16.4). Just before the devaluation, cereal prices were at the lowest point (considering only the two periods of the analysis) while after the devaluation prices increased gradually to the 1985 level. Peak prices just before the harvest occur irregularly (1984, 1996 and 1998) and seemed to be caused by realised harvest in the previous year as well as harvest expectations for the next season.

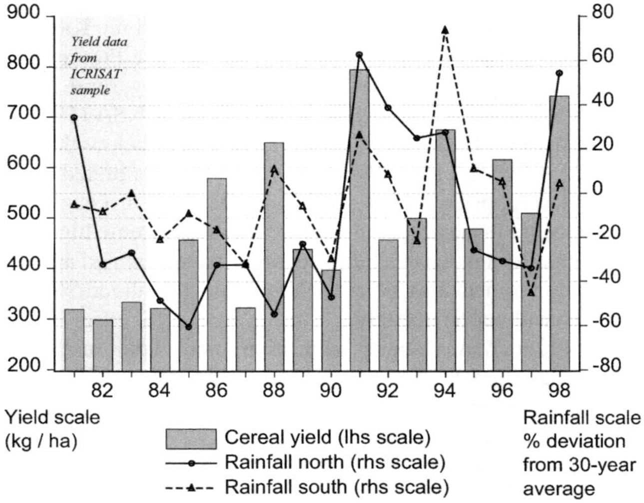

Figure 16.3. Northern Burkina Faso: Cereal yield and rainfall pattern (1981 – 1998).

Strategies for risk reduction to control the effects of climate variability within the context of available land, labour and other production factors have a major influence on land use patterns and crop yields. Table 16.6 presents a comparison of rainfall regimes, yield levels and variability during two periods. It is clear that quite fundamental changes have taken place in view of the near-average rainfall figures for both periods. Average crop yields improved by more than 100 percent (sorghum) and 200 percent (millet). Cereal yields per capita have increased from well below FAO minimum requirements (200 kg/capita/year) to levels just beyond the food security level. Another indicator of improved agricultural productivity is the reduction of the available area per capita. Due to processes of agricultural intensification and technological change, farmers proved to be able to increase their yields and food security under conditions of a reduced area. Also, though still small, the area of groundnuts has doubled. This crop is generally grown for local consumption, but also appears on the market as a crop specifically for women to sell. Its leaves are used for fodder as well. Lastly, there has been a dramatic shift away from millet towards sorghum, which is planted in more humid conditions (floodplains etc, see below) where yields are generally better. Certainly, the relatively more stable rainfall regimes during the second period favoured the adaptation of production technologies, particularly in the case of animal traction. On the basis of these figures alone, one could speak of an agricultural change which is unexpectedly rapid and which cannot be explained through the more beneficial rainfall in the second period of study alone.

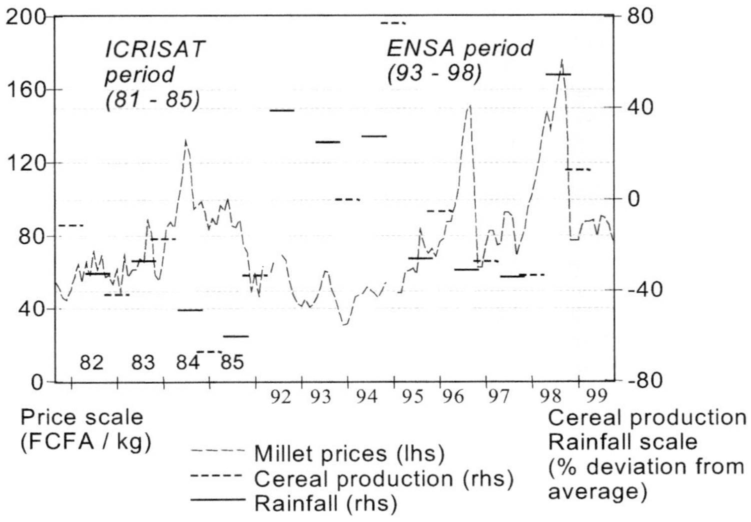

Figure 16.4. Northern Burkina Faso: Price trend, rainfall and cereal production. (1981 – 1985 and 1992 – 1998).

While most households maintained the dominant millet-sorghum based cropping pattern, in the margins they increased the area of maize, groundnuts and other minor crops. These commodities usually find important market outlets in growing cities. Most of this crop diversification takes place on plots located along the toposequence or across different soil types, but do not sufficiently reduce crop production risks (Carter 1996). Nevertheless, a long-term adaptation of plot positioning may have important implications for agricultural development. Households could abandon particular parts of the village territory in response to a sequence of dry years (as in the early 80s) or to a reduction of soil fertility (as in the 90s). The increase of rainfall during the latter period may also have incited households to leave the lower flood-prone plots uncultivated. These responses to change make clear that long-term productivity trends cannot be fully assessed without taking into account the quality of cultivated arable land. However, for lack of unambiguous indicators, data on the quality of fallow, bush and pasturelands is seldom included in rural monitoring programmes.

Table 16.6. Changes in agricultural production (1980s – 1990s).

	1981-'85	1993-'98
Rainfall		
Mean (mm yr^{-1})	547	605
Crop Yields		
Millet (kg ha^{-1})	232	763
Sorghum (kg ha^{-1})	293	679
Cereal produce per capita (kg)	120	292
Land use pattern		
Total area per household (ha)	9.8	4.8
Total area per capita (ha)	0.9	0.45
Millet (%) *)	63.3	35
Sorghum (%)	29.9	55
Groundnuts (%)	3.3	6.1
Input use		
Fertiliser use (kg ha^{-1})	4	2.9
Fertiliser coverage		
- % of farmers not using fertilisers	56	55
- % of farmers using less than 10 kg ha^{-1}	44	35
Animal traction (% of farmers)	12	25
Labour use (d ha^{-1})	94	n.a.
Marginal Factor Productivity		
(Cobb-Douglass production function coefficients)		
Land	0.79 (1%)	1.01 (1%)
Labour	0.03 (5%)	0.04 (n.s.)
Fertilisers	0.15 (10%)	0.02 (5%)
Animal traction dummy	-0.00 (n.s.)	-0.10 (5%)

Source: Brons et al., 2000b based on ICRISAT and ENSA surveys
Note : The ICRISAT sample includes a bias in favour of livestock farmers

Production technologies underwent considerable change during the two periods. Fertiliser use and animal traction seem to expand mainly in a horizontal direction, with more farmers applying the same technology. Important differences in input use are, however, registered between villages and regions. Fertiliser application during the 1980s was especially high in one of the villages in the Soudanian zone with most groundnut production. Looking at regional differences in fertiliser use for arable production, we noticed generally higher applications in *Bam* province (30-50 % of all cereal plots, compared to 5-10 % in *Sanmatenga*). Fertilisers are used on market gardens (around lake *Bam*) and in other areas in combination with small-scale irrigation, making use of intercropping and manure applications.

With respect to mechanisation, there is a clear tendency towards increased use of animal traction. During the early 1980s, animal traction was scarcely used in the Sahelian region compared to a higher application in the Soudanian region. The size of cultivated area is determined by the availability of animal traction and animal traction appears as a substitute for labour and external inputs. During the 1990s, an increasing share of the land was prepared using animal traction. As the area per capita has decreased, animal traction is also being used for timely planting and regular weeding, which are crucial in the short-cycle cropping system found here. It is therefore used mainly for land use intensification and not for horizontal expansion of crop production.

Finally, investments in land quality through erosion control measures provide an indicator for adjustments in farm management that took place in the period under study. Little information is available on the 1980s, but from secondary information it is evident that the diffusion of anti-erosion measures may not have been higher than a few percentages (Rochette 1989, IFAD 1996). However, in the late 1990s more than 30 percent of farmers' fields in any year used some type of soil fertility management practice, recording higher percentages on fields located nearby the villages. Moreover, application rates are substantially higher in *Bam* province compared to *Sanmatenga* province

Climate conditions are likely to influence the pattern of income composition, especially the balance between cereal consumption and sales, the contribution from cropping and livestock activities and the relative importance of off-farm and non-farm income sources. Since in both periods crop production was just enough to cover family food requirements, rural households necessarily maintain a diversified pattern of income-generating activities. The 1981-1985 data in particular reveals that non-farm and off-farm activities account for about 60 percent of the (monetary) household income or 35-40% of the total income (see Table 16.7). Recent income and expenditure surveys confirm a similar income pattern (Tabsoba 1998).

Table 16.7. Northern Burkina Faso: Structure of Income composition (1981-85 and 1993-98).

Period	1981-85	1998
Total (nominal) income per adult (FCFA) [a]	32,000	84,500
Income without retained harvest (FCFA)	20,000	48,000
Consumption of own crops	41	n.a.
Income composition (*% of total monetary income)*		
Crop sales	10	16
Livestock sales	25	23
Off-farm and non-agricultural revenues	65	61

Source: Brons et al. (2000b) based on 1981–1985 ICRISAT survey and 1998 Antenne Sahélienne survey

Notes: Data exclude remittances and pensions

a) estimated to include 292 kg of cereal production pc*average price of millet of 25 FCFA.

It is generally assumed that adjustments in market prices may partly compensate for changes in production levels. However, given the limited level of cereal sales, the extra income from higher cereal prices should not be overestimated, especially for the lower income categories. Only the wealthiest category of households is a net seller of cereals and presumably had the flexibility to adjust the timing of their sales to crop market prices. Though nominal incomes are higher in 1998, the general inflationary trend together with the devaluation of 1994 left the farmers with a limited increase in their real income. Although land use, yields and technology may have changed drastically, the net result hardly improved compared to the earlier period. The adaptation therefore succeeded in preventing a decline but has not been able to increase real incomes considerably.

Livestock is typically considered as a smoothing mechanism. Livestock incomes were most important during the dry year of 1983, particularly in the Sahelian villages and to a lesser extent in the poorer Soudanian villages. The revenues from livestock gradually decreased in the course of the 1981-1985 period, which suggests that the role of animals as insurance and a consumption smoothing device became more and more limited.

Revenues from off-farm and non-farm activities have become increasingly important. Parts of these revenues follow patterns similar to the agricultural activities (e.g. agricultural wage labour), but non-farm income patterns may be less related to the agricultural conjuncture. The relative importance of off-farm income sources is most notable in the Southern part of the research area, but is declining where market transaction costs increase. Hiring-in off-farm labour within the agricultural sector may be helpful to reduce labour constraints but mostly requires redistribution of labour within the supplying households. Incomes from non-farm activities are less co-variate with agricultural production.

Availability of assets and investments in land or livestock are usually considered as important risk-coping and consumption smoothing mechanisms (Deaton 1997, Udry, 1996). Therefore, attention is given to changes in the patterns of land and cattle ownership (Table 16.8).

Table 16.8. Changes in assets (cattle and land per capita) and adjustment of landholdings (% of farmers).

	1981-85	1993-98
Cattle per capita (number of heads)	0.7 [a)]	0.27
Land per capita (ha)	0.9	0.42
Increasing landholding (% of farmers)	10.8	7.8
Stable landholding (% of farmers)	81.4	80.6
Decreasing landholding (% of farmers)	10.8	11.6

Source: Brons et al., 2000b based on ICRSAT and ENSA Surveys

Note : a) The ICRISAT sample includes a bias in favour of livestock farmers

In the two periods there was no accumulation of wealth through increasing land holdings. Most adjustments in landholdings are likely to be occasioned by family lifecycle tendencies. Farms that decreased in size were generally found in the southern part of the area. However, average land per capita decreased substantially. The latter tendency may be related to the registered increase in yield levels (permitting family subsistence with less cultivated land) and to increasing importance of non-agricultural income sources.

Livestock holdings provide additional information on wealth distribution over time. During the 1980s, the Sahelian villages lost considerable numbers of cattle and small stock. In contrary, the Soudanian villages have seen their cattle herds grow – though they are still smaller than in the North – and experienced a reduction in small-stock. The general southwards trend in cattle ownership which was registered in other parts of West Africa as well is also apparent here. In this case, it is not due to the migration of cattle owners, but to differential wealth development. Although similar data is not available for the 1990s, it is generally understood that cattle ownership has become more concentrated in the latter period. Greater differentiation in livestock holdings took place, particularly among the sedentary farmers.

Rural households' investments of capital for improved crop and animal production reflect the use of external inputs for current land use activities, as well as for efforts towards soil conservation. As discussed before, efficient land use becomes more dependent on the purchase of external inputs (seeds, fertilisers, traction, hired labour). During the 1990s, public and private investments in soil and water conservation techniques increased considerably as well. Intensification of land use proved to be closely related to improved land management practices. In both cases, access to markets and institutions (credit, extension) appear as critical variables.

The production of cereals for home consumption seems to have increased considerably in per capita terms, which is impressive given the reduction of land holdings per capita. Rainfall indicators show that this could be due to slightly better environmental circumstances, but it is more likely that improved technology has facilitated this shift towards higher food security. Since fertiliser use has not increased significantly and manure applications have not increased either, yields will be threatened by nutrient depletion. Adjustments in crop choice, together with the selection of more fertile and humid lands near the village and the application of animals traction for better land husbandry have made it possible to increase yields. In recent years, higher rainfall has damaged these crops and sorghum yields have declined slightly. Increasing aridity has meant that sorghum will ultimately be replaced by millet.

The real value of cereal production in real monetary terms has not increased substantially, but the productivity increases enabled a rise in the share of crop sales in total income. Market prices of cereals have increased after the devaluation of early 1994, but long-term trends in cereal prices remain rather stable. This is a rather unexpected result of this devaluation and it made it difficult for farmers to generate income from their core business of cereal production. Livestock prices increased, but paradoxically the poorest households that need the smoothing effects of livestock sales most, suffer from these higher prices more when they need to purchase animals for substitution purposes. Non-farm activities have become a major income diversification device for wealthier households

Risk sharing is till a very important part of income generation and smoothing of the poorest group in society. It appears to generate almost a fifth of their income, and this during a period of relatively high productivity in cereal production. Or perhaps it should be stated that it is precisely because of this relative well being that it is possible to give and receive these gifts as, in times of drought, this source may disappear due to a general decline in incomes and risk-sharing possibilities.

The general conclusion is that long-term patterns present certain evidence of an important diversification strategy during the dry period of 1981 – 1985. The densely populated villages in the Soudanian zone proved to be most sensitive to the fluctuations in income from cropping activities while the villages in the North were able to rely on the sale of animals and income diversification. The 1993-1998 analysis shows a remarkable resilience of crop performance accompanied by important investments in soil and water conservation measures. This has led to higher yields and a stabilisation of incomes in real terms. Other investments included the purchase of livestock by the poorer households for future income smoothing, whereas the wealthier households seem to adopt alternative investment strategies outside of the agricultural sector.

4. CONCLUSIONS

The spatial analysis of development pathways (at village level) in Southern Mali and inter-temporal risk-coping strategies (at household level) in Northern Burkina Faso reveals a number of interesting tendencies regarding the interactions between climate variability, portfolio choice and livelihood strategies. Regional development policies for food security and natural resource management benefit from a more precise definition of local livelihood strategies based on multiple criteria. The typology currently used based on agro-ecological zoning and farming systems characteristics does not sufficiently respond to local variability in resource use options and strategies. The targeting of policy measures can be substantially improved if different responses to variability are considered (see Chapter 21).

Development Pathways

Multiple pathways are available to achieve adequate per capita productivity and income levels. Subsistence-oriented pathways face relatively low and stagnating income levels, due to the increasing scarcity of land and alarming levels of soil degradation. In other zones, negative effects of climatic variability and natural resource degradation are considerably reduced by diversification of cropping patterns through the inclusion of high value crops (rice, vegetables, fruit trees). In addition, diversification across production sectors (livestock, non-agricultural production) is shown to be an additional option to assure sustainable livelihoods.

The results of the principal component analyses show that specific types of inputs are related to different pathways. A diversity of development pathways results from various combinations of market development, population density and drought risk. Rural households exploit ecological niches through variability in crop choice (cash versus food crops), cropping intensity and livestock husbandry. These pathways result in different levels of economic development and natural resource sustainability. Three general tendencies can be observed: (i) labour use intensity increases with high cereal yield levels (rice and maize) and reliance on an external market, (ii) the use of traction equipment facilitates area expansion and extensive non-market related agriculture and (iii) the intensive use of external material inputs supports cotton cultivation for an external market. Knowledge of village resource endowments allows for strategic targeting of policy measures that respond adequately to observed variability and diversity. In addition, social welfare indicators (*i.e.* health, education) provide important information on village-level progress in domains other than primary production.

The observed patterns of portfolio choice indicate that diversity in land use and activity choice is closely related to rainfall variability. Commercially oriented cotton and rice production systems are concentrated in villages with higher and more stable rainfall and their development can probably be further reinforced through market-based policies (*i.e.* infrastructure, price information, *etc.*). Under more variable climatic conditions combined with relatively favourable resource endowments, either extensive cereal production or livestock farming become important options, the former combined with some engagement in non-agricultural employment. Improved risk management for these systems can be perceived through strategies of resource intensification (*i.e.* crop choice, soil moisture management and livestock feeding regimes or alternatively better marketing conditions for output, inputs and food). When highly variable climatic conditions coexist with an increasing pressure on the resource base, food security strongly depends on institutional support for human development (*i.e.* education, health care) and reliance on social exchange networks.

Livelihood Strategies

Livelihood strategies during periods of critical rainfall are characterised by a strong dependence on agricultural revenues, especially in the Soudanian zone. In the Sahelian zone, livestock holdings become more important. Increasing drought led to an increasing dependence on off-farm income sources. Changes in land use are almost exclusively determined by labour availability (Sahelian zone) or access to animal traction (Soudanian zone). The panel dataset permits to identify some important general trends. First, arable crop production remains by far the most important activity (in terms of time allocation) of rural households. Given the days of work per adult per season and the short duration of the planting season, the agricultural cycle leaves little room for diversifying the activity pattern during the agricultural season. Second, alternative revenues structurally compensate food deficiency in both zones, yet this source of income is less accessible to the poorest household stratum. Third, the relative stable income level during a period of climatic stress suggests that households successfully diversify their livelihood pattern, in particular in the Sahelian zone. Through this response, households showed a remarkable resilience. Fourth, a decrease in livestock holdings and food stocks, particularly in the Sahelian villages, gives reason for concern about the vulnerability for future income shocks when conditions deteriorate over a period of more than a few years.

With respect to the differentiation processes amongst households, the following pattern emerges. First, farm households follow two distinct crop production strategies: either land intensive or labour intensive. Both strategies can contribute to enhancing total crop production. Other income sources (*e.g.* animal sales and non-agricultural activities) provide an exogenous contribution to total income. Second, labour input (per adult) appears to be a major determinant of yields on the one hand and cultivated area per capita on the other hand. Since households are fully occupied during the agricultural season, progress can be made either by enhancing the efficiency of cultivation work (animal traction, tillage conservation techniques, and more efficient use of surface water) or by promoting dry season income-generating activities. Third, despite severe climatic shocks, no evidence is found of an increasing inequality of cropping income and land ownership. While crop production may be a volatile factor in the short run it appears to be a stabilising factor in the long run.

During the second period of more favourable rainfall conditions (1993-'98), livelihood strategies in the Kaya region of Burkina Faso experienced important changes. Increasing population pressure, infrastructure provision and market development led to adjustments in land use and resource allocation. In general terms, diversity between farm households tends to increase, but these are determined far less by agro-ecological conditions and depend more on access to markets (fertilisers, seed) and institutions (credit, extension). Natural resource degradation makes farmers increasingly interested in improved crop management practices and soil conservation control measures.

The increasing incidence of proximity to Ouagadougou is evident in the southern part of the area, particularly through the intensification of land use on smaller holdings and the decrease of available land per capita. At the same time, especially in the central area of both provinces and the northern *Bam* area, an improved technological package has been adopted based on high application of manure, fertiliser, intercropping, animal traction use and erosion control measures. High labour inputs, especially in the dry season, were an important factor in this context. The adoption of this package has led to a considerable increase in yields. In the northern part, better rainfall conditions contributed to improved yields as well.

The relatively high yields have been predominantly realised in the fields near the villages and on the better soils, rather than on the upper plains soils. Crop choice and management in this Sahelian zone seem to reflect similar decisions in the Soudanian zone during the earlier period, which indicates a flexible adaptation to changing climatic conditions in this northern zone. The wealthier households have reached production levels which are adequate enough to allow them to sell part of their cereal produce and buy other types of products instead, in particular rice which is preferred, has a higher price but cannot be produced everywhere due to lack of water.

Patterns of Change

The comparison of information derived from detailed panel datasets relating to two periods with rather distinct rainfall regimes provides useful insights into long-term development trends and farm household responses to changing agro-ecological conditions.

First, we noticed that average crop yields strongly improved and cereal production per capita is now well beyond FAO minimum requirements. Although land per capita has been reduced, yields could increase due to agricultural intensification and technological change. The most dramatic change in land use is the shift from millet towards sorghum cultivation, illustrating the tendency for preferences of less drought-sensitive crops. More favourable rainfall conditions enabled the adoption of more input intensive production systems based on (inorganic) fertiliser use and animal traction. Land productivity improved as well due to increased investments in soil conservation activities. Secondly, we registered a small (real) increase in incomes, basically due to better prices for crops and a stable and high contribution of income derived from non-farm sources. Third, important changes in landholdings took place in the 1980s and consequently the pattern of land holding and cattle ownership has become somewhat more concentrated in the 1990s. Farmers in the Soudanian zone are more sensitive to income fluctuations from cropping activities, while in the Sudanian zone livestock still functions as an important insurance device.

Chapter 17

SOME COMMENTS ON PORTFOLIO DIVERSIFICATION, RURAL PATHWAYS AND FARM AND HOUSEHOLD ANALYSES

Thea Hilhorst and Chris Reij

Abstract: At the ICCD workshop in April 2001, comments were made on the results of the ICCD analysis of patterns of change in Southern Mali (by Thea Hilhorst) and Northern Burkina Faso (Chris Reij), from a perspective of development practice. The emphasis is on the multi-locational (or even international) farming systems and the complexity of portfolios on the one hand (with the future for cotton production in some doubt) and on the impressive results of soil and water conservation measures in these areas, on the other. This chapter is an edited version of those comments.

1. PORTFOLIO DIVERSIFICATION AND RURAL PATHWAYS IN SOUTHERN MALI

The ICCD initiative was an important reflection on the possible impact of climate change on livelihoods and it has helped people to focus on solutions: what type of policies might help people in the Sahel to adjust more effectively to the impact of climate change. It has been a timely research programme, as global discussions on climate change are moving towards thinking on adaptation, given that some changes can no longer be avoided, whatever measures are taken.

It has also been important to emphasise a portfolio approach and thus look beyond farming practices. Farming and livestock rearing are still very important livelihood activities in the Sahel, but rural people are not only engaged in farming. They are engaged in a lot more activities and have many interests, some even beyond their village. This will affect their decisions regarding where to invest their scarce resources. They may invest in agriculture, but also in off-farm activities, in their village or elsewhere. They may also invest in social networks, through marriage,

A.J. Dietz et al. (eds.),
The Impact of Climate Change on Drylands: With a Focus on West-Africa, 293–300.

and other events. It is important for policy makers to acquire a better understanding of such wide-ranging livelihood strategies, since it will affect their priorities and possibilities.

Southern Mali is a very interesting region. As chapters 13 and 16 show, it is a region which, since about 1970, has experienced major changes in farming systems in the local economy and society as a whole. Malian farmers have proven to be very dynamic people, seizing upon new opportunities and adapting to new situations. In fact, many households in Southern Mali maintain an international farming system. These are extended households that grow cotton in Mali while also managing coffee and cocoa plantations in Ivory Coast. Over the last three decades, they have developed a system for rotating labour, money, and information. However, at present, they are experiencing a lot of insecurity, given that the future of these plantations is undermined by the troubles in Ivory Coast and the resulting uncertainty over land tenure. Households' portfolios have become quite complex, as the activities are many and may cover a large area. Households are confronted with many different types of insecurities: it is important to go beyond climate change and study it as part of a larger vulnerability complex. Looking back at the decisions taken by many Southern Malians in the last decades, it is clear that many have been engaged in migration, which was (and still is) seen as an opportunity and has become part of local culture: if you are young you have to prove yourself by going away for some time. Although some households or individuals may feel 'forced' to migrate because of poverty, this is certainly not representative. Involvement in off-farm employment is not only, or not even primarily, driven by poverty. The wealthiest households were also very much engaged in diversification and were actually better placed to develop new, profitable activities as they had the resources to invest in a mill, in a car, or in building a house in cities such as Koutiala or Sikasso. The perspective of climate change and people's behaviour during and after relatively bad rainfall years evoke too much of a suggestion that diversification strategies (migration, off-farm work) are poverty-driven or calamity-driven.

The study on portfolio diversification and development pathways, as reported in chapter 16, mainly uses data collected by the cotton company CMDT and the Malian research institute IER. A lot of effort (and money) has gone into the collection of this data and it is good to see that such good use has been made of it. However, these datasets are biased towards cropping systems and the field level data is very rich and complete. But the data on other elements of a diversified portfolio is far from complete. It is most likely that the datasets would not have captured the international dimension of farming systems if focused on a particular village. Given the flaws in these datasets, the results have to be interpreted very carefully and with the involvement of Malian experts and based on feedback sessions in the region. Without such precautions, conclusions and policy recommendations may be erroneous and flawed.

The study by Brons et al (chapter 16) gives a more varied picture of farming in southern Mali than is usually the case in existing policy documents. The researchers distinguish seven farming systems or components to characterise 43 villages in the sample. These components are, according to the researchers, different but not necessarily exclusive pathways for a village. Indeed, farmers in southern Mali are seldom specialised. A farming household may use some of its fields for growing cotton in rotation with cereals, while using some very distant fields, or fields of certain soil types, for extensive grain production. If they are lucky and live close to a lowland site which permits rice cultivation, and if they have managed to acquire access to a plot of land in these lowlands, they will grow rice. Taking a village as a unit of presentation (and as a target for specific agricultural policies) would ignore the considerable diversity between households which are, in fact, the decision-making units and it would also hide the differences between individuals within these household units of decision making. The danger is that village-level targeting will generate one-sided policy interventions, while the integration of many different aspects of the diversified portfolio should be a policy target.

For example, one of the farming systems which has been identified is a 'livestock pathway'. If that would mean that policy interventions in such villages should be restricted to livestock support, it would not be consistent with developments on the ground, in which livestock plays a role as one of the elements in a much broader livelihood strategy in these villages as well. The remarks on the livestock farming pathway are very confusing. It is stated that high numbers of cattle are found in the Northern zone and near Koutiala as well. To what extent does this system refer to a more intensive form of livestock rearing, linked to the only milking co-operative in the region? This would then also mean that the conclusions on a lower level of integration, related to livestock production, due to flexible and pragmatic management strategies, is very far from reality. It is certainly not proof of more transhumant forms of livestock holding.

Another example of questionable conclusions is the link made between high yields for cotton and rice production. There is no irrigated rice production in southern Mali, but there are a few flooded plains (Kléla, Niéna) where water is partly controlled by a system of bunds and dikes. Only rice is grown here, with maybe some vegetables in the dry season. It is an example of a well-exploited niche in the landscape. High cotton production in such villages is not, and certainly not directly, linked to rice growing.

Finally, the characteristics used for assessing 'institutions' are very meagre and possibly of limited value. The year of establishment of a village organisation or erosion control programme may be influenced more by CMDT than by the village. The authors should be more careful with the interpretation of the results and with their conclusions.

If climate change is progressing towards lower levels of rainfall and more variability, increased attention could have been paid to the northern, more semi-arid zone. These are areas where cotton is now disappearing (however: see chapter 10). Groundnuts are an alternative cash crop, but prices are rather low. This region is at risk. Village characteristics presented in chapter 16 indicate that the gross margin is lower and the percentage of households with a food deficit much higher when compared with the more southern areas. Research carried out by IER and IIED showed that resource degradation is a problem. It therefore becomes very hard for farmers to invest in fertilisers or organic matter production to maintain production and soil fertility levels. They do their best and may be more careful farmers than their neighbours in the south but soil fertility may still decline in the end on many of their fields. Due to increased population densities, farmers can no longer maintain a fallow system. For example, on average, a household in the village located in the semi-arid zone had about 16 hectare under cultivation and only 0.5 hectare fallow. Households in a village about 200 km to the North had about 25 hectare cultivated and 18 hectare fallow. Although rainfall was lower, these people were more food secure and they produced a surplus for sale. It is strange that the research did not find any evidence of a negative impact of soil fertility decline on crop yields. This is a very interesting point which requires further study: one of the first studies on 'soil mining' was based on datasets from this area. Is there a problem with data and with the construction of parameters? Or are policy makers too alarmist, assuming that all fields are degrading, while the problem might be restricted to only part of the fields?

The households in the semi-arid zone, in the north of Southern Mali are becoming poorer and off-farm employment will become a necessity. Maintaining the livelihoods (and levels of relative prosperity) is hardest in these 'post-cotton' zones. The question is whether this situation would become more widespread if the semi-arid zone were to expand towards the south as a result of climate change.

However, questions on the future 'after cotton' are not only relevant in areas where the cultivation of cotton is no longer possible because of rainfall deterioration. The cotton sector as a whole is suffering from the drop of cotton prices on the world market. Malian cotton farmers went on strike in the year 2000 and about half of them refused to grow cotton. The reasons were the low price offered by CMDT, problems of debts, poor relations and a lack of trust between farmers and CMDT, as well as the fact that CMDT had been involved in a major fraud scandal, which involved a lot of money disappearing over a number of years. Some former cotton farmers did well from a change to grains. In 2001, traders even came from Niger to buy these grains. The strike had major consequences for the economy in Southern Mali particularly the cities. Half of the cotton ginneries were closed thereby reducing opportunities for casual labourers, there was much less work in the transport sector and much less money to spend. This had a negative effect on local business. The cotton sector is now in turmoil and important policy changes now seem to be underway, such as partial privatisation. Some observers think that there is a trend towards only working with the bigger farmers.

The study was undertaken to assess the impact of possible climate change. The question is whether this study indeed provides some indications of trends and possible options which may help people in Mali to adjust more effectively to climate change? Most of the predictions as regards the effects of climate change in southern Mali appear to be negative: higher temperatures and less rain, and less reliable rainfall. What might happen to production systems and portfolios? Would cotton production, for example, still be possible and would it be possible to adapt? The next step should be to develop possible scenarios to show what might happen in the future regarding climate change and the related consequences. This will not be an easy exercise as climate change is not the only important factor. For example population growth, the impact of diseases like AIDS, as well as changes in population distribution (urbanisation), will also have an effect.

As the study rightly indicates, there will not simply be one policy which addresses every issue since the situation is very diverse. Ideally, governments should develop an enabling framework at national level, leaving more detailed policy making to lower levels of scale. It would therefore be interesting to see how decentralisation and the newly created communes and councils could play a role in this discussion. Until now, Malian people have proven themselves to be highly flexible and innovative and are prepared to try out new options, whether in agriculture or in other sectors. This can be facilitated greatly by an appropriate policy environment and by better prices for their produce.

2. FARM AND HOUSEHOLD ANALYSIS IN NORTHERN BURKINA FASO

In Chapter 16, the researchers have compared two databases for their study area in Northern Burkina Faso One was created by ICRISAT during a sequence of dry years in the first half of the 1980s and the other was created by ENSA between 1993 and 1998. This revealed average rainfall which was 10% higher than in the first half of the 1980s. The two databases refer to two different regions and in both cases the size of the sample is quite small. Furthermore, the researchers have reviewed a wide range of publications, making the chapter a rich source of information. Unfortunately, the researchers did not re-visit the villages included in the ICRISAT data base and hence they did not try to include the same 14-20 families per village in their study. This would have produced interesting conclusions and insights. It is a great pity that many years of development efforts in the study area did not produce a long sequence of panel data.

What is striking in this chapter is that it appears to be primarily a desk exercise. A lot of data for Bam province could be interpreted quite differently. The authors also treat the subject of soil and water conservation in more general terms than they should have done. To put it more bluntly, the authors ignore and underestimate the role of soil and water conservation in the process of adapting land use practices to climate change.

By the end of the 1970s, the northern parts of the Central Plateau in Burkina Faso were subject to serious land degradation. Despite relatively high population densities, there were no signs of agricultural intensification. Soil erosion, the destruction of vegetation, decreasing soil fertility and frequent droughts made many families decide to leave these areas. Land degradation and frequent droughts triggered farmers as well as NGO staff to experiment with soil and water conservation technologies, or better water harvesting technologies, and this resulted in improved *zaï* and in contour stone bunds. This, in combination, turned out to be very effective in rehabilitating strongly degraded land. On top of this, permeable rock dams proved to be very efficient in controlling gully erosion. In the beginning, farmers acted in many respects against the technical recommendations of experts and technicians and this proved to be a wise strategy.

These innovations started to spread quickly, particularly in the areas with highest population densities. This process accelerated with the arrival of major natural resource management projects in the northern part of the Central Plateau. Public investment funds were used to support farmers to create a soil and water conservation infrastructure in their villages. In several places the authors state that the implementation of soil conservation measures may prevent a yield decrease rather than bring about a significant yield increase. This is not the view of the farmers. In a recent detailed study of five villages in the northern part of the Central Plateau (Reij et al., 2001) it became evident that farmers have an integrative view of impact, while many researchers (including Brons et al. in Chapter 16) seem to have a reductionist view. Researchers tend to base their assessment mainly on the impact of soil and water conservation practices on crop yields. Farmers also take into consideration crop residues, grasses, agroforestry products, improved groundwater levels, etc. It is also difficult to understand the researchers' general conclusions about yield stabilisation, if one of their results is also that average crop yields improved by more than 100% (for sorghum) or even 200% (for millet) as a result of processes of agricultural intensification and technological change. Not enough has been done to explain farmers' behaviour. The authors write that there has been a dramatic shift away from millet towards sorghum, which is planted under more humid conditions (in flood plains). An obvious question is why farmers have shifted from millet to sorghum cultivation, when yields of millet have increased much more than yields of sorghum and when average millet yields seem to be higher than average sorghum yields.

Yield improvements can only partly be attributed to better rainfall conditions in the 1990s compared to the 1980s, and should largely be explained by technical breakthroughs in soil and water techniques and by substantial public and private investments in soil and water conservation, which have induced many changes. Many farmers did increase the quantity and quality of manure. In addition, in the province of Bam, many farmers are using chemical fertilisers. Three reasons can be given for this: intensive market gardening around two lakes (Bam and Bourzanga), the importance of cotton production and the impact of permeable rock dams which control gully erosion in valley bottoms and has made valley bottom land available for cultivation (both for cotton and for sorghum production). What was started as anti-erosion measures in fact resulted in water harvesting measures and made valley bottom cultivation more secure. It became a preferred strategy for many farmers with access to these lowlands. Farmers started to appreciate the soil and water conservation projects in the area and expressed an interest in becoming involved (as was required by the two major SWC projects in the area) and in the extension packages which were part of these projects.

However, it is not only the shift in field choice (a choice for more humid and more fertile humid lands near the villages; the *bas-fonds*) or the application of animal traction which created breakthroughs in yield levels. Water harvesting technology and a strong trend towards the rehabilitation of degraded land have contributed to increasing average yield levels and to improved household food production. Only a small fraction of the farmers have access to the village lowlands. Many farmers shifted towards the higher parts of the toposequence, since the adoption of better water harvesting techniques, in particular since about 1985. As a result, in the provinces of Yatenga and Zondoma, but increasingly in Bam as well, there has been an accumulation of wealth in land holdings and in livestock, based on a more secure and higher grain production.

As a result of the massive adoption of water harvesting measures (as 'anti-erosion', or 'soil and water conservation' projects) in different parts of the village territory, farmers increased their use of manure and fertilisers and this has led to an expansion of farmland (at least one hectare per family) and to higher and more stable yields and hence to improved food security at household level. It has also led to more investment in livestock and to changes in livestock management, which in turn has improved the manure situation and possibilities for animal traction.

Chapter 18

URBAN-RURAL LINKAGES AND CLIMATIC VARIABILITY

Annelet Broekhuis, Mirjam de Bruijn and Ali de Jong

Abstract: In the face of changing climate conditions, the expanding capital cities in the Sahel will try to keep their privileged situation with respect to the countryside and smaller towns. Tensions with respect to the use of land and water will increase. Whether the scarce water will be used for urban consumption and urban electricity generation or be used instead in irrigation schemes will largely depend on political choices and power arrangements. In any case, urban purchasing power will decrease even further and it will be impossible for large parts of the population to keep up their actual consumption levels (of food, fuel and water). As a result, urban poverty will increase and the discrepancy between poor and rich in the cities amplified.

1. INTRODUCTION

Rapid urbanisation has affected the whole of Africa during recent decades more than in other macro-regions in the world (Verkoren & Van Westen 2000). This process of urbanisation changes the relationship between rural areas and cities. Although almost all towns and cities have seen their populations increase, the large towns in particular have grown quickly. These large fast-growing African towns include many of the country's capital cities. These huge agglomerations accommodate a considerable part of the urban population, and often are typical primate cities. However, not only have the large towns grown rapidly, the regional towns and rural centres have also grown in number and have experienced population increases.

These processes are also evident in West Africa. The Club du Sahel mapped the changing spatial pattern of urban growth and spread for the period from 1960 onwards (WALTPS). The data for 1960 and the extrapolation for 2020 are shown in Figure 18.1. The growth of the

A.J. Dietz et al. (eds.),
The Impact of Climate Change on Drylands: With a Focus on West-Africa, 301–321.

capitals of the Sahel countries and their relatively important position within the urban hierarchy and within the zone is clear to see.

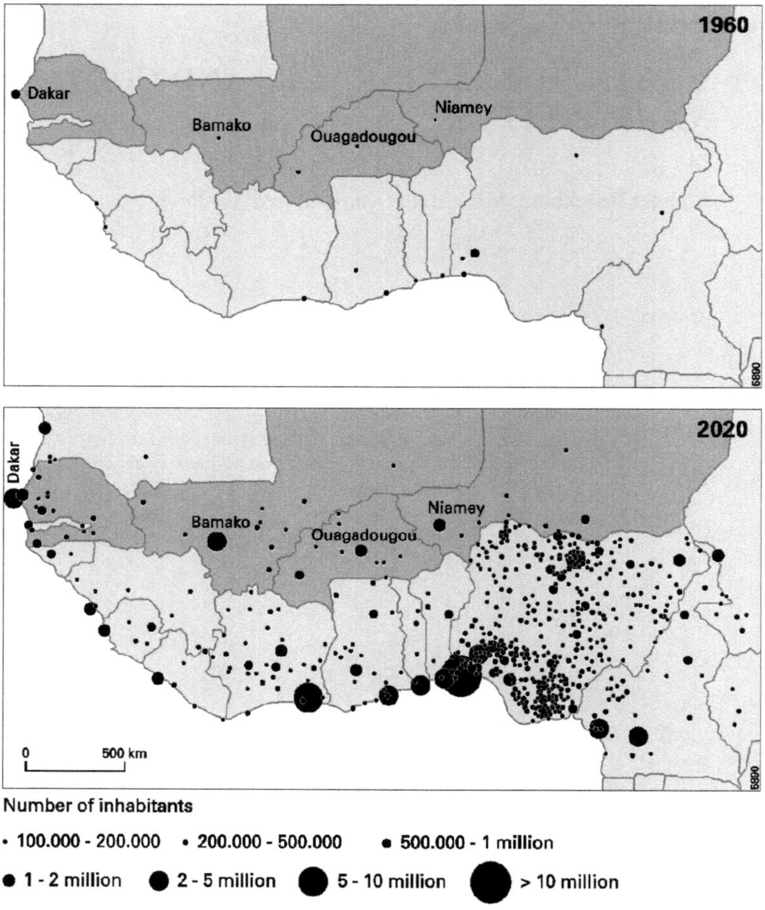

Map 18.1. Urban growth in West Africa, 1960-2020. Source: Club du Sahel, 1995, p. 14.

Cities can be seen either as islands of privileges or as concentrations of large-scale deprivation. Both images are true for the capital cities discussed here. On the one hand, the larger cities are privileged in terms of availability of important services such as schools and healthcare, markets and shops, electricity and piped water. On average urban populations score better in terms of wealth indicators. On the other hand, the real urban poor - often representing small percentages but yet large in absolute numbers - have to face miserable environmental living conditions, high expenditure on food, housing and transport, low wages and little access to formal work.

In many poor countries where agriculture is the dominant economic sector, rural-urban and urban-rural linkages are strong and important for both urban and rural economies and societies and the households and persons involved. These linkages have changed radically over the past decades. As Tacoli (2001:141) states: '...more than two decades of economic crisis and reform packages and, more recently, the internationalisation of trade and production have deeply transformed relations between town and countryside.'

In the Sahel, the most important cities are often located in the southern regions, along rivers, i.e. in the regions relatively well-endowed with water. Consequently, those urban centres have peri-urban surroundings, which are relatively favourable for biomass production (food and firewood). But as urban growth and expansion drastically transform the surrounding zone, the ecosystem in these regions comes under severe pressure.

In this chapter we present the first findings of research on these rural-urban linkages focusing on the effects of climate change and change in climate variability. Four fields will be treated more specifically, fields that are directly related to the problem of climate change: migration flows, food supplies to cities, biomass fuel, and water. Studies were carried out in the four capitals of Senegal (Dakar). Mali (Bamako), Burkina Faso (Ouagadougou) and Niger (Niamey), which are among the largest cities in the Sahel zone. The hinterland of all these cities is affected by changes in climate variability as is shown in previous chapters. Dakar is interesting because of its typical harbour situation. It is located at a peninsula and at the border with the sea. The other three towns are landlocked, but the aridity of the zone in which they are situated varies. Niamey is a typical Sahelian town in the sense that its hinterland is situated in the dry savannah zone. Bamako is situated in an ecologically better environment, the humid savannah zone, and Ouagadougou in a severely degraded environment on the border of the dry and humid savannah.

2. URBAN GROWTH AND DROUGHT MIGRATION

The history of the cities that were chosen for this study, show a rapid change from the beginning of the 20th century. The colonial rhythm of urbanization continued under post-colonial rule. As we will argue in this chapter, one of the reasons for this growth has been the search for shelter, food and security, as the hinterland no longer offers enough as a consequence of a changing ecological and economic situation (Amin 1995, Van Dijk *et al* 2001, Zaal & Dietz, 2002).

The last decades have shown a large increase in the urban population as is shown in table 18.1, an increase that is expected to continue, although on a somewhat lower level for the next 10 - 15 years. The table also marks the difference in urbanization level between the coastal and interior countries of West Africa. It should also be noted that the national urban hierarchies are dominated by primate cities: a relatively large part of the urban population is living in one (very) big city such as Dakar, Ouagadougou, Bamako and Niamey with a current size of nearly one million inhabitants or more.

Table 18.1. Percentage of the total population living in urban settlements

	Urban population as % of total population			Number. of inhabitants of capital city
	1975	1999	2015	
Burkina Faso	6	18	27	Ouagadougou:1 130 000 (2000)
Niger	11	20	29	Niamey: 660 000 (1999)
Mali	16	30	40	Bamako: 1 016 000 (1998)
Ghana	30	37	48	Accra: 1 900 000 (est.for 2000)
Senegal	34	47	57	Dakar: 2 079 000 (2000)

Source: UNDP 2001a, United Nations Statistical Yearbook 2000, Djibo et al 1999, Bonnard 2000; also see Zaal & Dietz, 2002.

Natural population growth currently plays an important role in this increase in urban size. During the first phases of the urbanisation process, when the urban settlements were still small, towns grew more as a result of immigration than of natural population growth. For the small and intermediate towns in the semi-arid regions in general this remark still holds: migration contributes substantially to the urban growth. Nowadays, the growth of the large capital cities is mainly determined by relatively higher natural increase rates caused by lower death rates compared to rural areas and still high fertility levels. There is evidence, however, that the third phase of the demographic transition - the decline of fertility rates - is starting to affect metropolises such as Dakar and Bamako (Diarra & Tangara 1999, p.45, DNSI/UNICEF 1996, p.10).

In addition, rural-urban migration, permanent or on a temporary base still contributes today to the growth of large cities and to the presence of a floating population in the cities. West Africa has a long tradition of circular migration (Mortimore 1998:1200). In the semi-arid zones of West Africa, long- and short-term circulation between home and (urban) places of employment or trading, is a form of adaptation to new opportunities in the rapidly urbanising economies, as well as a continuation of a historical tradition of spatial mobility. As a consequence, urban settlements can be characterised as important foci of in- and out-migration. Studies of small towns illustrate this point (e.g. the ICCD study of Douentza by Zondag 2001, cf. Baker 2001). These movements of people between rural and urban areas may be voluntary. In Africa, however, an increasing number of people have, in recent decades, been forced to move and search for a living in towns; their reasons being economic, political or ecological (Bascom 1995).

Regardless of the decreasing relative significance of migration to the overall growth of the larger cities, the national rural-urban pattern in Senegal, Mali, Burkina Faso and Niger is dominated by mobility to their national capitals. Migration takes place in all directions: from rural to rural areas, from rural to urban areas, from urban to urban areas and vice-versa. But the migration balance is always in favour of the capital city. In table 18.2 these trends are shown for Senegal and Mali.

Table 18.2. Population flows between areas during the years 1988-1992 in percentages persons older than 15 years).

Country	Capital > other cities	Capital > Rural areas	Other cities> capital	Other cities > Rural areas	Rural areas > Capital	Rural areas > Other cities	Total of flows
Mali	9.1	16.7	12.3	19.0	18.1	24.8	100
Senegal	13.3	19.6	16.5	12.7	22.2	16.0	100

Source: CERPOD 1995 in: Antoine et al 1998

During severe stress situations, such as drought and famine, population mobility intensifies. In the event of deteriorating rural living conditions, urban centres may expect increasing flows of immigrants who, eventually, stay or return to their home base when conditions improve. In this sense we can speak of forced migration and of displaced persons (Bascom 1995). In particular during the period of the severe drought in the early Seventies, all types of urban settlements in the Sahel have been drastically affected by migration. The populations of Ouagadougou, Bamako and Niamey grew by more than 10 percent in those years, rates which have remained unequalled ever since. Towns in the northern regions such as Tombouctou, Gao and Mopti, where food aid was provided, also experienced a large influx of refugees during the 1970s and early 1980s. Later on, this population growth slowed down.

During later drought years, for example the drought of the early 1980s, the influence of migration on the growth rate of certain cities weakened for several reasons. At first the population at risk during periods of stress decreased in relative terms as a result of changes in the general pattern of population distribution. There has been an ongoing movement of rural population from northern to southern zones and depopulation of regions with higher drought risks (Veldhuizen & Dietz 1999; see Chapter 8). Secondly, measures taken by national governments and international donor organisations to cope with future stress situations (establishing security stocks of food at national level and of cereal banks at village level; establishing early warning systems; see Chapter 22) enabled the government to respond earlier to food shortages and reduced the need of villagers to move away.

Table 18.3. Population size of the Sahelian capitals and yearly population increase1960-2000

Year	1960	1976	1987	2000
Period		1960-1976	1976-1987	1987-2000
Dakar				
Population size	374 000	799 000	1 490 000 (1988)	2 079 000
Population increase		4,9	5,3	2,8
Bamako				
Population size	89 000	404 000	658 000	1 016 000 (1998)
Population increase		9,9	4,5	4,0
Ouagadougou				
Population size	59 000 (1961)	173 000 (1975)	442 000 (1985)	1 130 000
Population increase		8,0	9,8	6,4
Niamey				
Population size	20 000	225 000 (1977)	398 000 (1988)	660 000 (1999)
Population increase		15,3	5,3	4,0

Source: Lahmeyer 2002: Recensement général de la population du Mali, 1987, 1998.

Many of those people who settled in town during drought years stayed. Because the influx had been so large and so rapid, the urban economies were not able to absorb all the newcomers: there was not enough work and housing. People settled in camps and in peripheral slum quarters and were engaged in all kinds of low-paid jobs. Drought migration was often related to poverty. The case of Niamey shows a process that may be labelled "urbanization of poverty". Niger is a very poor country with 63% of its population classified as poor (according to World Bank and PNUD definitions) while 43% are extremely poor (Niger is in this respect representative of the other Sahelian countries. Mali had 69% of its population below the poverty line of 1 US$ (ppp US$ 1993), Burkina Faso: 61.2 and Senegal 26.3%; HDR 2001). Many of the migrants who came from the rural areas left this area because they could no longer make a living there. These poor rural migrants came to town and live in peripheral quarters. In Niamey these quarters expanded especially during periods of drought and famine. For example, the drought of 1973 caused famine and very large migrations. Recent examples of drought are 1992-93, 1996-97 and 1999-2000. Their poverty often does not disappear when they arrive in the towns. There people end up just surviving. In town, their survival strategies consist of self-employed work or of casual labour; i.e. irregular and badly paid work particularly in small commercial undertakings, but also in agriculture (cattle herding), handicraft and other services, among which prostitution and begging are not the least important. The scarcity of land in towns leads to another strategy: namely land speculation. The poor squatters claim the land on which they live as soon as it officially becomes part of town and then they resell it to occupy another piece of land outside town, waiting until this also becomes part of the urban land (Gado 2000).

The situation of migrants in Senegal shows us a totally different picture. Although there is poverty and many migrants live on the margins of urban life, in Senegal opportunities seem to be more promising. In the past, the Mourides dominated a large part of the migration flow. They were a religious Islamic brotherhood which, from colonial times onwards, have played an important role in the production of groundnuts. With the droughts, the mining of the soils and the decreasing prices of peanuts on the world market, peanut cultivation became less attractive. Many Mourides went to town (in many cases from a small town, which was their first destination) and made a living in the informal sector, often in small commercial enterprises. Finally, some of them ended up in international commerce earning a lot of money, with which they were able to build up a luxurious life in Senegal and reinvest in their villages of origin and in the town where they live. They entered these informal urban and international networks through the strong social and religious networks of which they were part and which were also converted into a trade network (Fall *et al* 2000b). Recently, migration flows have changed and the peanut basin has again become an immigration zone. The capital, Dakar, still attracts large numbers of migrants (29%) from rural areas. Rural people leave for the city because of the agro-ecological degradation and because of better living conditions in town (UNDP 2001b).

Recent research on immigrants in Dakar and Bamako (Antoine et al 1998) indicates that their incorporation with respect to labour, houses and social life is not very different from that of the local population. Immigrants rapidly find jobs, although in Bamako to a larger degree in the informal sector than in the formal one, due to their somewhat lower education level. The conclusion of this research is that the cities of Dakar and Bamako are in a phase of stabilization. They are capable of absorbing the less numerous new migrants into the local population as a whole and the provision of work and housing has improved, while earlier migrants have been completely integrated. These research outcomes contradict the general idea that immigrants augment the number of the unemployed (Antoine et al 1998).

These relatively positive outcomes are confirmed by other research in the peri-urban zone of Bamako, where population growth rates were observed of up to 9%. The activities of the population in this zone, which included a large number of immigrants, were linked to the supply of the urban population with food, wood and charcoal. Rapidly expanding cities provide large markets for products of the primary sector and indeed enable the native population as well as the immigrants to earn a living. This change is not limited to the peri-urban zone. For example, a large proportion of the urban vegetable growers were immigrants with a rural background (Klompenhouwer & Van Soest 1999).

3. GOVERNMENT POLICY AND FOOD SUPPLY

Because drought periods in the past had such an enormous impact on urban and rural life, governments were easily convinced that policies with respect to agricultural production, agricultural marketing and storage should be changed. In the Sahel countries, structural adjustment measures, supported by the donor community, improved the functioning of cereal markets, extended local cereal banks and regional stocks (security stocks). The liberalization of the markets and the deregulation of trade stimulated the investment in storage facilities and transport by (small) private traders, as was shown by Dembélé & Staatz (1999) for Mali. In general, cereal trade became more liberalized and more efficient, contributing to some modernization of the production (this was especially the case for irrigated rice which shows interesting production increases in Burkina Faso, Mali and Niger and for the less important cereal maize, but this was however limited in the case of sorghum and millet due to the lack of high-yielding varieties adapted to the poor production circumstances).

The increase in production was rather spectacular in irrigated crops, i.e. the production of paddy, thanks to the improvement of existing schemes and enlargement with new schemes. In Mali, domestic production shot up dramatically, growing at an annual rate of 9% between 1980 and 1997, due to substantial yield increases in the irrigated area of the Office du Niger. As a result, rice production more than tripled between 1985 and 1998, from 214,000 m.t. to 688,000 m.t. (Dembélé & Staatz 1999:15). Mali is now capable of exporting rice to neighbouring countries. Mali exports higher quality rice, while importing lower-cost broken Asian rice for the low-income urbanites. According to Yade (1999 in: Dembélé & Staats 1999) this strengthens Mali's food security. Production of coarse grains also increased, above all by the extension of areas under cereals. Production size however, is very irregular and directly linked to climatic circumstances. The Famine Early Warning System Network (FEWS 2000 and 20002) for example noted record harvests in Mali, Burkina Faso and Niger for 1998 and 1999, and again for 2001 and 2002 thanks to favourable rainfall. In general, good harvests mean a better food supply and low prices.

Production size and producer and consumer prices used to be related although this relationship can be disrupted by government policies. The devaluation of the FCFA was one such disrupting factor. Cereal prices increased dramatically immediately after the devaluation in 1994 and caused difficult times for the urban population and hardship for the urban poor. However, prices settled and in Mali the real retail prices of cereals fell by approximately 20% to 30% during the period 1981/82 - 1997/98 due to increased competition between cereal traders (Dembélé & Staatz 1999). The prices of grains (rice, millet, sorghum and maize) have been rather low since 1998 due to the good harvests (exception is Niger's bad harvest in 1999/2000). However, in 2002, the prices of coarse grains on consumer markets were astonishingly high in Mali, Burkina Faso and Niger until March, in spite of (very) good harvests. The explanation can be found in the export to other countries, in the replenishing of the national security stocks and in the retaining of relatively large stocks at farm level (in the cotton zone this is possible because cotton revenues were quickly cashed after the harvest and farmers did not need to sell their cereals for cash) (FEWS 2002). The structural adjustment of the economies has caused a greater macro-regional integration, which means that international cereal trade has become much easier. This may be interesting for producers because they reach a much larger consumer market, but it puts the national consumers in a position of competition with consumers in the rest of West Africa.

Although the supply of the urban markets with food grains and also with vegetables and fruit has improved due to more liberal policies, this does not mean that food security has improved, neither in the short nor in the long run. In the short term the purchasing power of the urbanites is an important factor. This purchasing power has decreased during the period 1960-1994 for Africa in general and certainly for the Sahelian countries. Antoine et al (1998) states with regard to Bamako that there was a significant drop in purchasing power during the period 1962-1982. The structural adjustment measures, and especially the devaluation of the FCFA in 1994, again caused a fall in the purchasing power of the majority of the urban population. Low food prices due to good harvests improve the access of the urban population and especially of the urban poor directly (FEWS 2000). The improvement in the purchasing power of the urban poor, for example by stimulating small-scale enterprise, should be an objective of systematic government policy in the future.

Another aspect of government policy is also at stake. Capital cities accommodate a large number of public functions. The civil servants form the backbone of the urban economy. Their size, salary level and regularity of salary payment are elements affecting the urban economy and the purchasing power of other urban sectors. Before the *coupe d'état* of 1991, civil servants in Mali were not paid on time and this caused problems among traders and artisans who were dependent on their purchases. In Niger this kind of problems still affects the economy in the capital city where half of the population is directly or indirectly dependent on government salaries (FEWS 2000).

Although food production and food supplies to the cities have not presented major problems in recent years, production levels may become problematic in the long run. The production of rain-fed cereals is as dependent on rainfall as before. During recent decades no major droughts have taken place. Instead only local production deficits occurred which caused variable prices between zones. However, the effects of continuing climate change mean that droughts will be a serious risk to reckon with in the future. The expansion of irrigated agriculture is the only way to limit the effects of low rainfall.

4. FOOD SUPPLY CHAINS AND CLIMATE VARIABILITY

The relatively large capital cities of the Sahel obtain their food and other necessities of life from all over the national territory. Some regions are specialised in cereals, others in potatoes and tubers (southern regions) and again others in irrigated crops such as rice and vegetables. In the past, droughts have generally shifted the centre of gravity of food production to the south where rainfall was relatively less erratic. The extension of irrigation has, at the same time, retained and sometimes enlarged production in specific areas.

The extent and stability of the demand of the capital cities make their markets attractive for traders. Population growth, the presence of a large civil service with stable revenues and a concentration of the elite in these cities explain the growing demand in spite of decreasing purchasing power per capita. Traders are willing to invest in marketing chains directed towards the capital city. Dakar, for instance, accounts for 40 % of the total national demand for vegetables (Fall *et al* 2000a). Many of the marketing chains are short and contain only a few links: the produce is bought up in the producing areas, transported to the cities and traded by wholesale traders to retailers. Most of the small traders are implicated, especially in the collection and distribution of the products. In general, invested capital is limited and should have a rapid turnover. The cereal trade channels are sometimes more complex and longer and wholesalers (some of them are import-export traders at the same time) in the capital cities play an important role by injecting money into the buying up of cereals after the harvest.

The growing demand also exerts a strong influence on the immediate surroundings of the city and even on the wasteland within the city borders, even if provisions are coming in from the whole national territory and from outside. Production in these regions is adapting to the strong demand by way of the introduction of new varieties and products and by changing into more capital-intensive production forms. In the proximity of Ouagadougou Yonli (in: Rouers & Van den Bos 1999) noted an increase in yields and in capital investment in production for the urban market (sometimes by urbanites). In the region around Bamako, within a radius of 50 kilometres, new production forms were noted with respect to cattle breeding for meat and milk (Muis 1999; Gielen 2000) and with respect to the insertion of vegetable growing within a rotation with cereals. In the Sahelian capitals Dakar, Bamako, Ouagadougou and Niamey urban agriculture is important for local consumption. Vegetable growing has expanded in the cities themselves and in their surroundings.

Many immigrants who originate from rural areas find work in this sector (Klompenhouwer & Van Soest 1999; Tindano in Rouers & Van den Bos 1999, Fall *et al.* 2000ab, Gado & Salifou 2000).

Urban food patterns are different from rural ones, with less calorie intake but with a larger proportion coming from tubers, vegetables, meat and fat and a smaller proportion coming from rain-fed cereals such as millet and sorghum. During the drought periods in the past, the average consumption of the staple food (dry cereals) used to decline and imported cereals such as rice and wheat (imported by traders or in the framework of food aid) partly replaced millet and sorghum. Perhaps stimulated by these changes, and as a consequence of insufficient local production, imported cereals gradually replaced local cereals in the urban diet (in the form of bread, pasta and rice). This change in diet even took place in the countryside, albeit to a lesser degree. However, the devaluation of the FCFA has inverted this tendency and has been a stimulus for national food production. Imports of food grains have diminished in Mali, Niger and Burkina Faso and local rice has, to a large extent, taken the place of imported rice in the first two countries.

Climate variability is reflected in volatile food prices within and over the years. Because urbanites depend on the market for the largest part of their food intake they can be confronted with considerable price increases. Especially the poorer segments of the urban population may be hard hit. But government action in the form of food import and donor assistance in the form of food aid has up till now prevented large-scale hunger during drought periods. We expect that under more difficult climatic circumstances the combination of the large market and the presence of large traders will orient a significant part of national production towards the capital. According to Bonnard (2000) famine is rarely a threat in urban areas. When food is scarce and/or food prices too high social unrest will increase: the threat of famine in cities often results in riots and mayhem. The presence of the administration and of the many civil servants in the capital city will urge the national government to sustain food supply in the form of imports and distribution of food aid, possibly at the expense of rural and small town consumers.

5. BIOMASS FUELS AND CLIMATE VARIABILITY

Often the cutting of fuel wood is seen as an important cause of rural deforestation and land degradation in the semi-arid regions of sub-Saharan Africa. In line with this train of thought, the increasing urban demand for biomass fuel (wood and charcoal) leads to deforestation in ever-widening circles around urban settlements, particularly in countries where woodland is considered an open-access area, which could be mined for free energy. More recently, these visions (for comments on this kind of narratives see a.o. Benjaminsen 1993; Kristoferson 1997) have been adjusted somewhat and more recent views suggest that agricultural clearing is (sometimes) more important for the loss of woodland than fuel-wood cutting and that wood-fuels are supplied from dead trees, from agricultural land clearing or as a co-product of farming (Chomitz, Griffiths and Puri 1998). It is even stated that natural fuel wood production around (smaller) towns is often more than sufficient to meet demand and that, rather than investing in plantations, more rational harvesting and marketing methods should be adapted (CIRAD 2002). Here we will show that it is difficult to make generalisations. The cutting of wood is not the only cause of the deforestation in the urban hinterland, but in certain parts of the peri-urban zones (of certain cities) the wood used exceeds the natural production, while this is not (yet) the case in other parts.

In this subsection, however, the wood-fuel-driven degradation of the peri-urban zone is not the only central issue. At the end, we will also focus on the link between these aspects the other way around: what will the consequences be of climate change for the energy supply of the urban population in the Sahel.

In the semi-arid countries of West Africa, biomass fuels are the most important energy sources for households in both rural and urban areas as well for small-scale enterprises in the informal sector like, for instance, handicraft and roadside catering. Biomass fuel is also a key raw material for most income-generating activities carried out by women who process food or brew beer. About 90% of all households in Burkina Faso and Mali use biomass fuels for cooking and boiling water (Rouers & Van den Bos 1999; Ministère de l'Environnement, 1998); in Senegal and Niger these percentages are (somewhat) lower. Rural women depend almost entirely on firewood to prepare their meals, while a growing number of urban households, in particular in large cities such as Dakar, Bamako and Ouagadougou, have turned to charcoal instead of wood. The expectation is that in the year 2005 charcoal will be the most important energy source in 80% of households in Bamako (Ministère de l'Environnement, 1998). In Dakar, charcoal ousted wood as energy source many years ago (Sokona 2001; ENDA undated). In most African cities including Bamako, Niamey and Ouagadougou, alternative sources of energy (electricity, kerosene/paraffin, gas) are still used on a very modest scale even by richer urban families. Dakar and other Senegalese urban centres are exceptions (see also table 18.4).

As long ago as in 1974, the Senegalese government initiated a programme to substitute part of the charcoal consumption with liquefied petroleum gas (LPG). In addition, several types of LPG cookers were successfully introduced over time. This *butanization* campaign resulted in a diversification of cooking equipment among urban households and nowadays 85% of the households in Dakar own a cooker with gas bottle and use gas besides charcoal for cooking purposes. After its introduction in 1974, the LPG consumption in Senegal rose steadily from 3,000 to 100,000 tons and nearly half of this amount is used in the conurbation of Dakar (Sokona 2001, ENDA 1994). So, it would appear that the diffusion of alternative energy sources in urban circles is increasing, but that there is a huge difference between Dakar and the other capitals as regards the extent and pace of those changes.

Table 18.4. Energy consumption in Senegal in 1992 (tons).

	Fuel-wood	Charcoal	Gas
Rural areas	1,267,300	92,900	5,900
Urban areas	228,900	268,300	34,600
- Dakar city	7,700	137,300	22,500
Total	1,496,200	361,200	40,500

In rural areas, people (mostly women) go out to collect the wood they need for their household activities. An investment in time is all that is needed to obtain the quantity required. This makes firewood a more or less free good. Yet the amount of firewood consumed per person varies per region and location. Where wood is abundantly available, and consequently time investment is low, consumption per person is higher than in areas where wood is scarce and women have to invest more time to gather the required daily amount. Consumption is influenced by the necessary collecting time.

Urban citizens always consume less biomass fuel per capita compared to rural dwellers. Urban consumers are provided by the market and have to pay for wood and charcoal. Despite these costs, urban women continue to use traditional energy sources instead of using alternatives because they are accustomed to this type of energy and because of the relatively lower costs of biomass fuels (see table 18.5). Thanks to the open access to this resource for people making a living in this economic sector, biomass fuels are still cheaper than any other type of energy on urban markets (Kristoferson 1997; Chomitz, Griffiths & Puri 1998). In Burkina Faso, the use of wood was 40% cheaper than other energy sources despite taxes imposed on the cutting, transport and trade of wood (IOV 1992:101). Finally, some alternatives, such as electricity, are not available in remote and unauthorised urban quarters. This reduces the choices households have.

Table 18.5. Comparative cost of different cooking fuels with and without allowing for the cost of cooking appliances (fuel-wood = 1)

	Senegal		Niger	
	fuel only	*including cost of appliance*	*fuel only*	*including cost of appliance*
Charcoal	0.5	0.9	0.7	1.4
Kerosene	0.6	1.7	0.8	1.7
LPG	0.07	1.3-1.9	1	2
Electricity	1.3	3.3	1.6	2.8

Source: Anderson & Fishwick cited in Elkan (1988)

During the last few decades the price for charcoal and firewood on urban markets has gradually risen, in particular in Ouagadougou. This has forced the urbanites to reduce their consumption. For instance, between 1974 and 1992, the daily consumption per capita in Ouagadougou dropped from 2.8 kg to 0.5 kg. This reduction was obtained by an improvement in cooking efficiency. In 1990, 42 % of households in Ouagadougou were already cooking on improved stoves (Sokona *et al* cited in Rouers & Van den Bos 1999).

In 1992, the total annual consumption of biomass fuels of the population of Dakar was estimated at 820,000 tonnes (wood needed for charcoal) (ENDA 1992). In the mid-1990s, the populations of Bamako and Ouagadougou consumed 600,000 and 300,000 tonnes of firewood (charcoal included) respectively (Rouers & Van den Bos 1999; Ministère de l'Environnement, 1998). Considering the population size of these cities, the figures indicate a higher consumption among *Bamakois* compared to the inhabitants of Ouagadougou and Dakar. This difference in consumption reflects a difference in scarcity of wood and in population density of the city's hinterland. Around Bamako the countryside is less populated and more wooded and more stocks of trees and shrubs are available, while in the more densely populated surroundings of the capital cities of Burkina Faso and Senegal stocks have diminished substantially during the past decades (Sène 1993; Rouers & Van den Bos 1999, Ministère de l'Environnement, 1998, Sokona 2001).

Donkey carts and trucks are used to transport fuel-wood and charcoal from rural areas to the cities. According to some older sources (Dankelman & Davidson 1988: 67) this traffic was already causing a deforested 'ring of desolation' around Ouagadougou, Dakar and Niamey more than 10 years ago. Rural women normally collect or cut only dead wood for their own consumption, because it is easier to carry and it burns better. This type of collection is considered as harmless as the amount of collected firewood doesn't exceed the natural production. On the contrary, firewood collection destined for urban markets is often blamed for the depletion of rural wood supplies (see Ministerie van Buitenlandse Zaken 1991: 34), among others because of the urban preference for charcoal (turning from wood to charcoal aggravates the environmental degradation as more wood is needed to provide the same energy) and because of the tendency to cut whole trees instead of dead wood. We found some evidence to support the view that urbanisation contributes to deforestation in the direct vicinity of the city, as will be discussed below for Bamako.

The more the population of large cities increased, the more the supply region extended and the longer the distances across which wood and charcoal were transported. Nowadays the biomass fuels for Bamako and Ouagadougou are brought in from a distance of 200 kilometres, while during the 1950s these cities were supplied from an area within a radius of only 20 kilometres. Today in Dakar, charcoal is even brought in from as far as 600 kilometres away (Sokona 2001; Rouers & Van den Bos 1999; Ministère de l'Environnement 1998; also see Post & Snel 2003 for the impact of recent decentralisation policies of forest management in Senegal). The extension of the supply area of Bamako and Ouagadougou does not mean that all the wood collected in this area is consumed in the capital city. Around Ouagadougou only 17% of the collected wood is commercialised (Rouers & Van den Bos 1999), indicating that the members of the rural population themselves use by far the largest amount of the firewood collected. In the more wooded surroundings of Bamako the proportion of the wood sold is higher and is estimated to be 30% (Ministère de l'Environnement 1998). For charcoal alone, compared to biomass fuel in total, the picture is somewhat different. This product is destined more for the urban market, as illustrated quite clearly by the situation in Senegal (table 18.4)

Research around Bamako irrefutably showed that the more closely located sub-districts, in particular those situated along the main roads to the capital city are providing the greatest share of the total biomass fuel supply. In addition, the degree of commercialisation of the collected wood is much higher compared to the above-mentioned averages. In those areas the collected quantities exceed the natural production substantially, while in more isolated sub-districts the natural production is more than sufficient, compared to the quantities cut or collected (Ministère de l'Environnement 1998). The total supply region of Bamako still has a surplus production, but deforestation takes place in the sub-districts near the urban fringe and soon those sub-districts will be faced by severe deficits. The situation in the sub-district of Kalabankoro, which surrounds the territory of Bamako District, illustrates this tendency quite clearly. The yearly Kalabankoro wood production is calculated at 53,610 tons. For the consumption of the local inhabitants, 29,710 tons is needed and 95,066 tons is exported from this area to Bamako, with half of this amount being in the form of charcoal. The result is a deficit of 71,166 tons of wood every year (Ministère de l'Environnement 1998: 26). Given the rapid population growth in this sub-district (9 % per year between 1987 and 1998), woodcutting and the clearance of woodland is certainly taking place for other reasons too. Land is steadily being transformed into housing areas and agricultural land also is expanding.

- the substitution of wood fuels by (subsidised) gas or kerosene (Senegal)
- reforestation, promotion of village woods (Burkina Faso, Mali)
- the establishment of plantations (Mali, Burkina Faso, Niger, Senegal)
- the diffusion of (subsidised) high-efficiency household wood stoves (all countries)

- the introduction of better carbonisation techniques (Senegal)

All these efforts had limited success for several reasons. In Senegal the only successful measure was the promotion of subsidised butane gas as a household fuel (Sokona 2001). Improved stoves often proved unattractive to housewives, while adoption of the stove resulted in rather modest fuel savings (Barnes *et al.* 1993, Wilton 1996 cited in Chomitz, Griffiths & Puri 1998). Plantations turned out to have high exploitation costs and yielded high wood prices (Kerkhof 1990), to mention only some of the bottlenecks observed. But the most important constraints on the implementation of these policies were the lack of financial means and of follow-up actions and, above all, the low costs of wood-fuel itself. Highly efficient - in terms of costs for the consumer - informal structures supply the urban markets with biomass fuel at low costs.

During the last decade many Sahelian governments have rewritten their forestry laws, but despite the new more community-based approach advocated herein, the relationship between the commercial and domestic use rights did not change substantially. This means that production and marketing permits are still under Forest Service control and the vast majority of the rural population did not acquire any new rights to forest resources. The regulations still favour powerful urban-based merchants with permits for charcoal production and the trade of wood and charcoal (Ribot 2001). However, recently some positive results have been achieved in Niger and Mali due to the devolution of responsibility of forest resource management to rural communities and the introduction of a differential tax system. In Niger the government decided to encourage conversion of open access areas to community managed organisations and to impose a tax on wood-fuels produced from so-called unmanaged areas (CIRAD 2002; Ribot 2001).

What can be deduced from this information for the future wood-fuel market in urban areas in the event that climatic conditions change and further degradation of the area takes place? The first observation is that the demand for wood and charcoal will continue to grow. Population growth rates and urbanization rates are still high and a fast, short-term decline is not very likely. More people means a higher demand for energy. In the light of relatively high prices for alternative fuels (oil, gas) on the world market this signifies a demand for cheap wood-fuel. Secondly, because of the weak economic conditions and labour market situation in the countries concerned the wood-fuel sector will be attractive as provider of employment and as an important cash income opportunity for villagers and migrants to peri-urban zones. Thirdly, as long as wood is regarded as a free access good and no policy changes for managing natural resources are implemented, the current cutting and collecting practices will persist. Finally, we may expect that in case of increasing drought as a consequence of climate change, the biomass fuel supply will first grow due to the increasing availability of dead wood. In the long run, though, the demand for biomass fuel will exceed the natural production to a great extent, including in regions where this is not yet the case at the moment. As a result, wood-fuel will become scarcer and consequently more expensive in urban areas where consumers have to pay for it. Urban households will increasingly have to turn to other energy sources.

6. WATER SUPPLY

Deteriorating climatic conditions may contribute to growing water shortages in urban areas, in particular in the dry season. In general, urbanisation contributes to higher water consumption due to better water supply facilities in urban centres compared to services in rural areas. In African cities the consumption is still limited because of a low level of public water coverage in terms of a citywide network. The cities studied fit into this picture, except for Dakar (see table 18.6), where a large majority of the households have an in-home water connection. In Bamako and Ouagadougou, on the other hand, four out of every five households depend on standpipe operators and carters for their water supply.

Each of the cities studied has its particular constraints regarding the availability and the quality of water due to its specific (physical) geographical location. Hydrological conditions are problematic for various reasons. Dakar has problems with salinity. As one of the largest African cities with a coast location, its water table is vulnerable to seawater invasion. Over time existing boreholes in the heart of the city have gone brackish. Ouagadougou, Bamako and Niamey face low yields from underground water sources in their geological crystalline ground layers. For this reason their water producers must make use of surface water resources, which require large infrastructure investments. Dams and reservoirs to supply Ouagadougou and treatment stations for river water in Bamako and Niamey (Collignon & Vézina 2001:8).

Although substantial numbers of urbanites still have no access to clean drinking water, in absolute numbers a growing number of households are connected to the water reticulation system or use water from bore holes. In Bamako, for instance, the number of households with access to piped water or bore hole water rose from 21,840 to 34,797 and 456 to 953 respectively between 1990 and 1996 (ODHD 1998: 81). In 1996, it was estimated that 70 % of the population had access to water from these facilities. For Ouagadougou this percentage was more or less the same.

Table 18.6. Access to drinking water services in Bamako, Dakar and Ouagadougou per water source for household use (percentage of households) 1999.

	Bamako	Dakar	Ouagadougou
In-home connection	17	71	23
Standpipe water	19	14	49
Fetched by household (*fontainiers*)			
Independent providers or traditional sources	64	15	28

Source: Collignon & Vézina (2001): 5

Dakar has the best public water service compared to the other three cities in terms of coverage percentage and as a consequence water consumption is relatively high. In 1999, the water agency SdE served 71% of the households by direct connections and 14% via standpipes. Per capita consumption of the water service customers is estimated to be 76 litres per day (Collignon & Vézina 2001). In Bamako – where only 18 per 1000 residents are supplied by the water agency EdM - the population consumes 54 litres per capita per day, more than twice and a half as much as a villager elsewhere in Mali who consumes only 20 litres a day (ODHD 1998). For Ouagadougou the picture is somewhat different. Although in Burkina Faso the urban population also uses more water than the rural one, the consumption per capita in Ouagadougou decreased substantially from 57 litres daily in 1978 to 39 in 1986 and 26 litres in 1993 (Rouers & Van den Bos 1999). The reason for this drop in consumption was malfunctioning waterworks, which were not able to supply the growing number of affiliates sufficiently. Moreover, n Bamako the water pipe system does not function properly. Here water is lost due to technical problems such as leaks estimated at 45% as opposed to only 20% in Dakar (Collignon & Vézina 2001). Despite these enormous losses, there are as yet no shortages due to scarcity in Bamako.

Ouagadougou suffers from water shortages during the dry season, particularly after a bad rainy season. It is estimated that the available water amount during dry years is a third less than during a year with average rainfall. Due to the decreased per capita consumption the available water amount is still sufficient (Rouers & Van den Bos 1999).

In Dakar, Bamako and Ouagadougou, as in many other large African cities, the water supply network is run by a monopolistic operator, often a public enterprise. There are also other independent water providers. They distribute the piped water by reselling it in various ways. Either they deliver the water directly to households by cart or truck, or they sell it from fixed locations such as standpipes or cisterns. These water providers operate in particular in peripheral urban quarters and in spontaneous settlements, which are not connected to the water network. In general, they provide the poorer members of the urban population which consequently face higher water prices compared to richer families with an own in-home connection, see table 18.7 (Collignon & Vézina

2001). In Ouagadougou the price of water from the independent resellers even fluctuates during the year. During the dry season the water price is many times higher than during the rainy season in particular in periods of severe drought (Campaore 1992 and Jaglin 1995 cited in Rouers & Van den Bos 1999).

Table 18.7. The water supply in Bamako, actors and cost of water.

	Households served	1000 cubic water sold / year	Unit price (FCFA)
City water agency (EdM)	18 000	30 000	55
Independent providers	92 000	2 400	580
- private wells	50 000	100	500
- small network operators	2 000	100	400
- standpipe operators	35 000	2 000	400
- carters	5 000	200	2 500

Source: Collignon & Vézina (2001)

In all four cities studied, small (private) water points provide for the water demand of many families. People use small boreholes pumped by hand (Ouagadougou and Bamako), small private wells (Dakar, Bamako and Ouagadougou) or river water (Bamako and Niamey). Much of this water is not clean enough for human consumption and is used only for washing and bathing. By using this water, the poor urban people can limit the purchases of water and thereby the costs of living.

In the dry season, the demand for water by the urban population has to compete with the demand from the agricultural sector. The water demand of large and small irrigation schemes for food production in the urban surroundings is growing and will continue to do so as the transformation of the agricultural sector in this area continues. Urban market gardeners, even in Bamako which is located in a sub-humid zone, are already experiencing problems with their water supply during the end of the dry season due to the low ground water level (Klompenhouwer & Van Soest 1999).

7. CONCLUSIONS

1. The capital cities of the Sahel countries are primate cities with a large proportion of the urban population within their boundaries (Burkina Faso is more or less an exception with a relatively large second city: Bobo-Dioulasso). Their growth is mainly based on a relatively high natural population increase, immigration from the countryside being of decreasing importance. The growth rate of these cities is diminishing.
2. Mobility is an aspect of the strategy of rural households and individuals designed to help them to survive in periods of stress. This mobility, which is often oriented towards the

cities, may cause people to end up in the very poor population groups in the capital cities, although it may also cause an improvement in status.

3. In the past, urban growth rates were affected by droughts through huge numbers of refugees from the countryside flocking in to the cities looking for food aid. During the last few decades no important droughts have occurred. National governments have taken different measures to cope with food scarcity due to insufficient agricultural production. Examples are the creation of security stocks, the stimulation of cereal banks at village level, the liberalisation of cereal markets etc. However, future droughts as a consequence of climate change may stimulate urbanisation and may also cause a population shift to more southern regions inside and outside the countries concerned.
4. The food needed for the population of these growing cities comes from all over the national territory. However, production in the immediate surroundings of some of these cities has been influenced by the huge, concentrated and stable demand: new, more capital intensive production forms are developing, directed to urban consumption patterns. Agricultural production in the surroundings of cities and in the cities themselves has relatively become more important.
5. With climate change and increasing drought the food supply of the cities will come under pressure and this will be reflected in the first place by increasing prices. The urban poor will be the first victims.
6. The fuel consumption in the capital cities is largely based on biomass production in the surrounding rural areas. In these areas, woodcutters and collectors have, in practice, free access to the natural vegetation. The demand will determine the radius around the cities within which cutting and collecting is a remunerative activity. Climate change may extend this radius or intensify cutting in nearby areas contributing to further deforestation.
7. The water supply of the capital cities in the Sahel is already problematic. It is not the availability of water that presents the problems for the moment, but the lack of organisation and investment in waterworks and drainage systems. Large parts of the urban population do not have access to safe water and have to use wells, often contaminated, or have to buy drinking water, which is relatively expensive.
8. Climate change will restrict the availability of surface water and will lower the underground water level, thereby raising water costs and limiting water use by urbanites. The lower

ground water level will have a negative effect on vegetable growing in the cities.

9. The urban poor form the most vulnerable group of urban citizens during periods of environmental deterioration. First of all, their intake of food is insufficient in terms of quantity and quality and the money they spend on food often accounts for the bulk of their meagre cash income. This group in particular will suffer in the event that food prices rise during periods of severe crop failures in rural areas. Secondly, the poor have less access to safe drinking water because they lack provisions like piped water or bore wells. They have to rely on surface water, which is often polluted, or they have to buy their water from commercial vendors at prices far in excess of those of piped water. Thirdly, if fuel costs rise due to increasing scarcity, the poor will be unable to afford this energy source and they will have to switch to more uncooked meals.

In the face of changing climate conditions, the capital cities in the Sahel will try to keep their privileged situation with respect to the countryside and smaller towns. Tensions with respect to the use of land and water will increase. One wonders, for example, whether the scarce water will be used for urban consumption and urban electricity generation or be used instead in irrigation schemes. Much depends on political choices and power arrangements. Anyway, urban purchasing power will decrease even further and it will be impossible for large parts of the population to keep up their actual consumption levels (of food, fuel and water). As a result, urban poverty will increase.

Chapter 19

CLIMATE DETERIORATION AND REGIONAL SPECIFICITY

Regionally differentiated portfolios of options for drylands

Ton Dietz

Abstract After the regional case studies this chapter provides an overview of the main findings and a comparison of portfolios and adaptation strategies that Sahelian people have used to cope with climate change and variability. It then puts the Sahelian experience in a wider perspective of (sub-)tropical drylands as a whole.

1. SUMMARY OF FINDINGS

In the first phase of the Dutch National Research Programme on Global Air Pollution and Climate Change (NOP-I) the emphasis was on global modelling, on a natural science-based approach, with more detailed studies focusing on the Netherlands/Western Europe. In 1995, it was decided that in NOP-II (1996-2001) more attention should be devoted to the impacts of climate change not only on natural systems but also on societies. This meant including a social science perspective and addressing problems in developing countries. This project (NOP-ICCD) was one of the results of this change of focus. It brought together groups of scientists from an agro-biological and climate modelling background with those from an anthropological, geographical and economic background. It also linked existing research efforts and expertise from a Wageningen-based consortium of natural scientists and economists ('DLV'), RIVM-Bilthoven and the national research school for resource studies and development CERES (for this project adding geographical and anthropological expertise). The research team also succeeded in involving researchers from Burkina Faso, Ghana, Mali, Niger and Senegal, partly financed by funds from the Dutch participating universities. The research team also succeeded in involving not only senior scientists but quite a number of postdocs and PhD and M.A/M.Sc.

A.J. Dietz et al. (eds.),
The Impact of Climate Change on Drylands: With a Focus on West-Africa, 323–339.

students as well, both from the Netherlands and from Africa. NOP-ICCD has worked as a catalyst for the initiation of a number of new (and still on-going) projects. We knew in advance that it would become a complex, rather pioneering effort. Looking back, we can say that we have not yet fully succeeded in bridging the large disciplinary and location-specific gaps in approaches and styles of doing research and of reporting. However, some bridges have been built and they seem strong enough to use for further attempts to cross boundaries. We feel that the research field that we have tried to cover deserves greater scientific energy and funding investments.

The research effort started with a geographical inventory of all tropical and sub-tropical drylands in order to get an idea of the diversity in aridity, land degradation, population densities and the urbanisation of the world's drylands and to put the drylands of West Africa in perspective. It also guided the choice of in-depth study regions within West Africa. An analysis was made for West African drylands as a whole and for case study regions of the climatological evidence for changes, looking at the 1960-2000 data and comparing that with 1930-1960. What was evident from these data was:

- a rather drastic deterioration of the rainfall/evapo-transpiration situation between 1930-60 and 1960-1990 and an overall southward move of the boundaries between humidity and sub-humidity, sub-humidity and semi-aridity and semi-aridity and aridity. Everywhere in dryland West Africa the risks of agro-climatological droughts had increased; the start of the monsoon season had become less predictable with the period generally starting later as was the case with most of the other drylands in the world;
- large fluctuations in rainfall were observed, including a number of years with extreme risks of agro-climatological droughts;
- an improvement of the rainfall/evapo-transpiration situation after about 1990, but not to a level of the years with relatively abundant rainfall in the 1950s;
- at national levels a rather high, but by no means complete, correlation between average annual rainfall situations and average annual production figures for grains;
- at local levels (and in the sub-humid zones in particular - Northern Ghana was studied in depth) only very minor and sometimes even a negative correlation between rainfall and crop yields (and for all regions a lot of doubts about the reliability of both rainfall and crop yield data and about the usefulness of the 'agro-climatological drought risk indicator' that was developed for this research).

The scenario analysis, using different global circulation models, shows a wide variety of outcomes for 2050, but with rather strong suggestions that most of dryland West Africa is expected to become a lot dryer (due to higher temperatures and lower rainfall levels). The consequences of these projections are an increase in high-risk environments for agriculture and a further southward shift of the arid and semi-arid zones. Changes in rainfall distribution could mean additional stress on agricultural production in these areas. Simulation studies clearly reveal a shift in the onset of the growing season and lower yield levels.

From a global perspective, West Africa is probably one of the exceptions, with most of the world's tropical and sub-tropical drylands becoming wetter (increasing temperatures more than compensated by increasing rainfall). Hence, extrapolation of the rather unique situation of sub-Saharan West Africa to other drylands may not be a very useful exercise. In addition, the driving forces of Africa's agricultural production performance are so much more influenced by a very problematic macro-economic and macro-political context, and by failure to catch up with the positive side of globalisation, that this is another reason to be very cautious when extrapolating findings for Africa to the world's drylands as a whole.

To understand farmers' behaviour in West African drylands in preparing for ('insuring') dryer conditions and agro-climatological droughts, in coping with droughts and adverse production conditions and in adapting to changed conditions afterwards, it is wise to look at their performance before, during and after drought years in the past. In the case of West African drylands, there were really dry years in the early 1970s and early 1980s and those experiences live on in people's memories. However, just before the field research started farmers in many areas had experienced rather poor years (1996-1997) which could also provide some insights into their behaviour. The method that was developed to study people's 'pathways' and relate those to life cycles and external shocks (droughts, but also economic and political turmoil) looks promising. It also proved very valuable to assess people's behaviour based on the realisation that they use a combination of options to secure a living, many of which are unrelated to agriculture. This was studied with an eye for contextual differences at village/area level (as in Southern Mali, the Koutiala-Sikasso area), with an eye for differences between aridity levels (as in comparisons between the semi-arid north - Douentza in Mali and Kaya and Gorom-Gorom in Burkina Faso - and the sub-humid south - as in Koutiala and Northern Ghana -), with an eye for the growing importance of major urban centres (as research around Bamako and Ouagadougou shows) and with an eye for differences in endowments at household level and resulting classifications of vulnerability (as shown by household and plot-level research in Burkina Faso). However, a major problem of doing research on long-term changes in regions such as West Africa is the lack of reliable statistics and the lack of long-term panel data, despite decades of foreign-funded regional development projects and programmes. Our original optimism of being able to use some long-term datasets almost always turned out to be a deception. Any attempt to get a better grip on changes in order to arrive at a better understanding of farmer's/people's behaviour needs to be supported by genuine attempts to build up monitoring data in a variety of chosen areas and with enough attention for regional and social detail. Reconstructing climate, agricultural and socio-economic histories became a puzzle and it was rather difficult to interpret the multitude of existing studies from our perspective.

However, some tentative conclusions can be formulated which are not as negative as the commonly used 'picture of doom' for Africa. West Africa's shock experience in the 1970s and 1980s did result in the region being much better prepared for possible new drought shocks and in improvements in its agricultural production performance in the 1990s (when rainfall became considerably better):

- there was a major increase in the importance of non-agricultural sources of livelihood, both locally and (mainly) as remittances from places of work elsewhere;
- there was also a major increase in urbanisation and in the urban demand for food, water and energy; this improved the options for farmers in the surroundings of (big) cities and those for farmers much further away (charcoal and non-perishable food came from areas hundreds of kilometres away);
- in the countryside there was a significant increase in the importance of cash income from higher-rewarding elements of agricultural production: cotton, vegetables, livestock, milk; and of charcoal;
- in a number of places (*e.g.* most of Burkina Faso) there has been a major investment in soil and water conservation, which not only slowed down erosion but also improved the water retention capacity of the soils;
- in a number of places there were considerable investments in irrigation, and a movement to the hitherto hardly-used riverine and marshy areas (partly enabled by the adoption of oxen ploughs, and of ploughs drawn by other animals);
- all over dryland West Africa, but in the north even more than in the south, there was a strategic attempt by many farmers to develop a multi-locational, and multi-sectoral household economy, both in agriculture and outside agriculture;
- in some areas livestock became increasingly important as a drought-risk buffer as did the grains-for-livestock trade, which made use of (very) positive caloric terms of trade for farmers who sell livestock; in areas with high population densities livestock management became more intensive,

manure management improved and animals were integrated better into crop agriculture;

- old socio-cultural dichotomies in livelihoods (*e.g.* between the agricultural ethnic groups like Bambara and Mossi, and the livestock groups, like the Fulbe) are gradually disappearing, but cultural identities and identification processes are still important and still have political significance at national and local levels; the decentralisation processes of government power and the growing importance of non-governmental agencies can be expected to have a major impact on the culturally diverse ways of coping with adverse situations and on access to natural resources and to livelihood options in general.

Our rather positive findings may well be based on the rather positive rainfall situation of the 1990s and on the lessons that people have learned from the droughts in the 1970s and 1980s (individually and collectively and based on assistance from foreign donor agencies which have played quite important roles in West Africa). If the 1990s prove to be a temporary upswing in a climate cycle that is basically tilted downwards (as expected), government and donors are advised to be more cautious and to prepare for more difficult periods to come. Research efforts are needed to support this caution.

If indeed there is going to be a decreasing rainfall trend in most of the Sahel, the livelihood system might collapse under stress. A lot will depend on the severity of relatively bad years and the lengths of the recovery period in-between. A lot will also depend on the resilience of local social security systems, in combination with external assistance. Although the continuation of foreign development aid will remain important in a considerable part of the Sahelian area, the financial support coming from the coastal economy will probably be more important. This partly depends on the political willingness of governments to use coastal wealth (e.g. oil in Nigeria) to invest in the north, but also on the continuation of massive private streams of remittances from the coastal areas to the Sahelian source areas of migration. Remittance investments in the Sahelian urban areas might trigger a further development of circum-urban agricultural, water and energy resources. With adequate land and water policies this might be feasible in ways which are ecologically and socially sustainable. A gradual depopulation of the most marginal areas will then be (temporarily?) compensated by concentrations of people in and around urban zones of expansion. The future for the Sahel is not necessarily gloomy. However, system breakdown can occur during droughts. One fear might be that in those situations religion will be used as a major catalyst for political support to exclusionist claims (Islam versus Christianity and religious sub-groups versus sub-groups) and may result in massive violence and rapid deterioration of all local livelihood options, resulting in a large death toll and mass migration southward and overseas.

2. A COMPARISON OF PORTFOLIOS AND ADAPTATION TO CLIMATE CHANGE IN THE CASE STUDY AREAS

Five rural and two urban case study areas have been studied in depth, as part of the ICCD project, embedded in a more general analysis of primary and secondary data, for Mali, Burkina Faso, northern Ghana and - for the urban chapter - Senegal and Niger.

The rural study of Kaya, and the urban study of Ouagadougou were chosen as examples of the drier (semi-arid) parts of the Sahel, with relatively high population densities and relatively high levels of land degradation. The rural areas of Douentza in Mali and of North-eastern Burkina Faso were chosen as examples of semi-arid fringe areas with - on average - relatively low population densities, but high levels of land degradation and a precarious situation for farmers who have to rely on crop cultivation.

The rural study area of Koutiala-Sikasso in the southern part of Mali was chosen as an example of a sub-humid area, with a relatively low population density and relatively low levels of land degradation, but with a rapid expansion of cotton production over the last thirty years. The urban area of Bamako was chosen as an example of a rapidly expanding city in the sub-humid zone, but which is surrounded by a rural area with relatively low population density and low levels of land degradation. On the other hand, the rural area of North-eastern Ghana was selected as an example of a sub-humid area with a high population density and a relatively high level of land degradation.

In all study areas the impact of the periods of droughts in the 1970s and 1980s has been considerable, with long-lasting effects. Especially in the northern part of the study area, the massive death of cattle and other stock, and the collapse of the grain and livestock markets created a famine that resulted in death and a massive south-ward migration of people who had lost most or all their assets, built up during the relatively favourable 1960s. However, the situation improved everywhere after about 1987 and the 1990s can be regarded as a relatively good decade, with strongly improved conditions for crop cultivation and livestock, compared to 1970-1985. If we interpret the long-term climate data in terms of climate change, there is some empirical evidence of rising temperatures (and hence evaporation levels; see chapter 12 about northern Ghana) and we might conclude that long-term rainfall fluctuations are tilted downwards if we compare the 'ups' of the 1950s-60s and of the 1990s with the 'downs' of the 1920s-40s and of the

1970s-80s. However, the recovery of the 1990s is such that the results are not conclusive and, on the other hand, drought conditions prevailed in some years during the 1990s. Farmers' shock experiences of the 1970s and 80s also resulted in many innovations and a much better preparedness for another period of relatively low rainfall. Farmers all over the region perceive the long-term climate change not only in terms of lower average rainfall, but also as fewer rainy days and shorter rainy seasons (in Kaya - see chapter 14 - also as more intense rains, with higher erosive power), as less reliability as regards the start of the rainy season (in northern Ghana farmers say they can no longer rely on the 'traditional signs' - see chapter 12), decreasing water availability in rivers, dams, and pools and lower groundwater levels. In some areas farmers have pointed out the gradual disappearance of characteristic species (like the *Dawadawa* tree in northern Ghana and some perennial grass species in Kaya) and farmers everywhere have adjusted their farming system (and broader: their livelihood system) to the likelihood of problematic periods in the years to come. Despite the fact that the large majority of the inhabitants in the study areas are relatively young, the memories of the disaster years 1970-75 and 1983-87 still form an important framework of reference.

Ecological differentiation

In most study areas, a more detailed geographical analysis shows important differences within the study area and highlights the importance of attention being paid to inter-village and intra-village differences in ecological and socio-economic conditions. During the bad periods, access to lowlands proved to be rather crucial. Especially in the Kaya case study (chapter 14) being a 'local', with established rights of access to preferred areas (valley bottoms, marshy areas), proved to be important. Being a 'stranger' generally does not mean that people do not acquire access to any land, but it does mean that their choices are much more restricted and that they are much more vulnerable. Recently, the decentralisation policy and the growing authority of village-level decision-making agencies, means a formalisation of these rules of inclusion and exclusion, and in some places (e.g. in Koutiala, see chapter 13) this has led to the exclusion of the itinerant population, like the transhumant Fulani. All the case studies make it abundantly clear that generalisations as regards study areas, as a whole, can be very deceptive both geographically and socially. Chapter 16 has shown that even an area that is often depicted as rather homogeneous (the 'cotton zone' of southern Mali) can be differentiated into at least seven completely different 'village types'. Unfortunately, even the existing datasets at village or household level do not give reliable information about long-term change at the various levels of scale, let alone for the area as a whole.

Expansion of the cultivated area and the importance of lowlands

One might well ask what has happened to agriculture during the last thirty years. The areas of land being used for cultivation increased significantly everywhere and the lower areas in each of the village territories have all become intensively exploited. In some of the most densely populated areas (most of Kaya, the Bandiagara area in Mali and parts of north-eastern Ghana, all with more than 100 rural inhabitants per cultivable km^2) a clear 'saturation of available space' has taken place with the complete disappearance of fallow practices. With the exception of the built-up area, sacred grooves and graveyards, all the land is now continuously being used for crop cultivation. Almost everywhere, sorghum and millet dominate the food crops. During and after the drought periods, short-cycle millet and sorghum became more important and there was a shift towards millet in the driest areas (e.g. in Kaya). During the 1990s, sorghum and in the southern areas (like Koutiala, but not so much in north-eastern Ghana) maize became more important again. Wetlands were often transformed into rice growing areas (partly for sale). Many lowlands became accessible for farming after the successful eradication of tsetse fly and other disease vectors in the 1970s and 1980s. Some of these areas became official settlement areas with government projects for recruited farmers (e.g. the Volta areas in south-western Burkina Faso). Other areas became more spontaneous settlement zones (e.g. the Nasia Valley in northern Ghana). Everywhere, the construction of dams and other sources of permanent water (boreholes, rock catchments) resulted in a rapid increase of horticultural production ('market gardening'; tomatoes, onions and in some areas irrigated forage production as well!). Commercial fruit trees were also on the increase and provided a growing opportunity for many farmers to earn cash. Near the big cities, the expansion of commercial fruit and vegetable production has resulted in a major rural enterprise. However, in more isolated areas as well (like Douentza, chapter 13), the production of onions for far-away markets has also increased. Traders from as far away as Niamey and Bamako come to Douentza and traders from Accra come to southern Burkina Faso and northern Ghana to buy horticultural produce. An infrastructure for trade in fruit and vegetables also meant increased employment for local employees in the trade, transport and banking sectors, often located in small rural towns.

Grain yield levels

During the 1990s, average crop yield levels generally increased. In northern Burkina Faso, average sorghum levels improved from 400 to 500 kg/hectare from the 1980s to the 1990s and millet levels from 300 to 400 kg/hectare. In Bandiagara, extensive grain production yields up to 400 kg/ha and more intensively managed fields up to 800 kg/ha. In the southern areas, production levels of 900 kg/ha for sorghum and maize and 800 kg/ha for millet (and 1800 kg/ha for rice) are the norm. During the 1970s and 1980s, farmers were often forced to rely on a few crops, with fast-maturing varieties. During the 1990s, many farmers succeeded in diversifying crop production and putting more of an emphasis on pulses.

Expansion of cotton and other forms of cash crop farming

The expansion of cotton cultivation has long been regarded as a major success story in Southern Mali. The Koutiala study area underwent rapid growth as regards cotton production in the 1970s and 1980s and was the basis for a regional boom economy during the decades of drought. However, in the 1990s, the cotton yields deteriorated, partly as a result of a negative price development between cotton and inputs like fertilisers. In the late 1990s, the slump in world market prices for cotton added to a crisis situation which resulted in the 'cotton strike' in 2000. The future for cotton is quite unclear and long-term climate change might indeed result in a disappearing cotton economy in the current cotton belt of southern Mali. In northern Ghana, the 1990s were characterised by an increasing cotton economy, particularly in the less densely populated areas of Northern Region. However, major parts of the sub-humid area of the Sahel have never developed a cotton-oriented economy. In some of these areas groundnuts rather than cotton were the preferred cash crop (for example in some parts of north-eastern Ghana and in Kaya). In other areas, grain production for home production expanded to include some commercial outlets as well. For instance, in some parts of Douentza, niche areas of successful grain production succeeded in providing the more northern areas around Timbuktu with commercial grain. One of the side effects is that many farmers no longer store surplus food of good years as an insurance for bad years but sell their surplus harvests for money. Often they buy livestock as their store of assets, but a growing number of farmers buy relatively luxurious goods (bicycles, radios), partly as an 'investment' in social relations with their adolescent children. Many farmers hope to convince their children to stay, or stay a bit longer, in order to assist with farming instead of 'going on migration' (see later).

Agricultural innovations

There are all kinds of farming styles and a wide range of opinions on the adoption of agricultural innovations. In general, animal traction has increased (not only using oxen, but also donkeys or camels) and this made the cultivation of lowlands easier. It has also solved some of the problems related to increasing labour shortage in areas with a lot of youth out-migration (e.g. in Kaya). Some agricultural innovations have been adopted by almost everybody (e.g. treatment of seeds). Researchers have differing opinions on other innovations depending on their views or areas of study. Some regard the anti-erosion and soil and water conservation measures as a major success and the yield improvements of the 1990s as being mainly due to the improved water harvesting technology applied by farmers (some of it successfully introduced by government extension agencies or by non-governmental agencies, see chapter 12 and chapter 17). Others admit that soil and water conservation measures have been important in some areas for some farmers, but that its adoption is not very widespread (chapter 14 and chapter 16). Some authors see a major breakthrough of 'organic' agriculture (in some areas taking over from chemical fertilisers which have become too expensive) and a major integration of crop cultivation and livestock keeping. Others regard this as 'patchy' at best and not enough to counter the rapid disappearance of fallow systems (and the change of 'bush farms' into 'compound farms', as in northern Ghana). Some authors point to the adoption of more 'northern' farming styles in more southern areas, mainly by immigrants from these northern areas (e.g. Burkinabé in northern Ghana, or in Ivory Coast), while others remark that these immigrants had to accept the worst pieces of land where they can only farm using methods which make the best of badlands. In Koutiala, millet production by immigrants from the Dogon areas near Bandiagara, or by sedentarising Fulani, makes use of the driest and shallowest soils (see chapter 13). Of course, compared to the situation which caused many authors in the 1970s to be appalled by the rate of erosion and soil degradation in the densely populated parts of the study regions, the stabilisation of the soil conditions is a good result. However, some authors suggest that there has been a reversal of soil degradation, as a result of massive adoption of appropriate measures. Even if this can be proven, it remains a matter of interpretation as to whether it can be regarded as the result of improved farm management, or of generally improved rainfall and overall production conditions in the 1990s (see chapter 17 for an example of a SWC viewpoint).

Kin-based multiple farm enterprises

The most successful farmers in the study areas have often succeeded in setting up multi-locational farming enterprises which include international elements. During the 1990s, Kaya-based farmers tried to acquire (coffee, or cocoa) farms in immigration areas in Ivory Coast, next to their farms in their home area. To interpret these developments it might be better to look at the situation from a perspective of kin-group related corporate property, in which a relatively large group of kin-related people has established a multi-spatial and multi-unit livelihood 'system'. This can be linked to another observation of growing 'atomisation' of individual farm units, whereby the labour involvement in lineage (family) fields dwindles relative to household fields, and particularly individual fields, of each of the men and women who are involved in farming. Localised labour pooling is gradually being replaced by a new type of security pooling, for those who can afford to do so. These more successful entrepreneurs are the ones who succeed in getting a clientele of foster children, of labourers (but often not paid in the form of wages) and of 'friends' who rely on them. It appears to be a very volatile system, with a lot of mobility and with fluctuating 'unit' composition. Those kin-organised farm enterprises, which have invested a lot in their Ivory Coast farms, are now being threatened by the political and economic turmoil in that country, a situation which might develop into a massive crisis for the Malian and Burkina Faso societies. If the home areas are hit by a period of drought again (and they are bound to be), the fallback option of Ivory Coast might have disappeared and the currently rich farmers will be faced with major livelihood threats. The same might happen to the formerly successful cotton farmers in the Koutiala area.

Changes in the livestock economy

There have been rapid changes in the livestock economy as well. During the 1970s and 1980s there was large-scale destruction of cattle wealth and a major movement of herders and their remaining animals to the south. In many areas the livestock-per-capita situation never recovered (e.g. see Chapter 16 for northern Burkina Faso). Shifting caloric terms of trade led to an undermining of the commercial position of herders who had to pay much more for grains. This created a situation which lasted for about fifteen years and which only started to improve after the late 1980s. It also meant a southward shift of ethnic groups, with the Tamachek moving towards Douentza and northern Burkina Faso and the FulBe or Fulani moving towards Koutiala and northern Ghana. Quite a number of former herders had to start some form of crop cultivation, or they became herders for other people's animals. However, there are still groups which can be regarded as specialised pastoralists (see Chapter 15). Major changes took place in the traditional relationship between livestock-keeping groups (like the Fulani) and cultivating groups (like the Minyanka, around Koutiala, or the Dogon, around Bandiagara or the Moose, around Kaya and Ouagadougou). Many farmers started to increase their numbers of animals (partly for traction). Many farmers no longer gave their animals in the custody of the herding groups. Tensions increased between (transhumant) herders and farmers, whose fields had increased partly in wetlands which used to be dry-season grazing areas for herders. Crop residues became important for the farmers-cum-livestock owners and had to be protected from herders. In the most densely populated areas, farmers started to keep animals near their homes and collect, or even buy, stock feed. Many farmers also diversified their livestock and some became successful pig farmers (as in northern Ghana) or commercial chicken and guinea fowl producers. Livestock trade expanded and also gave women an opportunity to expand their economies. Goats and sheep generally became more important than cattle and camels and this also gave women a better stake in the local economy.

Non-agricultural enterprises and services

With the exception of the big towns, industrial production is very limited in the Sahel. Moreover, in smaller towns like Koutiala, or Kaya, let alone Douentza or Bolgatanga, there are very limited industrial employment opportunities. However, the informal, small-scale non-agricultural economy has rapidly expanded during the last three decades. The development was largely related to the expansion of the urban economy in countries like Mali, Ghana, and Burkina Faso, with a strong increase in demand for (fire) wood and charcoal. Supply areas around big cities like Ouagadougou and Bamako (and even more so in the case of Dakar) cover thousands of square kilometres and reach even the most remote parts of our study areas. Some other rural non-industrial jobs can be found in niches of mineral activity, with an important option of small-scale gold digging in some areas of northern Burkina Faso (see Chapters 14 and 15). The commercialisation of the rural economy has also increased the opportunities for small-scale businesses (basket, and hat weaving, cloth production, pottery, butchers, tanners, etc.), while the demand for 'modern' goods like bicycles, or radios, or watches has also created a demand for bicycle repair, radio repair, and watch repair enterprises. In general, there are clear differences between off-farm income opportunities for men and women. In some areas (e.g. Oudalan and Soum in northern Burkina Faso; see Chapter 15) artisanal activities are primary activities for a substantial minority of women. Here and there, tourism is beginning to create a market for a variety of rural artisanal and service products (e.g. attracted by rural architecture, as in the Dogon areas of Mali or in northern Ghana). The gradual increase in local government functions, health care and education (although this is rather limited everywhere compared to other parts of the world) has created opportunities for educated people to find jobs in the service sector. The growth of the non-governmental sector and the support from foreign donor agencies has also resulted in quite an extensive sector of specialised 'development services'. In some areas the growing activities of religious agencies (a variety of Christian churches, Christian break-away groups, and Muslim organisations) meant an increase in job opportunities as well. Chapter 13 portrays interesting individual 'pathways' of successful Muslim scholars in Mali, while church-based evangelists and development workers in northern Ghana, and parts of Burkina Faso have also become important.

It is evident from the various case studies that off-farm and non-agricultural income (excluding migrant remittances, see next section) has become an important element in livelihood portfolios. The household-level data for northern Burkina Faso, as presented in Chapter 16, even shows that around two-thirds of all monetary income came from these revenues, both in the crisis years of 1981-85, and in 1998, followed by livestock sales (a quarter) and by crop sales (10% in the early 1980s, and 16% in 1998).

Labour migration

The growth of a multi-locational economy, with a very mobile (labour) population has become one of the most remarkable facts of the Sahel. In most areas it was started as a colonial attempt to recruit labourers for plantations and military service, or as a way to get tax money. Through (forced) labour it gradually changed to become a fact of life, at least among the cultivating groups (among the pastoralists a 'cultural aversion against migration' was still found among the Fulani of northern Burkina Faso; see Chapter 15). First almost all young boys wanted to 'go on migration', to get away from home, to earn some money and to 'get to know the world'. For a long period this led to considerable tensions between fathers and sons. The droughts of the 1970s and 1980s resulted in a necessity for many to escape from poverty and hunger and resulted in a depopulation of some of the most northern Sahelian areas (see Chapter 8; although some return migration was noticed during the 1990s, e.g. in the area around Gorom-Gorom, see Chapter 15). Gradually migration became a general element of people's livelihood. Not only young men, but also young women and families with children started to participate in long-distance migration. It was no longer a seasonal phenomenon or one that lasted for only a couple of years. During the 1990s, a lot of labour migrants succeeded in acquiring their own farms in the rural areas where they had worked as farm labourers, or they succeeded in starting a business in one of the towns where they had first worked as (casual) labourers. The Sahelian economy has become very much intertwined with the Coastal economy, with some areas becoming dominated by northern migrants. Migration often took the form of relay migration, where certain immigration areas became settlement outposts for northern villages, linked with family and lineage ties. In some areas it would become the root cause of major tensions with the 'autochthonous' population (like in Ivory Coast, or parts of south-western Burkina Faso, or parts of northern Ghana) who, in some areas, became minorities in "their own" land. Coupled with religious differences (Muslim versus Christian) this might undermine the successes of the 'multi-locational lineage business economy', which is one of the most characteristic outcomes of post-drought Sahel. A major forced return migration is not very likely and if it happens it will not result in a stable local economy in the 'home areas'. It is more likely that a further crisis in the immigration areas will result in the growth of overseas' migration, following the example of for instance the Ashanti of Ghana, or the Yoruba of Nigeria. The next round of drought years will probably result in the first waves of massive long-distance out-migration towards more prosperous areas in Africa, or outside Africa.

Changes in social security

The importance of lineage networks (and ethnic allegiances) also results in major social security buffers. The diversification of the Sahelian livelihoods, and its geographical and sectoral spread, functions as a fallback cushion during adverse periods. Widespread cultural practices of fosterage (with important functions for the mother's brother sister's son-relationship; see chapter 14) and life-long economic friendships within and beyond lineage and ethnic linkages do form an ingenious insurance 'system' in the vulnerable economies of the Sahel. We should not overestimate its reach, though. Among the poor, the lack of relevant network partners is often one of the main reasons why they are persistently poor and why, according to some researchers, their poverty has increased (as in the relatively 'rich' area of Koutiala, see chapter 13). Some of those poor people can fall back on alternative systems of social insurance, like the ones provided by the church or Islamic charities or by food aid provisions organised by government, foreign donors, or non-governmental agencies. These agencies sometimes also succeeded in building up local credit and mutual assistance provisions. Some authors hope that the new institutions, which are the result of decentralised government (e.g. the *associations villageoises* in Mali), will also become local insurance agencies, with emergency activities during droughts, preventive functions before droughts and mobilisation and reinvestment functions after droughts. Others regard these village-level agencies as power bases for the village elites and as a return to pre-colonial conditions of servitude, or even slavery, for the poor and the 'strangers'. Occasionally they might succeed in functioning as fallback institutions in cases of ideosyncratic risk. As soon as village-wide calamities strike with a collective risk, they might function as agencies for redistributing assets from the relatively poor and marginal villagers, to the relatively prosperous village elites. Agencies with their power bases elsewhere (central governments, NGOs, foreign donor agencies, religious centres) might be necessary to redistribute wealth the other way around.

3. WHAT ABOUT OTHER DRYLANDS?

It is questionable whether the research findings for dry-land West Africa have any relevance for other drylands parts of the tropics or sub-tropics. The general expectation in the regionalised climate change extrapolations for the year 2050 is that overall, the dry-lands of the tropics and subtropics will become wetter, not dryer, although the increase in temperature will play a moderating role. It is probable that for many dry-land areas in the world the P/ETP situation, and hence aridity, will not worsen. In many of the different climate change extrapolations dry-land West Africa is an exception: here it is expected that the area will become dryer, although we have seen that for the eastern part of dry-land West Africa, the situation is predicted to be far less gloomy than for the western part, and will probably resemble most of the world's tropical dry-lands.

However, if we compare the 1960-1990 aridity situation with the 1930-1960 situation that formed the basis for the UNESCO aridity classification for all dry-land areas in tropics and sub-tropics, the number of dry-land 1°×1° cells which experienced a deterioration of the average rainfall situation (that is from sub-humid to semi-arid or from semi-arid to arid) was far higher than the number of cells which experienced an improvement (from semi-arid to sub-humid, or from sub-humid to humid). The global distribution is shown in Table 19.1 and in Map 19.1.

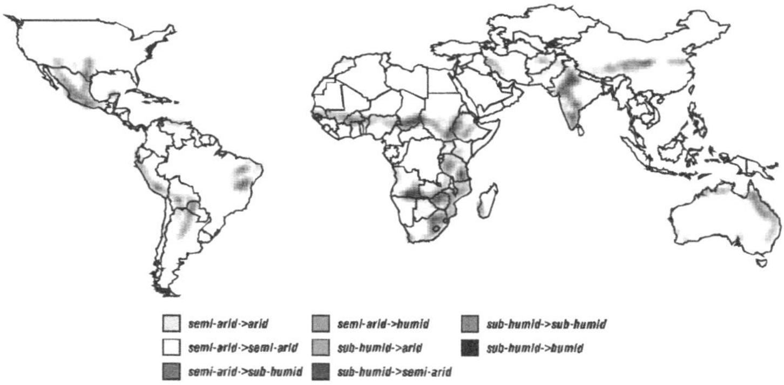

Map 19.1. The world's drylands: changes in aridity classification 1960-'90 compared with UNESCO's classification (ca 1930-'60). source: IIASA (1930-1960 Global Climate Database), CRU/LINK (1960-'90 Global Climatology) *(See colour section, p. 465).*

It is clear from this overview that UNESCO's well-known and much-used aridity classification was made using rainfall data which give a rather 'rosy' picture of the world's recent aridity situation. Out of 1844 dryland 'cells' in this comparison (semi-arid and sub-humid combined) 542 cells, or 29%, experienced a change to a dryer class and only 125, or 7%, a change to a more humid class.

West Africa belongs to the areas which only experienced deterioration at levels which were almost the same as the average for the world's drylands. However, West Asia and the Indian sub-continent experienced a much more severe deterioration in relative terms. The drylands in China are a big exception. They experienced a major improvement.

West Africa's dryland experiences (a negative trend, droughts, a more variable and less predicable climate) in the years between 1960 and 1990 were shared with many of the world's drylands. The severity of its impact (in particular during the drought of the 1970s) was a result of West Africa's relatively weak economic structure and its reliance on local subsistence-oriented agriculture and on a few world-market crops (cotton and groundnuts) of which the relative terms of trade deteriorated. Both the transport network and the trade structure were weak in West Africa. There were very few opportunities to revert to other, non-agricultural livelihood resources. In addition, the state structure and performance was rather dismal. Added to these negative characteristics is the fact that West Africa's drylands have a uni-modal dryland climate (with only one major harvest in a year; unlike the bi-modal climate situations in *e.g.* most of dryland India and parts of dryland Eastern and Southern Africa). West Africa's irrigation infrastructure was also very weakly developed, with the exception of the areas next to the Niger River in Mali.

Table 19.1. Changes in aridity classification: 1960-1990 compared to UNESCO's (ca 1930-classification, world's drylands, number of 1°'1° cells.

World region	Deterioration	Improvement
USA	14 17%	0 0%
South & Central America	97 24%	32 8%
North Africa	11 25%	0 0%
West Africa	37 27%	0 0%
Southern & Eastern Africa	192 33%	44 8%
West Asia	35 58%	0 0%
Indian subcontinent	85 47%	8 4%
China	0 0%	39 46%
Australia	71 26%	2 1%
Total	542 29%	125 7%

Source: Veldhuizen & Dietz, 1999.

West Africa's shock experience in the 1970s and 1980s did result in it becoming much better prepared for possible new drought shocks and in considerable improvements in its agricultural production performance in the 1990s (when rainfall increased considerably) :

- a major increase in the importance of non-agricultural sources of livelihood, both locally and (mainly) as remittances from places of work elsewhere;
- a major increase in urbanization and in urban demand for food, water and energy;
- in the countryside a major increase in the importance of cash income from higher-rewarding elements of agricultural production: cotton, vegetables, livestock, milk; and of charcoal;
- in a number of places (*e.g.* most of Burkina Faso) a major investment in soil and water conservation, which not only led to a slowing down of erosion but also an improvement in the water retention capacity of the soils;

- in a number of places, investments in irrigation and a movement to the hitherto hardly-used riverine and marshy areas (partly enabled by the adoption of oxen ploughs and of ploughs drawn by other animals);
- a strategic attempt by many farmers to develop a multi-locational, and multi-sectoral household economy, both in agriculture and outside agriculture;
- the increasing importance of livestock as a drought-risk buffer, and of grains-for-livestock trade making use of (very) positive caloric terms of trade for farmers who sell livestock.

Many of the world's drylands have developed in the same ways during the past few decades. This is partly a sign of real change, but also a result of improved measurements and broader research and policy interests. The big difference between most of sub-Saharan Africa's drylands compared to most of the drylands in the rest of the world is the fact that the above-mentioned changes took place in a macro-economic, macro-political, and security situation that was pathetic in most countries and with relative demographic increases higher than anywhere else. Outside sub-Saharan Africa, climate deterioration in the course of the late 20th century took place in a situation of improved economic performance and relatively diminishing demographic growth. This created opportunities for net out-migration from drylands to more humid areas and it created rather fast changes in the rural-urban and agricultural-non-agricultural balances in a situation of overall improvement of income and wealth levels. This enabled higher investments in social and physical capital, partly through the government but increasingly through private initiatives, both by commercial enterprises and by non-governmental institutions. In sub-Saharan Africa net out-migration only occurred in the arid areas, not in the semi-arid and sub-humid drylands. In sub-Saharan Africa the drastic changes in the rural-urban and in the occupational balances took place in a situation of overall economic deterioration and in a situation of relative weakening of Africa's participation in the world market.

Chapter 20

THE IMPORTANCE OF SOCIO-CULTURAL DIFFERENCES AND OF PATHWAY ANALYSIS FOR UNDERSTANDING LOCAL ACTORS' RESPONSES

Mirjam de Bruijn and Han van Dijk

Abstract: The enormous diversity of responses to the drought conditions in the last thirty years makes it difficult to formulate general conclusions about people's responses to climate change. It is important to study the pathways of decision-making units at the micro-level and even at individual level and to emphasize the socio-economic differences in changing patterns of responses and the gradual changes in people's 'habitus'. To understand the options available to people it is wise to focus on the technological changes in land use, the changes in the control over resources, migration and mobility, the trends of livelihood diversification and institutional change

1. CLIMATE VARIABILITY

The climate in the Sahel with its low and erratic rainfall is the main problem farmers and livestock keepers have to deal with. The amount of precipitation, rather than its distribution in time and space, is the main structuring factor for decisions with respect to the use and management of natural resources and the allocation of labour (Mortimore & Adams 1999).

Risks resulting from climate variability have an impact that goes far beyond the domain of agricultural production alone. Given the enormous fluctuations in agricultural production, market prices react sharply to rainfall variability and even magnify its effects (Swift 1986, Hesse 1987, Davies 1996). However, climate variability is not the only important factor in decision-making. A whole range of exogenous factors (international markets, international and national policies related to

A.J. Dietz et al. (eds.),
The Impact of Climate Change on Drylands: With a Focus on West-Africa, 341–362.

agricultural development, laws) needs to be addressed to analyse local-level strategies aimed at dealing with climate change. Obviously, these high risks have an impact on the way in which people organise their lives. It has been shown that in the Sahel livelihood strategies, laws and institutions, moral codes, social security mechanisms, rituals and understandings of their environment have emerged out of the interaction between local actors and their environment while handling these high risks. Global circulation models predict greater aridity in parts of the Sahel, implying that the variability and unpredictability in the timing and spacing of rainfall will also increase (Dietz *et al.* 2001, Van den Born *et al.* 2000). This will make it even more difficult for local actors than at present.

The high risks at stake compel people to adjust constantly to variable conditions of all sorts and to preserve a large degree of flexibility. The strategies that local people have developed over the years are the result of their interaction with climatic conditions and other contingent factors and the sequential adjustments they have made. The key to an understanding of the strategies people develop under conditions of climate variability is to focus on the daily decisions they take to mitigate climate variability and a host of other factors to ensure their subsistence and survival. This approach also allows a better appreciation of the role of individual differences in resource endowments and in social and cultural backgrounds in moulding the distinct paths actors seem to follow under high-risk conditions.

An important part of the research effort within the ICCD project was thus devoted to an investigation of the role of socio-cultural differences for the understanding of the evolution of specific pathways followed by local actors to mitigate climate variability and climate change. After a short introduction to the ways in which the research was executed (Section Two), we focus on the methodological aspects (Section Three) and the results of the analysis of local-level responses based on field research conducted within the framework of the ICCD project and earlier research by the authors and others (Section Four). This is done in relation to contextual factors and climate variability. We also examine the significance of socio-cultural differences in explaining the reasons for choosing specific responses and their relation to climate change scenarios and modelling exercises (Section Five).

2. INVESTIGATING DIFFERENCES

2.1 Methodologies for Investigating Differences

Broadly speaking there are two strategies for investigating socio-cultural differences in relation to actors' responses to climate variability and climate change. The first involves formulating hypotheses and investigating these by means of a statistical analysis of a broad range of qualitative and quantitative data in relation to each other, to see whether these cluster and whether they indicate specific responses. The second method is more inductive and is based on focused fieldwork leading to the description of responses and an in-depth analysis of these responses compared with a broad range of contextual factors. Both methods have specific advantages and disadvantages as is shown by Chapters 9 and 16, which all lean more towards the first strategy and Chapters 12-15, which are more typical of the second strategy. The main bottlenecks of the first strategy are the selection process of the data and the relevant scale level of analysis for decision-making units and the reliability of the data used. The data gathered under the second strategy are in general better validated but a major drawback is the specificity of the data, which renders generalisation of the conclusions of analysis more difficult

These difficulties are greater in a situation of uncertainty of some of the most basic contextual parameters such as rainfall, market prices, and the political situation. The flexible responses, which are consequently developed by local actors and other decision-making units, further complicate research. One way in which many actors respond is to look for better opportunities to earn a living elsewhere. This means in the first place that, for a statistical analysis, the farm or village level is insufficient and that data at a regional (Chapter 9 and to a lesser extent Chapter 16) and even sub-regional level are needed. The difficulty is to relate this back to the socio-cultural characteristics of the actors involved.

To investigate the role of socio-cultural differences in explaining differential responses to climate variability, the following general research domains can be formulated:

- a description of the major responses developed by local actors to mitigate climate variability and climate change across different agro-ecological regions, focusing on the relationship between rainfall variability, crop and technology choice, on/off-farm resource allocation, and social and cultural factors;
- a description of current trends in order to review potential response reactions (e.g. conditional chances) of different categories of local actors and higher level decision-making units to selected scenarios of climate change;
- a construction of a household typology/stratification based on the analysis of local actors' responses which integrate both quantitative and qualitative dimensions and the integration of this typology into up-to-date simulation models; and
- the creation of a generic methodology, applicable to West Africa, aimed at producing insights for policy makers on the

basis of an analysis of responses by local actors under conditions of climate variability and climate change.

The first two domains were covered by field research and bibliographic surveys focusing on the description of a variety of responses. For these domains, the results of past field research were used (see De Bruijn & Van Dijk 1995, Breusers 1999). In addition, new field research was carried out in relation to a number of selected topics (see below).

The third domain proved the most problematic because it was difficult to come up with sufficient quantitative indicators for a household typology. The range of these indicators within selected clusters of respondents was vast. This was caused by extreme variations in climate, soil properties and other indicators (or a combination thereof) from year to year and even within one year as the result of an intra-seasonal variation of rainfall in time and space in one case-study area. These variations interfered with clustering into specific types. In Koutiala, more quantitative data was available but was concentrated on just one group of actors. Longitudinal data on other groups was not available, so only partial analyses could be made.

The fourth domain was aimed at creating a methodology able to deal with the complexity of the process of construction of responses. This methodology had to differentiate between climate and other physical factors on the one hand and institutional, economic and socio-cultural factors on the other. Given the variability of the main factor to be analysed (climate), some of the other factors involved (markets, political situation), and the complex interaction between the relevant institutional factors, a decision was taken to focus the methodological exercise on decision-making as a process and on individual actors and larger decision-making units. The reason for this was that all the relevant factors to be taken into account converge in the decision-making process.

2.2 Research Activities

Initially the following research activities were planned:

a) an analysis of the quantity and quality of farm household resources and their fluctuations over time (land, water, labour relations and allocation, the households' relations with the institutional and market environment and social and cultural factors);

b) an in-depth analysis of different farm household risk-taking and/or avoidance strategies (implicit time horizon, reference position, relative importance attached to different farm household objectives e.g. food security, reproduction, accumulation, maintenance of resource base, etc.) and the bio-physical, socio-economic, cultural and institutional variables that contribute to the selections of these strategies; and

c) a review of the potential impact of specific scenarios of climate change on the adjustment capacity of resource use and food security.

Field research was envisaged in three areas of the West African Sahel: Kaya in Burkina Faso, and Koutiala and Douentza both in Mali. Later on, Bolgatanga in northern Ghana and Gorom-Gorom in northern Burkina Faso were also added. Reports on the results of these case studies can be found in Chapters 12-15. In this chapter we mainly discuss the contents of Chapter 13 and 14 and consider the results of the case studies in Douentza, Koutiala and Kaya.

Bibliographic Surveys

An analysis of secondary data was made for these three research locations. This analysis was carried out by Mark Breusers for Kaya, and by Han van Dijk for Douentza, both with a broad disciplinary outlook. For Koutiala, Mirjam de Bruijn focused on the anthropological and economic literature.

The main constraint was the availability of literature. For the Douentza sub-region there is little quantitative information on the technical, agro-economic aspects of land use in the region. Furthermore, the northern and western part of the region is quite well described but there is hardly any information on the southern part. In the case of Kaya, the bibliographical survey lacks information on pastoral strategies. Past research in this area focused mainly on the Mossi, who are described in the literature as sedentary agriculturalists, and their survival strategies. There is no data on the mobile and pastoral aspects of the economy and land-use systems in the Koutiala region. In general, anthropological literature on this area concentrates on the sedentary Minyanka (Jonckers 1987, 1995) and is difficult to access.

Field Research

In collaboration with various university departments (the Department of Agronomy and the Department of Tropical Nature Conservation at Wageningen Agricultural University; the Department of Cultural Anthropology at Utrecht University and the Amsterdam Research Centre for Global Issues and Development Studies at the University of Amsterdam), several field research activities were organised. The following projects were carried out:

- land-use strategies by sedentary farmers in Koutiala (2) and Douentza (3) (Nijenhuis 1998, Nikiéma 1999, Maas 2001, Brandts 2002, Griep 2001);
- mobility strategies of semi-nomadic pastoralists around Koutiala (1) (Van Steenbrugge 2001);
- the dynamics of a small rural centre in Douentza (1) (Zondag 2001);
- informal trade systems of cereals in Douentza (1) (Rutgers van der Loeff 2001); and
- vegetation dynamics in Douentza (1) (De Boer 2001).

A PhD project on the dynamics of entitlements to fallow land in Douentza and Koutiala started in 1999 and is still continuing. This project has resulted in several papers (Nijenhuis 2002, forthcoming).

3. PATHWAYS AND HABITUS: TWO CONCEPTS FOR THE ANALYSIS OF DECISION-MAKING IN HIGH-RISK ENVIRONMENTS

3.1 Pathways

The methodology for the analysis of decision-making in high–risk conditions was geared towards analysing the dynamic interaction between individuals and groups and their environment. This methodology, labelled ‘pathways’, focuses on decision-making. Pathways can be conceptualised as the strategies arising out of the decisions actors, households and groups take in order to deal with all kind of risks in an unstable environment. A pathway is different from a strategy because a pathway is not designed to attain a pre-set goal after a process of conscious and rational weighing-up of the actor’s preferences. Instead it arises out of an iterative process in which goals, preferences, resources and means are constantly reassessed in view of new (unstable) conditions with which the decision-maker is confronted. In this process a wide range of past experiences are at the basis of decisions rather than a sharp vision of the future, while these recollections of the past depend to a great extent on the intellectual concerns of present recollections (Ortiz 1980: 80). Knowledge of these unstable conditions and how to deal with them is gathered in an incremental learning process.

The following assumptions underlie the concept of pathways:

- the environment of decision-makers is inherently unstable;
- decision-makers proceed on a step-by-step basis in a high-risk environment and decision-making is an iterative process with the resulting pathway not necessarily having an intrinsically planned or rational character or following a logical order;
- past decisions have to be taken into account because they influence the pathways and the condition of the decision-maker and his/her mental state in the present;
- decisions are made within a specific context by decision-makers with a specific history and a variation in decisions therefore need not to be based on the synchronic attributes such as resource endowments but can also arise from life history; and
- decision-makers coordinate their decisions explicitly and implicitly.

This methodology does not aim to create idiosyncratic descriptions of actors' and collectives' decisions but rather to analyse the dynamics underlying these decisions and to pinpoint the opportunities and constraints which cause a specific type of actor or group of actors to be likely to follow specific pathways to mitigate instability. This may result in the formulation of a number of 'rules of the game', which may be fed into formal decision-making models as treated in other chapters.

In the study of pathways, special attention is given to moments in their evolution when the environment is marked by crisis or a situation of shortage. It is during these periods that we find extremes and an accelerated pace of change in the decision-making strategies of the actors. The options open to actors during these times vary according to the actor but still we expect regularity in options and in decision-making strategies to be detected which relate to the typology of actors according to the varying constellations of their capitals.

A typology of pathways can only be made for the level of the individual or household. Cultural capital appears to have an important influence on the way decisions are taken. People refer to who they are with reference to rules and norms of behaviour. For instance, a cultivator from a Dogon culture will consider different aspects of his habitus when deciding whether or not to use new technologies than a cultivator within a Fulbe cultural environment. The outcome may be the same but in most cases it is not.

Another important element is economic capital which is also used in the linear models developed by economists. It is clear that the possession of tools, money, etc. is very important at a given moment and heavily influences the decision-making process. Another prominent form of capital in the typology is social capital, i.e. social networks, social care relations, labour relations, etc. Knowledge can be an equally important capital. These capitals define the access people have to social and ecological resources and to institutions. They also define perceptions of the environment. A Dogon cultivator with a long history in the region and who sees him/herself as an autochthon will have a different outlook to a cultivator who was once captured as a slave by the Fulbe.

3.2 Habitus

The data requirements for an analysis of pathways are enormous. A strict application of the concept would compel us to assess an actor's perspectives and attributes, amounting to an enormous number of relevant contextual variables, and the social and cultural factors involved at any moment in time. Therefore, we need an intermediate concept to economise on data and one that can be used to link up 'pathways' to contextual factors.

This intermediate concept is labelled 'habitus' to denote the habitual aspects of many of the concepts, devices and perspectives local actors have and use to interact with ecological and other environments. The concept was defined by Bourdieu: 'Systems of durable, transposable dispositions, structured structures predisposed to function as structuring structures, that is as principles which generate and organise practices and representations that can be objectively adapted to their outcome without presupposing a conscious aiming at ends or an express mastery of the operations necessary in order to attain them. Objectively "regulated" and "regular" without being in any way the product of being of obedience to rules, they can be collectively orchestrated without being the action of the organizing action of a conductor' (Bourdieu 1990: 53). 'Habitus' can be defined as the way in which the environment is interpreted and used by local actors. Its properties can be abstracted from the decisions they take in dealing with this environment and the opinions they have about it, discourses and the cultural means they develop for reflecting on these properties. The habitus is the sum of 'cultural understandings' of the environment (Croll & Parkin 1992). The habitus is embedded in the institutional environment, creating and reshaping this institutional environment and through this the physical and social environment. The latter in turn define the availability and the nature of social and natural resources and the possible modes of use.

Habitus denotes both the constructed and objectified nature of the socio-cultural devices people use to interact with the environment and the more permanent character of these devices without implying that they are consciously constructed. Only when confronted with an unexpected situation, e.g. a drought, severe flood or other disaster, will it change form and content. These transformations arise out of the most recent interactions, reflections and people's changing opinions of their environment when they attempt to give meanings to the events that caused these changes and the changes in the natural and social resources perceived and used by them.

For instance, perceptions of the environment change drastically because of recurrent drought. Once seen as the basis for existence, climate variability has become a major concern for people who have lost most of their assets. Though the situation may return to normal, the effects of such events may be permanent. The crisis moulds the relations to which people direct their actions. Rules, norms and ideologies change. With respect to Sahelian droughts, the religious community has become more prominent as a focus of solidarity (Niezen 1990, De Bruijn 1994, 1997) and the role of ritual changes (De Bruijn & Van Dijk 1995). Kinship ties and village solidarity become weaker and people retreat to smaller units such as the 'hearthhold' and individual enterprises (De Bruijn 1997, Van Dijk 1994).

3.3 The Decision-Making Unit: Actors and Groups

The decision-making unit or actor that follows a distinct pathway is a self-reflective agent, taking decisions based on the available information in the face of constraining factors and with the assets at his/her disposal. A relevant decision-making unit can be an individual, a family or a household or any other social unit that acts in a corporate manner.

High-risk environments require specific decision-maker dispositions and organisational set-ups to deal effectively with risk and to ensure continuity. All the environmental domains are defined in relation to decision-making units. They become relevant only because of the decisions taken by these units acting upon one or more attributes, functions or roles of a specific environmental domain. While taking these decisions, the relevant characteristics of an environmental domain have to be defined by the decision-making unit. In turn, the decision-making unit changes as a result of its interactions with the environment and has to be defined as a highly flexible unit. The environment itself, as well as its relevant characteristics, is transformed as a result of its interactions with the decision-making units and processes within this environment itself. In order to explain human behaviour in high-risk environments we need to deal with this two-way interaction.

The survival options of each decision-making unit depend on the possibilities a unit has to interact with its environment. This in turn depends on the degree to which a decision-making unit is able to appropriate and/or incorporate and consequently make use of elements from its environment, which then become 'resources' or 'capitals'. The following types of capital can be distinguished: (i) economic capital: access to technology, tools, crop varieties, equipment, knowledge, labour, cattle, land, water, cash etc.; (ii) social capital: social security networks, family, friends, neighbours, marriage relations, village, lineage, composition of the household; (iii) cultural capital: religion, knowledge, skills, educational level; and (iv) political capital: status, ethnic identity, position in local hierarchies, relations with government and development organisations.

3.4 Related Concepts

These concepts can be used in isolation. However, all kinds of related concepts can be employed to differentiate decision-making units and position them in the various contexts in which they operate. The idea that actors take the various forms of capital at their disposal into consideration in the process of decision-making does not seem problematic. However, in the institutional environment, all kinds of constraints exists for specific types of actors when they interact with their environment. People may not be equal players in the various capital markets such as those for land, credit, employment and political support.

They are ranked in hierarchies that influence the range of choices open to them and they may be excluded from certain types of resources and capitals.

These processes of exclusion (or inclusion) have predominantly socio-cultural and political dimensions. They concern, for example, issues of identity (who belongs to a particular group and is therefore entitled to a specific form of capital and who does not) and status (who is expected to perform a given type of labour because he/she is expected to do so on the basis of a specific status). Actors are not necessarily equally predisposed to taking similar decisions when facing similar conditions. Neither are they equally vulnerable to the impact of climate variability and climate change. They occupy different 'risk positions' (cf. Beck 1992).

4. SOME RESULTS

4.1 The Past in the Present

One of the results of the study indicates that there is a certain congruence between ethnicity and social and political status and strategy because ethnic and social groups have already followed a specific pathway in becoming what they are. They have done so under the pressure of past conditions. For example, under the impact of slave raiding by the Fulbe, the Dogon in the Douentza area are organised around a small strip of land located on the Bandiagara escarpment, whereas the Fulbe organised their existence around the possession and management of livestock on the plains.

Once the political situation changed during the colonial period, this division of resources also changed. The Dogon now occupy more and more land on the plains in their search of fertile cropland and organise new villages around this resource. Their strategy is different from the cultivating Riimaybe who are the former slaves of the Fulbe. They stay where they are since they have never owned the land they cultivated. The noble Fulbe owned that land. As a result they still cannot claim land in Fulbe territory because they do not dispose of cultural capital in the form of genealogies (i.e. they cannot say 'this land belongs to my kinsmen'), for they were part of the family of their masters. Instead these Riimaybe organise themselves around new resources such as aid and development initiatives. A similar development can be observed on the Bandiagara plateau, where the Dogon do not have sufficient land. They are also turning to new resources such as labour migration and aid.

Pathways are at least partly shaped by this historical legacy. Even today Fulbe pastoralists tend to decide more often to follow a pathway, in which livestock keeping plays a prominent role. Likewise Minyanka,

Mossi and Dogon farmers tend to invest in cereal cultivation as their main subsistence strategy, and Riimaybe pathways are more often marked by a more diversified pathway, going for cereal cultivation in combination with the gathering of wild food grains.

These patterns persist when representatives of these groups decide to move to new contexts. A Mossi farmer does not usually suddenly change from cereal cultivation to livestock keeping when moving to the southwest of Mali. He may, however, adapt his choice of crops, add cash crops such as cotton or other cereals like maize and rice, adopt new technology and may possess a herd of cattle tended by a Fulbe herdsmen. However, he will remain a farmer. Likewise, the Fulbe have a strong preference for occupations related to livestock, such as herding or the livestock trade. They only invest in cultivation and land when they have disposed of their herd.

4.2 Technological Change in Land Use

Under the impact of the droughts of the 1970s and 1980s, the partial recovery of rainfall conditions in the 1990s, the integration of the study regions in the world economy and population growth, tremendous changes have taken place in land-use strategies during the period under study (1960-2000). However, in the sub-regions, changes were very different due to the differences in weight of the various contextual factors. In the Kaya sub-region, population growth coupled with poor soil fertility and declining rainfall have driven Mossi farmers to cultivate heavier soils in the valley bottoms, which were hitherto left uncultivated because of the technological difficulties involved. This development has also had an impact on the strategies of Fulbe pastoralists in the area. With increasing rainfall, the Mossi may be expected to return to their former land on the slopes of the hills that dominate the landscape. Mossi interest in livestock increased during the drought years and many of the cattle they own are entrusted to sometimes impoverished Fulbe herdsmen who, in this way, see their liberty of movement further restricted.

In the Douentza area, rainfall is clearly the most limiting factor for agricultural production. Over the past decades, a technological revolution has occurred in response to the increasing variability in rainfall conditions. Due to the droughts, the number of livestock in the area has declined dramatically. As the manure of these livestock was used to restore soil fertility after cultivation, productivity from cereal cropping has also declined. Dogon farmers and Fulbe herdsmen compensated for this decline by investing in ploughs and animal traction, an innovation that has allowed them to cultivate more land with the same amount of labour. Strangely enough, camels are now being used as draught animals in the north of this sub-region.

In the Koutiala area the main driving force has been a state-run cotton development scheme. This agricultural intensification project has managed to encourage cereal farmers to devote one third of their land to commercial cotton growing. This authoritarian agricultural intensification model worked exceptionally well as long as the cotton prices on the world market guaranteed a profit for the Cotton Company and the farmers, and as long as there was enough wasteland to put into production. The latter was essential because, despite the use of inputs, productivity failed to grow at the same pace as production and profits were being made at the expense of soil fertility. The result has been an almost total transition from hoe to plough-based cropping systems. Small and marginal farmers and villages have, however, been ignored. This success story came to an halt at the end of the 1990s when the cotton price dropped, farm-gate prices were reduced and the financial stability of cotton development came under threat. Rampant corruption at all levels of the organisation also played a part in this downturn of fortunes.

For each of the areas, there seem to be critical factors of a political, economic and climatic nature that shape decision-making strategies with respect to land use. In the Douentza area, declining rainfall might become the most critical factor. Cereal cultivation here is at the margins of feasibility. A further deterioration in average rainfall and an increase in its irregularity might drastically reduce the production potential of the region and induce numerous people to opt for something else in the future. A similar reasoning might be envisaged in the growing of cotton in the northern parts of the Koutiala area.

For the Kaya area, the story is somewhat more complicated given its intermediate position and the possibility of substituting one kind of land for another. Both modern and endogenous technological innovations can be observed. However, the evidence of improvements in productivity is contested. The increase in productivity cannot be directly attributed to the changes observed and relatively few investments are being made to improve the performance of local cropping systems. Local farmers and herdsmen do not invest in landed resources but rather in capital that can be moved from one location to another, such as livestock or social capital in the form of access rights to land in a variety of locations (see also below). Other factors such as rainfall patterns, the increase in livestock numbers and the relative importance of infields versus outfields might be more important factors on which the observed increases in productivity are based.

4.3 Control over Resources

Maintaining control over resources is pivotal to the continuity of a decision-making unit and for the pathway it follows. It is clear that units that have preferential access to land in a specific area are more inclined to invest in a land-based strategy such as the cultivation of cereals and cash crops. These units are in general more sedentary than others. Poor households of Dogon are moving to the south (Nijenhuis forthcoming) and individuals from the Kaya area who do not belong to the core lineages of villages and are at the bottom of the pile when land is being distributed often decide to move to other areas where land is still available. In the past, they moved into areas in the Kaya zone which are less densely populated. Nowadays they are moving to the southwest of Burkina Faso and Côte d'Ivoire where population densities are relatively low and agro-ecological conditions are better.

Likewise, control over cattle is an important factor in understanding the development of pathways of livestock-keeping Fulbe families. When they have control over their cattle they are more inclined to engage in small-scale movements in their home areas to look for the best pasture and watering areas and the optimal place to market livestock products (milk). When not involved with livestock they devote themselves to cereal cultivation and tend to live an economically marginal existence in the northern Sahel or a socially marginal existence in the southern areas (the Sudan) where the conditions for cultivation are better. Alternatively, they may end up as wandering paupers, relying on temporary activities such as guarding someone else's herd, wage labour and/or religious services.

The case studies also show the importance of non-material resources and social capital in the form of social relations that enable people to access specific types of labour, support and income. Cultural and political capital, such as religious knowledge or affiliation to specific status groups may have material consequences in the form of access to land and/or access to sources of income in the form of gifts and/or payment for specific services rendered. Affiliation with the founding lineage of a village or belonging to the autochthonous lineages is still an important criterion for gaining access to land. The position of 'strangers' is by definition much weaker. In Douentza, members of the noble groups among the Fulbe remain in firm control of the land whereas the Riimaybe are still extremely dependent on the nobles.

4.4 Population Movements

Mobility is not only a part of daily life and a form of crisis management. It has always been an integral feature of life in unstable West African climatic conditions as the prime strategy for coping with instability (Adepoju 1995). The history of the Mossi, for instance, is marked by the expansion of their kingdoms (often they are still considered aggressive expanders), and by forced labour migration to Côte d'Ivoire. The history of the Dogon is one of expansion in the 20th century with the pacification of the area by the colonial powers. The Fulbe herdsmen have a specific history of movement and are regarded all over West Africa as the people who come from elsewhere, as strangers.

An analysis of these histories and actual patterns of movement have shown that the conclusion Gallais (1975) drew, based on the pre-drought situation, that people in the Sahel and Sudan have to move to earn a living to accommodate climate variability is still valid. The events during the drought years of the 1970s and 1980s only confirmed this, although new patterns of activities have emerged. Current trends in population mobility will certainly become more articulated with increasing climate variability and climate change and this will have far-reaching consequences for policy formulation and recent trends in urbanisation and population growth in more southerly areas.

In all the studies undertaken, mobility emerges as an important aspect of people's pathways. This aspect of life goes way beyond an individual level and may be designated a group strategy. Being mobile has various causes, among which climate change, variations in rainfall and conflict appear as the most important. Mobility refers not only to the population of the areas themselves but also to others migrating to the study areas. These movements have all kinds of consequences in the area from which people originate and in the regions to which they migrate. Migration to a specific place may channel the movements of those who decide to move later. For example, people from the same ethnic background and even from the same micro-unit (village, lineage) tend to cluster in specific places.

The consequences for those who remain behind depend on the characteristics of the people who migrate. For Fulbe women, the elderly and children, the massive outflow of young men, coupled with the failure of these migrants to contribute substantially to the welfare of their dependants who remain behind, is a severe problem (see De Bruijn 1998). Among the Dogon and Mossi, ties with the migrants remain close and their decision-making units may more aptly be called multi-spatial livelihoods (cf. Foeken & Owuor 2001).

The form this mobility takes and the way it is used depend on the personal characteristics underlying a specific pathway and the perception of the environment as comprised in the habitus. For instance herdsmen perceive their environment as food for their animals and behave accordingly. For them, transhumance is an important part of their mobility. And if their usual transhumance routes are no longer accessible – maybe because of drought – they will search for another option. This has led in the Douentza area to a shift of transhumance towards the west (in the Seeno to the west of the Daande Seeno and from the Bandiagara plateau to the Inner Delta of the Niger).

Another important strategy for nomadic people is to flee the regions where they anticipate problems, i.e. conflicts or drought. They simply move to the south. Their perception of the ecological environment is much more as a space to feed their animals than as a space to produce. Cultivators will appropriate and exploit their environment with the idea of settlement in mind and are mobile if mobility serves this objective.

Nevertheless, the Mossi and Dogon also have their rural seasonal migrations. They move to locations in which they expect to obtain better results. This is, however, limited and is related to their political position (dominance over other groups) and to the availability of open space. In the southern research area, Koutiala, it was clear that this way of exploiting the environment had come to an end. There, even the herders (who often came from the north) had no further possibilities to move freely with their cattle, leading them to flee to relatively empty zones in the border region between Mali and Burkina Faso.

Exceptional conditions lead to more articulated forms of population mobility. An on-going development that can be currently observed and which can be attributed to climate change or at least to successive drought years, is the expansion of the Dogon and Mossi towards the south and within their home areas. Coupled with the introduction of the plough and camel or oxen traction, they are able to occupy more land as a consequence. This process started in the northern areas of Douentza and Kaya but the same pattern is now apparent in the southern area of Koutiala.

Mobility also links the different agro-ecological zones in which the various case studies were carried out. It appears that the geographical borders drawn around these areas had nothing to do with social borders or ethnic relations. This linkage between the regions could even be seen as multi-spatial land use and production units. Access to various natural resources in different agro-ecological zones is crucial for the survival of some families and individuals. They are individually mobile between the zones, or they spread different family members over the zones to assure the survival of the family. Mossi, for example, try to maintain rights of access to agricultural land in various climate zones, using an elaborate kinship network. In this way they can play with rainfall conditions and adapt their farming strategies to various phases in their life cycle. Concrete examples of this were studied among Fulbe herdsmen and Dogon cultivators (Steenbrugge 2001, Brandts 2002). One can also find examples of Mossi and Fulbe outside their home area co-habiting with Senoufo and other ethnic groups in the southwest of Burkina Faso.

In their search for land Dogon tend to become more mobile while the Fulbe, in their search for cereals, tend to become less mobile (since their principal source of labour, the Riimaybe, have disappeared). Likewise, the Mossi frequently move their fields and compounds, both within and beyond village boundaries. Mossi are moreover heavily involved in long-distance migration to Côte d'Ivoire where over the last thirty years a shift in their main occupation can be observed from mostly wage labour to more entrepreneurial activities (e.g. share croppers, cocoa or coffee plantation owners). Others from all groups tend to go into non-agricultural activities (wage labour, trade).

Urban migration has received relatively little attention in this study (see Chapter 16 this volume). For Dogon and Mossi cultivators, migration to towns is a seasonal activity, accepted as an alternative way of earning additional income. Fulbe herdsmen may also move to towns. However, for them, the town is a last resort, only to be considered if their main resource (their herd) is depleted.

Finally at the opposite end of the scale is sedentarity. The studies done for this project show that both mobility and sedentarity are a reaction to the same changing circumstances. Technological change may lead to the settlement of herding people who are adopting similar technologies to the Dogon and Mossi. The Riimaybe, the former slaves of the Fulbe, have become less mobile as a reaction to climate changes. They have become active in appropriating development projects propagated by non-governmental organisations (NGOs), which has stabilised village life for them.

4.5 Diversification

Trade

One of the most interesting aspects of the changes over the last 40 years is the enormous increase in the amount of trade and the number of people (partly) dependent on this activity. The town of Douentza has, in particular, changed from a dusty administrative centre into a lively market town. The area has also become an important transit route between cereal-producing areas in the south and the cereal-deficit areas to the north. In the Douentza area this has given rise to a big increase in the number of small millet traders based in the villages. This evolution has been promoted by the liberalisation of the cereal trade, the introduction of donkey carts, the end of political unrest in the north of Mali and the availability of capital from wage labour in town and abroad. Personal characteristics of people engaging in trade are very important, for example whether or not they use their capital to buy a donkey cart or to invest in merchandise, and whether there is someone at home to replace them during periods of prolonged absence.

In Koutiala, trade is certainly booming, though less so than in Douentza. Here, imported goods have acquired a certain importance as people seek to spend the money they have earned from growing cotton. Even more importantly, earnings are being invested in livestock. The cotton-growing zone has become the main area for rearing cattle in Mali. Cattle are a good investment and fit into the farmers' strategies of doing something about declining soil fertility. However, little is known about who manages and owns the cattle presently in the area, and for what purposes their products are used.

In Kaya, trade is less obvious. Incomes are not as high as in Koutiala in Mali and the need to trade is less urgent than in Douentza since large numbers of people have settled in other areas. The ties maintained with these people often serve as the basis for alternative sources of income, or for acquiring access to other resources.

Wage Labour and Services

Though trade is also a poor man or woman's occupation as a coping mechanism, wage labour is one of the principal forms of crisis management. The diversity and forms of wage labour are too numerous to discuss here in full. At the local level, wage labour plays a limited role. Sometimes poor people work for others on the land during the rainy season. However, in general the possibilities for earning a wage in one's home area are limited, especially in Douentza and Kaya. This is not only due to a lack of employment but also because labour shortages are often covered by exchange between individuals and households.

People have been going to towns and to other countries to look for work since the colonial era. The period of time people are away ranges from a couple of weeks to several years without interruption. Earnings from wage labour are mainly used for consumption purposes, for example, to solve the cereal gap after a bad harvest. Direct investments in agricultural equipment are limited to oxen or camel teams and donkey carts, which can also be used for trade.

The kinds of activities and their locations vary enormously between ethnic and status groups but also within these groups depending on gender, age, knowledge and status. Social networks play an important role as they often provide entry to specific activities. A distinction must also be made between people who remain in the countryside and those who move to town. Young Fulbe herdsmen, for example, often take up salaried herding in the south of Mali, northern Côte d'Ivoire or southwest Burkina Faso. They may also settle and become agricultural migrants. Among the Dogon and Mossi this kind of rural-rural migration is quite common, though these groups are not looking for work with livestock but for work on the land. In areas like Koutiala with its cotton industry, the prospects for employment are somewhat better than in other areas. Not surprisingly, many Fulbe and Dogon from Douentza and Bandiagara can be found in Koutiala.

Cultural preferences also play an important role in the choice of labour. The Fulbe avoid manual labour as far as possible and stick to herding and trade. Dogon and lower status individuals from Fulbe society are much less sensitive in this respect. There has been a striking increase in the participation of women in temporary migration, for example as domestic servants for rich people in town and among the Mossi they join their husbands in agricultural enterprises in Côte d'Ivoire.

The delivery of services like healing and counselling is also a way of earning an income. The Fulbe, for example, are regarded as Islamic clerics by non-Islamic groups and are reputed to be potent healers with the help of Koranic texts. The Dogon have a reputation for being skilful herbal healers and are often consulted in their home areas as well as further afield.

4.6 Habitus, Institutional Change

Another question is, of course, what happened to the institutional environment in relation to climate variability and economic policy and political changes at higher levels and whether these institutions have been affected by the evolution of the pathways as sketched above. In the first place such an analysis must distinguish between effects caused by global and national economic and political change and those related to on-site (i.e. regional and local) factors. For reasons of space we limit ourselves here to the habitus as perceived by the local people. Higher order changes have been treated in Breusers (2002), Van Dijk (forthcoming) and Brons *et al.* (Chapter 16 this volume).

A major institutional change that has had a great impact on the construction of pathways concerns social relations, especially relations between generations. Among all the groups studied for the micro-level studies it is noticeable that the younger generations make different use of the opportunities offered by the outside world and have developed a different perspective on their future. For them, their careers are not limited to farming or herding. Instead they orient themselves towards wider social, economic and ecological environments. They clearly perceive the need for outside income and have different ambitions in terms of employment and consumption but also in social relations and their view of the world. Strikingly, education is not seen as a way of realising these ambitions, at least not by most of those originating from the countryside.

The role of patron-client relations is changing. For example, in the past the Dogon were subordinated to the Fulbe but now impoverished Fulbe herdsmen are becoming the clients of Dogon cattle owners and if they migrate to the south, of Minyanka and urban cattle owners. The Riimaybe are increasingly gaining independence as they move to town and acquire independent sources of income, whereas their patrons are becoming increasingly impoverished.

Though no research was conducted in this domain, these changes have certainly been promoted by the amount of information available about the outside world by means of modern communications such as (local) radio, telephone, television, which is penetrating the countryside, and the experiences of returning migrants. Better and cheaper means of transport and improved infrastructure have also helped to open up people's worlds.

Thirdly, the ways in which resources are being appropriated and distributed through society have changed enormously. As a result of droughts, property relations with respect to livestock have changed fundamentally, making it less attractive for young people to remain in the livestock economy. There is increased competition over land and water due to growing scarcity but changes in land use have also promoted these changes as well as the fact that an ever-increasing number of people are settling or establishing production units outside their home areas. Administrative interventions, the decentralisation of administrative power, environmental and land-use planning projects, legal reforms of land tenure and forest management have opened up opportunities for local people to use other means of acquiring access to resources. As a result, tenure systems have hardened and the number of conflicts has increased.

Fourthly, the image of the state has changed fundamentally. From an oppressive all-powerful colonial state and its successor, it has changed into a weak, corrupt bureaucracy. The credibility and legitimacy of the state is declining, a development that has gained momentum with structural adjustment programmes. Administrative decentralisation and democratisation have not repaired the damage because these changes have only confirmed the image of weakness and have primarily promoted the decentralisation and democratisation of corruption, while not (yet) having led to the full participation of the population in politics. Informal political hierarchies have remained largely intact. State influence differs dramatically across the study areas, from a quite active (and in the past activist) state in Kaya to a weak bureaucracy in Douentza, and to the tight control of the para-statal cotton development company (CMDT) in Koutiala.

Fifthly, an important change has occurred in the way that urban areas and small rural centres are perceived. They are naturally the centres for all kinds of commercial activities, which are increasing in number at an amazing speed. At the same time they are the nodes for the exchange of all the kinds of information mentioned above. The dynamics of these centres are perceptible even in the most remote corners of the study areas though the impact on the economy is not yet being felt in the more inaccessible places.

The increase in the number of NGOs in the region is an effect of the semi-arid climate and its variations. The droughts of the 1970s and 1980s attracted a large number of aid agencies (bilateral, multilateral, NGOs) to the Sahel. A strange kind of symbiosis has developed between the development sector and target populations. In some cases the courting of aid agencies has developed into a substantial aspect of a household's income-earning activities, as described in some cases of dam building on the Bandiagara plateau (Van Beek & Peters 1999). In some Riimaybe villages a sequence of projects is providing a substantial part of the income of the population in the form of food for work, salaried jobs and the like, whereas no substantial improvement of income-earning capacity can be observed. Part of the reaction of aid agencies can be related to the specific position of their target populations. The Riimaybe present themselves as former slaves and thus as hard-working people (also in their own cultural definition). The Dogon on the rocky Bandiagara plateau are famous for their artwork and seem to be engaged in a heroic fight to survive in their unfriendly ecosystem as a result of tremendous physical effort. Development agents see them as such and therefore like working with them.

Lastly, perhaps the most important innovation is the way in which social relations are used over distance. As we have seen, mobility is one of the most important characteristics of strategies to accommodate change. This has led to an enormous broadening of the life-world of Sahelian populations. There is hardly any individual who does not have a distantly related kinsman living somewhere beyond his/her immediate surroundings. This geographical expansion of the habitus and the new meaning and content that are given to these kinship relations are extremely important for an understanding of the evolution of pathways. Hardly any research has been carried out into this domain.

4.7 Pathways

Though individual pathways are extremely varied and are almost unique inventions by individuals, families and higher-level social organisations, some general patterns can be detected in the three research areas and a characterisation of pathways can be made. This typology will naturally not be exhaustive and no claim can be made that these pathway types will be the most dominant in the future as the conditions under which they have evolved may change rapidly. Pathways also have a dynamic of their own and as we will see some are dead-end streets and will disappear over time when conditions are (un)favourable. Several activities may be combined simultaneously or sequentially as options open up or disappear.

A pathway evolves over time as a combination of contextual factors, the way in which the social actors perceive these factors (habitus) and the cultural and psychological predispositions and assets owned by the actor. In addition it is clear that there is a considerable degree of chance, arising from variations in climatic and economic conditions. An evaluation of pathways and decision-making by actors and other decision-making units needs, therefore, to be carefully evaluated against the background of all these factors.

A classification of pathways must distinguish first between those that are more and those that are less mobile, though the degree of mobility may change in the course of their evolution. Cultural predispositions may provide part of the explanation for the amount of mobility and the choice of activities. Fulbe herdsmen tend to be more mobile than farming Riimaybe, Dogon and Mossi, though the Kaya case study clearly shows that this distinction is not watertight. Likewise, people with a herding background tend to refrain from manual labour and opt for herding, trade, and services. People from a farming background seem to invest more in access rights to landed resources than people with a herding background. Urban pathways are present in both groups.

Another distinction must be made between winners and losers. One must not forget that the people we observe today are those who have survived, in other words have been able to cope with climate variability, drought, and economic trends. Closer inspection of impoverishment pathways reveals that the difference between failure and success is often minimal. Several cases discussed in De Bruijn & Van Dijk (2001) and Breusers (2002) show this all too clearly, with some ending in extreme poverty, chronic illness sometimes with psychological origins, psychological problems and even an untimely death (Chapter 13 and 14 this volume).

Another important feature of the evolution of pathways is the emergence of what has been labelled multi-spatial livelihoods (Foeken & Owuor 2001). Risks are not only spread through a diversification of activities but also by geographically dispersing the members of decision-making units. Mossi farmers in particular, as well as representatives of other ethnic groups, have created complex livelihoods where the boundaries of the decision-making unit or livelihood are sometimes difficult to draw because people are moving around the various parts of the unit in a very flexible manner. The multi-spatial units formed in this way can be quite large and are based on a variety of social relations, of which kinship remains the most important.

This extension of pathways over multiple locations seems to run counter to another observed trend, namely, the increasing fragmentation of decision-making units. However, this contradiction is only apparent. Fragmentation indeed occurs in most places under the impact of problems related to climate variability and the ensuring of survival. The departure of members of decision-making units causes a lot of this fragmentation. Moreover it seems that risk avoidance puts a premium on investing in small-scale units because of the co-variance of the consequences of climate events (the case of drought in Douentza) and market fluctuations (the case of cotton in Koutiala).

4.8 Conclusion

The studies on decision-making have made it clear that wide-ranging factors have to be taken into consideration when investigating pathways of specific individuals and groups. Studies of decision-making tend to be limited to on-site variables such as land, availability in the household and personal assets. The circles of relevant spaces have to be extended enormously. The variables taken into account by decision-makers cover a far wider range than the household or the village or even the district, By following people from one area to another, a more complete analysis can be made of responses to climate variability and possible reactions to climate change.

Since the research executed within this sub-project was limited to a small number of settings and social groups, it is obvious that not all the underlying patterns have emerged at this stage. More carefully contextualised and focused field research is needed to make up for this lack of data. Economic data will have to be built into these efforts from the start but the integration of socio-cultural and economic approaches can only be successful when non-linear elements are incorporated into simulation models and wider sets of data covering not only on-site conditions can be used.

People make use of a variety of resources and networks to earn an income and to survive calamities. Choices made about undertaking a particular activity are never permanent. People shift regularly from one activity to another in a variety of locations. Pinpointing who chooses which pathways and under which conditions is important for any projection of people's responses to climate variability and climate change. The methodology has demonstrated its ability to make more sensitive analyses of these processes of decision-making. It has to be supplemented by more quantitative assessments of the economic behaviour of people, and more basic and original data gathering and field research. The linkages between the individual level and higher order phenomena at village, regional and national levels still require more attention.

PART C

CONSEQUENCES FOR POLICY

Chapter 21

CLIMATE VARIABILITY, RISK COPING AND AGRARIAN POLICIES IN SUB-SAHARAN AFRICA

Ruerd Ruben, Arie Kuyvenhoven, Keffing Sissoko and Gideon Kruseman

Abstract: In this chapter we use a bio-economic model to assess the influence of market and structural policies on farm household resource allocation in different rainfall conditions. We present the simulated effects on income level, soil nutrient balances, food security and labour use for three types of households with different farm resource endowment. The model simulates production and consumption decisions taking into account weather conditions, off-farm employment opportunities, market functioning, risk motives and saving and investment behaviour. The results show that a changing frequency of dry years hardly affects farm income because major decisions are based on dry-year conditions. Output price policies particularly enhance the income of the more wealthy households but have a negative effect on soil nutrient balances. Input price subsidies bring about a more sustainable land use. Structural policies that reduce transaction costs generate the largest income improvements while, particularly during dry years, poor farmers benefit relatively more than wealthy farmers. Market reform programmes therefore continue to be a relevant instrument for reducing poverty and enhancing food security.

1. INTRODUCTION

Agricultural development in semi-arid regions of Western Africa takes place in a context of extreme variability of rainfall and poor capacity of the soils to retain moisture. Consequently, farmers' livelihoods and the local economy are highly vulnerable to shocks that simultaneously affect production, consumption and exchange conditions (Davies, 1996). Farm households rely on different strategies for coping with these risks, trying to satisfy their consumption requirements through careful management of natural and human resources.

A.J. Dietz et al. (eds.),
The Impact of Climate Change on Drylands: With a Focus on West-Africa, 365–384.

During the last decades, ample attention has been paid to the design and development of appropriate land use systems and technologies that would enable farmers to achieve better risk management in agricultural production (Breman and Sissoko, 1998). In arable cropping, soil conservation measures have been promoted to maintain water balances and control run-off, while mulching of crop residues is used to improve soil organic matter content. For livestock production, the emphasis is primarily on improved pasture management and the production of fodder crops to enhance feeding practices. Mechanisation is sometimes promoted to reduce labour demand for land use activities. Better integration of cropping and livestock activities is an important device for integrated nutrient management. In principle, all these technologies contribute to improving the sustainability of the natural resource base, but the implied high input demands could outweigh potential welfare gains (Kruseman, 2000).

Far less attention has been paid to the role of economic policies in mitigating risk and enhancing farm household adaptive resource management in conditions of rainfall variability. Most studies on exogenous shocks focus on external events like changing commodity prices and their impact on macroeconomic balances (Collier and Gunning, 1996). In analysing the performance of the domestic economy, 'average' conditions are commonly used as a benchmark, even though actual conditions may be strongly diverging.

In this chapter we analyse the effectiveness of various policy instruments as regards influencing farm household resource allocation strategies in different rainfall conditions. Therefore, we assess farmers' supply response to selective economic incentives on input and output markets. Instead of using average conditions, we explicitly account for variability in rainfall as a major factor that influences farm households' resource allocation decisions. These policy simulations are based on an integrated bio-economic modelling approach originally developed to assess the prospects for agricultural intensification in sub-Saharan Africa (Ruben *et al.*, 1998; Kruseman and Bade, 1998, Kruseman, 2000).

Basic data for the simulations are derived from different sources. Information on farmers' resource endowments and their stratification is based on multi-period farming systems surveys conducted in the Southern region of Mali (Brons *et al.*, 1994; CMDT, 1994). Technical coefficients for major cropping and livestock activities are based on agro-ecological simulation studies (Hengsdijk *et al.*, 1996). Farm household consumption behaviour is analysed through a review of cross-section budget surveys (DNSI, 1991). Statistical sources are used for information on input cost and output prices.

Climatic conditions in Southern Mali are typical for semi-arid regions, with an average rainfall of 780 mm/year and a very high potential evapotranspiration.. Rainfall has clearly declined from more than 1,110 mm during the 1950s to less than 800 mm during the early

1980s. Three clearly distinguished types of rainfall regimes can be identified, viz. dry years (average rainfall 694 mm), normal years (855 mm) and wet years (1,129 mm). During the last three decades, 10 % of the time was known to be dry, while wet and normal rainfall conditions occurred in 45% of the period under review.

The findings reveal the existing options for the short-term adjustment of farm household resource allocation to reinforce livelihood security. Attention is primarily paid to two different dimensions of *adaptive livelihoods*, i.e. (i) adjustments in land use systems and technologies (e.g. crop choice, production technologies) and (ii) changes in labour allocation (e.g. engagement in farm and off-farm activities). Model simulations are presented that provide insight into the simultaneous implication of economic policy incentives on production, consumption and natural resource balances.

The remainder of this chapter is structured as follows. In section 2 we discuss economic policy instruments for mitigating climate variability, and their potential impact on farming systems and the household economy. Section 3 provides an explanation of the modelling framework used to assess farmers' responsiveness to policy instruments. Section 4 introduces the characteristics of different farm types and their ability to cope with climate-related risk. Section 5 illustrates adaptive land use strategies with regard to different rainfall expectations. Section 6 discusses the policy framework, paying attention to different types of instruments to enhance a more secure livelihood. In section 7, policy simulations are conducted for selected economic incentives in different rainfall conditions. Section 8 presents conclusions and policy implications.

2. CLIMATE VARIABILITY AND ECONOMIC POLICIES

The analysis of interaction between climate variability and economic development involves a number of consecutive steps. First, the impact of climate on agricultural production has to be considered, making use of available models for water-limited and/or nutrient-limited production. Second, farm household coping strategies should be taken into account, providing insight into adaptive behaviour. Third, aggregate effects at the level of the local economy are included which consider the role of public policies in mitigating drought risks.

Climate variability is defined as the annual variation in the level of precipitation, which may fluctuate around particular trends or cyclical long-run patterns (Hulme, 1992; Climate variability during years is a more subjective concept that refers to the difference between farmers' *expectations* of rainfall compared to *actual* rainfall. This is usually addressed with probabilistic approaches.). This implies that in certain years rainfall supply is above or below specified average conditions. Climate variability becomes critical for agricultural production when moisture availability drops below the optimal level required for biomass growth during different stages of the agricultural cycle, resulting in reduced yields. Climate variability becomes critical in economic terms when a shortfall in water availability is beyond the recurrent seasonal rainfall pattern that farmers consider as benchmark for selecting their cropping system and activity pattern.

Rainfall variability is considered to be one of the major causes of risks that threaten livelihoods in semi-arid regions where farm household income is strongly dependent on rain-fed and labour-intensive agricultural production. Given the stochastic nature and the high co-variance of weather risk, options for local insurance are rather limited (Our analysis does not include the implications of idiosyncratic risk related to individual events - illness, death, marriage, etc - that can be locally insured or managed). Implications of rainfall variability become visible at three different level (Hardaker *et al.,* 1998; World Bank, 2000): (i) changes in yield, (ii) changes in prices, and (iii) adjustments in (disposable) income. The latter effect refers directly to the availability of purchasing power to ensure food security. Changes in yields and prices can partly compensate each other.

Farmers can rely on a particular set of cropping and land use practices to manage yield risk. The usual distinction is between *ex post* coping strategies and *ex ante* adaptive strategies (Davies, 1996). Coping strategies refer to built-in mechanisms that enable rural households to overcome chronic and transitory food shortages (e.g. cropping and planting practices, maintaining stocks, borrowing), whereas adaptation means a permanent change in the activity mix and/or adjustments in the local exchange system. , Our analysis is limited to those coping strategies related to productive choice, resource allocation and market exchange that enable farmers to recover from adverse rainfall conditions.

Risk-coping behaviour comprises a wide variety of procedures that farm households use in their attempt to mitigate shortfalls in income and consumption. Five major adjustment strategies are distinguished (Corbett, 1988; Parry, 1990):

1. changes in land use (for arable cropping: cropping area, crop types and crop location; for livestock: fodder crops and stocking rate);
2. changes in farm operations and resource management (input use, pests and disease control, soil erosion control measures, feeding practices);
3. changes in assets (livestock, food reserves, hire, mortgage or selling of land; savings);
4. changes in labour allocation (seasonal and gender division of labour; off-farm employment, migration, leisure);

5. changes in market relationships (market exchange, local exchange, reciprocal claims, input purchase).

In practice, these strategies tend to be interrelated. Adjustments in land use require changes in resource management and labour allocation and will have repercussions for asset management and exchange relationships. Different types of farmers having unequal access to markets and institutions can only rely on a specific number of options. Wealthy farmers operating within a relatively stable environment are able to use strategies for crop diversification to reduce exposure to risk, while resource-poor households located in a more vulnerable setting can only draw on assets ad reserves in their effort to manage risk and maintain food security.

Economic policies can be a useful instrument to mitigate income risk (Ellis, 1982). Three different types of policy interventions related to rainfall variability can be distinguished. At the agricultural sector level, price policies for inputs and outputs can be used as a device for income stabilisation. Price support is, however, frequently rather pervasive and can lead to undesired distributive effects. At the regional level, public investment in physical or social infrastructure is usually considered to be a suitable device for market stabilisation. This implies that market exchange intensifies and transaction costs for buyers and sellers are reduced. Finally, macroeconomic policies can be used to reduce potential spill-over effects of yield or income shortfalls into the national economy. Exchange rate policies, interest rate regulation and public expenditures are important instruments for maintaining macroeconomic stability. Here we focus our attention on price policies and public investment as major instruments for income stabilisation at farm household level. Benson and Clay (1998) present an analysis on the impact of macroeconomic and structural adjustment policies for drought mitigation in sub-Saharan countries. They conclude that the macroeconomic impact of drought is less in economies that still maintain a high level of self-provision, due to limited spill-over effects. Otherwise, in more complex economies, structural adjustment policies allow a better containment of the economy-wide impact of drought shocks.

Economic policies for managing climate variability focus on regional measures. Early warning systems are considered highly relevant for improving access to information, enabling better targeting of support measures (e.g. promoting drought resistant or early maturing crop varieties; food aid; stock building). Impact assessment is usually based on appraisal of yield and income loss due to drought or excess rainfall (Dinar *et al.,* 1998). Davies (1996) calls for a close linkage between early-warning systems and monitoring of farm household coping strategies to enable a systematic analysis of adaptive behaviour.

More detailed studies on the effectiveness of agrarian policies for enhancing risk management in rainfed agriculture demonstrate considerable scepticism regarding generic policy instruments. Scoones *et al.* (1996) present a lengthy review of common drawbacks and failures in sector and macroeconomic policies for dealing with drought risk in Zimbabwe. Most promising policy alternatives include land titling, local research and extension systems, non-agricultural employment and safety nets. Davies (1998) strongly advocates decentralised policy-making to enable a more flexible response to local conditions. Sub-national contingency plans should include provisions for asset creation, access to financial and labour markets, and relief aid in crisis years. However, there seems to be a general understanding that economic policies that effectively address risk management and livelihood vulnerability should be based on three major pillars:

- Flexible production technologies that enable farmers to adjust land use and the timing of resource allocation;
- Adaptive economic instruments that allow farm households to absorb or even benefit from climate variability (resilience);
- Access to markets and information to reduce the likelihood of adverse shocks and diversify exposure to risk.

Reliance on such a policy framework implies implicit acknowledgement of the fact that farmers exhibit a wide variety of responses to interventions. Diversity in farming styles and operations can be considered the result of differences in farm resource endowments, in the availability of assets, and/or diverging household objectives and risk perception (Steenhuijsen Piters, 1995). Therefore, we should rely on an analytical framework that accounts for multiple actors and diversity in response to policy incentives.

3. ANALYTICAL FRAMEWORK AND MODEL APPROACH

Different approaches can be used to analyse interaction between climate variability and farm household behaviour. Studies that address the technical and economic sensitivity of farming in varying climatic conditions pay attention to changes in yield and net revenue due to adverse weather conditions. Within this tradition, a difference can be made between (i) *production function* approach and (ii) *Ricardian* approach. The production function approach estimates crop yields derived from agro-ecological simulation models, which it incorporates into an economic model of farmer behaviour (Rosenzweig and Parry, 1994; Kaiser *et al.*, 1993; meta-modelling approaches have been developed that allow the direct estimation of production functions based on technical coefficients derived from agro-ecological simulation models; see Ruben and van Ruijven, 2000). The Ricardian methodology is based on econometric procedures for the analysis of pooled panel data regarding farm performance across different climate regimes, explicitly accounting for adjustments (e.g. crop mix, planting and harvest dates, input use) that farmers make in response to varying agro-ecological conditions (Dinar *et al.*, 1998). The production function approach focuses more on short-term response, but tends to overestimate damage caused by climate variations by failing to account for substitution effects. The Ricardian methodology is more appropriate for long-term analyses of climate effects, but cannot fully disentangle the adjustments made in response to climate and those induced by other processes.

Social science approaches for the analysis of interaction between climate variability and farm household behaviour rely strongly on the concepts of *livelihood security* and *entitlements* (Davies, 1996; Ellis, 2000; Toulmin and Scoones, 2000). This framework pays attention to the available resources as well as the rights that enable households to cope with income shortfalls. Moreover, possible adjustment strategies related to engagement in labour and capital markets are explicitly included. Income derived from off–farm and non-farm employment is considered to be an increasingly important stabilisation device (Reardon *et al.*, 1988; Bryceson, 1997). In a similar vein, farm household assets (including cattle) and precautionary savings can contribute to consumption smoothing (Deaton, 1992; Udry, 1990). Consequently, different coping strategies can be distinguished based on either internal adjustments in resource management procedures or external modification of exchange mechanisms. In recent debates, attention is focussed on the prospects for adapting farm households' to vulnerability and enabling *sustainable livelihoods* in both biophysical and socio-economic terms.

We integrated both frameworks and used a bio-economic model that accounts for simultaneous adjustments in production and consumption behaviour in conditions of climate variability. Therefore, on the production side, attention is given to a wide range of agro-ecological land use coefficients for cropping and livestock activities that are based on local farmers' experience and expert systems. Farm household preferences regarding consumption are derived from cross-sectional expenditure surveys that provide information regarding the utility derived from major consumptive categories (cereal, meat, milk, non-agricultural commodities).

The modelling framework is based on the functional integration of different separate modules (see Figure 21.1) for (i) expected prices and transaction costs, (ii) farm household resource endowments, (iii) technical coefficients for cropping and livestock activities in different weather regimes, (iv) consumptive expenditures and (v) savings and investment. Information for the specification of these modules is derived from different sources: local farming systems surveys, market studies, expenditure surveys, regional statistics and a technical coefficient generator for input-output data (Kruseman *et al.*, 1997; Hengsdijk *et al.*, 1996). The data sources are combined in a multi-objective optimisation procedure that provides information on the selected pattern of land use and resource allocation as determined by farm household welfare objectives.

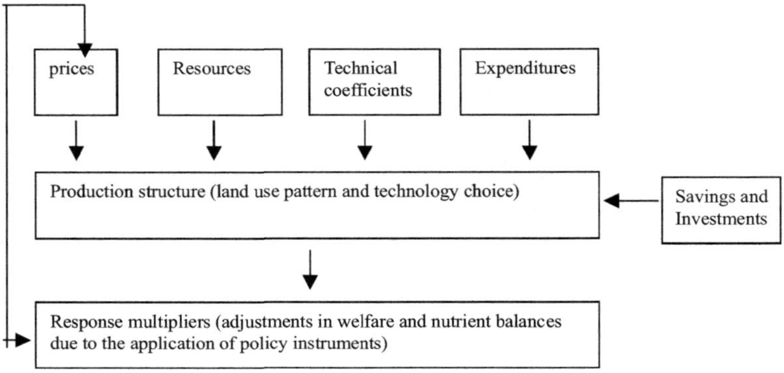

Figure 21.1. Modular structure of the bio-economic farm household model (Specific goal weights are attached to the objectives of net revenue and sustainability criteria (i.e. nutrient and organic matter balances). The model is then calibrated through a comparison of the simulated production structure with the actual land use pattern).

The modelling approach includes a number of behavioural mechanisms that enable adjustments to income shocks. Farm households may be engaged in off-farm employment or (temporary) migration to generate additional income that can be used for consumption or investment purposes. Moreover, factor exchange between farmers is allowed to ensure access to e.g. animal traction for less wealthy households in exchange for labour services provided to wealthier households. Market imperfections for exchange on commodity markets are approached through the inclusion of a mark-up for transaction costs that represents the difference between buying and selling prices. These price bands can become prohibitive for market exchange and explain major differences in supply response between net buying and net selling households.

The bio-economic model includes an algorithm for savings and investment behaviour, making a difference between permanent savings (e.g. assets accumulated from the past, calculated as the value of livestock in average rainfall conditions) and transitory savings (resulting from income fluctuations between average rainfall conditions and low or high rainfall years). While permanent savings represent the available options for accumulation, transitory savings are required for stabilising consumption in adverse weather conditions. Access to rural financial markets is linked to permanent savings that provide the necessary collateral for borrowing. Risk behaviour is household-specific and modelled through the specification of different time discount rates (see section 4).

An important distinguishing feature of the model refers to the non-separability of production and consumption decisions. In conditions of market imperfections, consumption and production behaviour are likely to interact and recursive optimisation procedures are used to link technical options with behavioural preferences. Dynamic properties are included in the model through the specification of (past) savings and investment behaviour. Moreover, local resource constraint can be circumvented through mutual exchange and interlinked transactions. Relations with the non-agricultural sector are considered through the incorporation of incentive goods in the expenditure system and the specification of opportunity costs of labour for migration and engagement in non-farm employment.

The bio-economic modelling framework is particularly useful for the analysis of farm household reactions to adjustments in market and/or weather conditions. It is designed to address short-term responses to changing production conditions and does not consider the impact of consecutive periods of a-normal years. The model has been used earlier to discuss the effectiveness of different policy instruments for improving simultaneously farmers' welfare and the sustainability of the resource base in Southern Mali. Sissoko (1998) applies the model to identify interventions in the output price for cotton, fertiliser price support (combined with cereal market liberalisation) and a reduction in transaction costs as the most effective means for enhancing agricultural intensification. Kruseman (2000) concludes that structural policies to reduce transaction costs are generally more efficient compared to input price subsidies. Both applications rely, however, on average rainfall conditions as a benchmark for policy simulation and disregard climate variability.

4. FARM HOUSEHOLD STRATIFICATION

Farm households can rely on a wide variety of adjustment strategies for coping with climate variability. Therefore, bio-economic models should regard diversity in supply response as a major feature. Technically, homogeneous farm stratification is also required to reduce aggregation bias. Different types of farmers will typically exhibit specific response reactions to changing weather conditions depending on their objectives (i.e. risk behaviour), their resource endowment (i.e. assets and savings), and their access to markets and institutions.

The economic behaviour of farm households in the Southern region of Mali is strongly determined by the availability of family labour and animal traction. Given the short rainfall season, timely preparation of land for arable cropping activities (cotton, cereals) is of major importance. During the last few decades, cotton revenues have been increasingly used to purchase livestock. The availability of animal traction (and later mechanisation as well) permitted an expansion of the cultivated crop area at the expense of pastures. The latter tendency is further aggravated by immigrated labour from other regions. Current stocking rates (0.3 TLU/ha) are far beyond the carrying capacity of rangelands (Bosma *et al.*, 1993). Intensification of land use is nowadays considered a major imperative (Breman and Sissoko, 1998).

Basic information on farm household resource endowments and time preferences is provided in Table 21.1. Three different types of farmers are distinguished (Struif Bontkes, 1999):

A. Wealthy extended-family household with sufficient access to land who has a substantial herd of cattle to afford high savings;
B. Medium-size farm household that maintains a balanced resource base and an average savings capacity;
C. Small nuclear household with a more fragile resource base and little livestock and draught power, hence it hires in additional equipment from other farmers.

Table 21.1. Southern Mali: Farm Household Stratification.

	Farm Type A	Farm Type B	Farm Type C
Resources			
Family size (persons)	25	12	9
Labour force (persons)	12	6	4
Land (ha)	18	10	6
Livestock (TLU)	25	4	1
Oxen (number)	6	3	1
Ploughs (number)	4	2	1
Ratios			
Dependency ratio	2.1	2.0	2.3
Land/labour ratio	1.5	1.7	1.5
Oxen/labour ratio	0.5	0.5	0.3
Coefficients			
Savings coefficient (%)	20	10	5
Time discount rate (%)	5	10	18
Number	9,100	7,900	2,400

Source : based on CMDT (1994) and Kruseman et al., (1997)

These different farm household types exhibit specific patterns of land use and rely on particular livelihood strategies. Moreover, input provision and extension messages are oriented towards specific farm types. Farm type A is able to cultivate relatively large areas of food and cash crops for both consumption and commercial purposes. These farmers use crop residues for animal feeding, and have access to both chemical fertilisers and animal manure to maintain soil fertility. Farm type B is smaller is size but still manages to cultivate land rather intensively with available resources. Production technologies are similar to farm type A, but with a lower share of cotton in land use. Farm type C relies more on manual field operations and hires in oxen in exchange of labour for land preparation.

Analyses of long-term tendencies of farm evolution in Southern Mali illustrate most likely development pathways (Struif Bontkes, 1999). Distinctions between farm type A and B are rather flexible since the number of type A farmers might decrease due to farm fragmentation (*eclatement*) whereas type B farms could be transformed into type A farms due to livestock accumulation. The number of type C farms will initially increase due to immigration and settlement, but in the long run lower cereal prices limit the amount of farm households within this farm type. Finally, farm households below the size of type C are bound to disappear and will become involved in agricultural wage labour and non-farm employment.

5. ADAPTIVE BEHAVIOUR

In order to meet their goals and aspirations as effectively as possible, households use the resources available to them when responding to changes in the circumstances they face. Climatic change can be regarded as a change in the resource endowments of the household that are outside their direct control. The household model can be used to explore the possible reactions of farmers to changes in rainfall regimes. Although farmers do not have perfect foresight, they may have certain expectations about rainfall levels. Figures 21.2 and 21.3 presents the adjustment in income and organic matter balances of different farm household types under various levels of expected rainfall. The horizontal axis shows a decreasing probability of the occurrence of dry years (towards the right).

The results show that the response to higher certainty regarding expected rainfall is generally low, although somewhat stronger for wealthier households. The main reason for this remarkable stable pattern is that - although the rainfall expectations become more secure - there is still a (decreasing) chance that rainfall levels will be very low. Since farm households are generally risk aversive, this implies that they will seek to garantee household food security before perceiving higher profits. The relatively low reaction in terms of income and organic matter balance is especially apparent in resource-poor households. They are willing to forego higher average incomes in good years to prevent shortfalls in bad years.

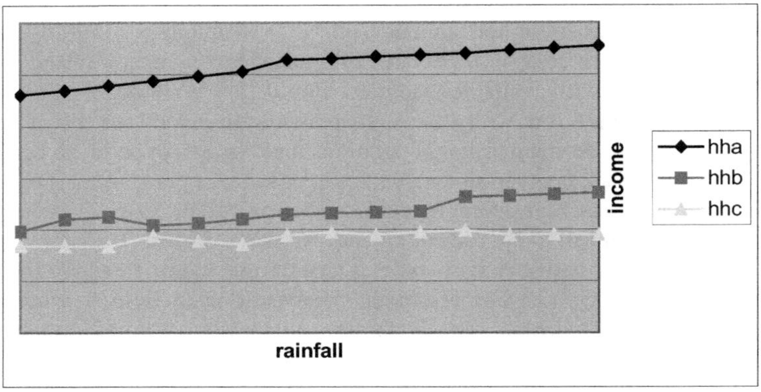

Figure 21.2. Relationship between rainfall and income.

The risk-aversive strategy becomes evident when we analyse the adjustments in land use and cropping patterns under changing expectations with regard to rainfall.

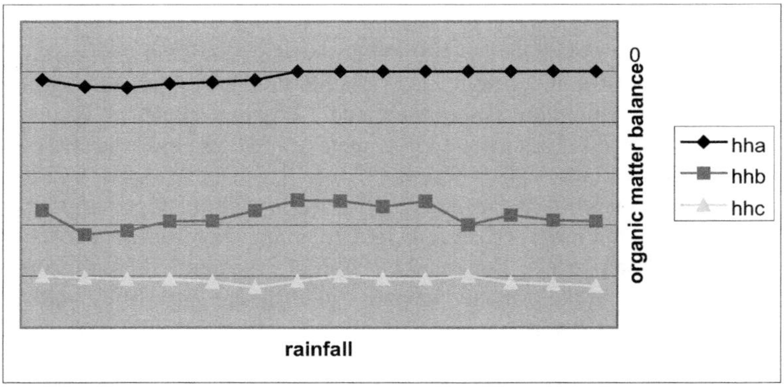

Figure 21.3. Relationship between rainfall and organic matter balance.

In Figure 21.4 this is presented graphically for the poorer type C households. With decreasing rainfall the area under more resistant millet and groundnut crops increases and the area devoted to cotton decreases, especially when rainfall reaches critical levels. Households can attain their food security goals using a variety of cropping patterns (*e.g.* substituting sorghum for millet) without sacrificing much of their welfare (Kruseman, 2000). This explains the shifts in cropping patterns observed in many parts of Africa. Differences in threshold rainfall levels at which crops still perform or outperform alternatives mark the adjustments in land use. This implies that variability in cropping patterns continues to exist but that, as rainfall regimes change, an overall shift in production structure is to be expected towards crops better adapted to these circumstances.

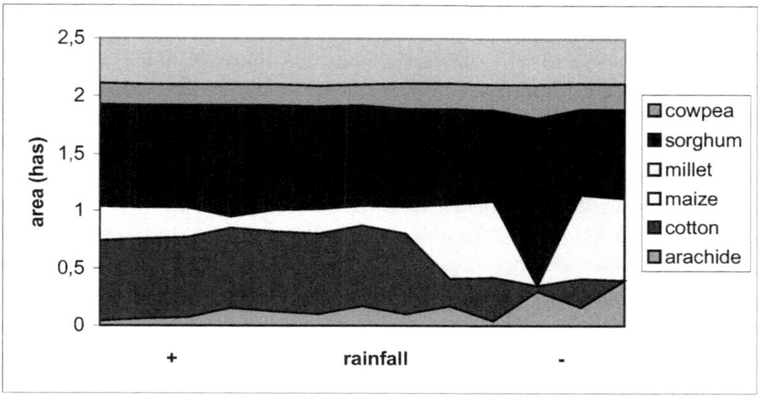

Figure 21.4. Cropping pattern of Household type C in different rainfall conditions *(See colour section, p. 465).*

6. POLICY INSTRUMENTS

Different types of policy instruments can be used to influence farm household behaviour and resource allocation. Economic incentives for land use intensification can be divided into two types of instruments (Ellis, 1992):

a) market policies related to adjustment in prices of production factors, inputs and marketable output;
b) structural policies that look for a reduction in transaction costs through institutional reforms of delivery networks or (public/private) investment in infrastructure provision.

Market and price policies generally aim to improve resource allocation and the level or distribution of (real) incomes. Selective price incentives usually modify production conditions through changes in the terms of trade for certain commodities and thus encourage farmers to make adjustments in land use or technology choice. On the other hand, structural policies are aimed at lowering transaction costs and reducing price bands (de Janvry *et al.*, 1991). Improving access to factor and commodity markets and delivery networks enables farmers to specialise and increase their involvement in market exchange. Although this may lead to a higher exposure to market risks, it also increases the possibilities for insurance. Consequently, structural policies are considered most important as a risk mitigation device (World Bank, 2000).

Both types of incentives generate simultaneous adjustments in farm production and household consumption behaviour, since revenues are related to factor rewards. On the production side, substitution effects occur on the input side as well as in the output mix. Moreover, the distribution of resources between farm and off-farm activities may be subject to change. Otherwise, positive income and consumption effects are expected when exchange conditions improve and transaction costs are lowered. Farm household welfare eventually depends on the balance between income and substitution effects. Although this balance may be negative for selective price policies, both effects can operate in the same direction when markets adequately perform their allocative functions.

Within the framework of Structural Adjustment - favouring market liberalisation and institutional reform – some regard market policies with scepticism. Moreover, recognising widespread institutional failure, structural policies are increasingly executed by private agencies. With most attention focussed on macroeconomic policies for maintaining fiscal and foreign trade balances, sector and regional policies generally lost ground. However, it can be demonstrated that enhancing sustainable land use usually requires direct interventions (Kuyvenhoven *et al.*, 1999). As policy measures often have different distributional implications, effective targeting warrants selective interventions (van de Walle & Nead, 1995). Consequently, more attention is given to the identification of policy instruments that simultaneously support farmers' welfare and sustainable resource management practices and generate effective supply response from the farmers' community.

Scoones and Toulmin (1999) provide an overview of major factors that influence farmers' behaviour regarding soil fertility management. Three types of parameters are considered most important for explaining different levels of farming systems intensification, viz. (a) biophysical factors (e.g. climate, soils, cropping and livestock systems), (b) socio-economic factors (e.g. population, market access, livelihood strategies) and (c) institutional factors (e.g. macro-economic environment, tenure regimes, marketing chains). Diversity in supply response to economic policies can be attributed mainly to selective market access and institutional constraints. Favourable or adverse biophysical conditions do not necessarily entail more or less productive and sustainable systems. Instead, access to resources and exchange conditions that influence farmers' willingness and ability to invest are recognised as factors that contribute to higher welfare and sustainable land use. Public intervention is considered appropriate to reduce market imperfections and to account for externalities, but only in conditions that (i) positive short-term and long-term income effects are ensured and (ii) limited options are available for (farm or off-farm) factor and/or output substitution.

In a similar vein, Sanders *et al.* (1996) discuss the prospects for agricultural technology development in sub-Saharan Africa, focussing on the requirements for input intensification in water and nutrient-limited situations. Major policy options include (i) the introduction of subsidies for critical inputs without viable short-run alternatives (e.g., chemical fertilisers) and (ii) the development of effective public-private partnerships to guarantee input provision and output marketing. Both measures lead to higher revenues and less price variability thus enabling farmers to replace essential nutrients and to invest in their farms. Regarding market failures and demand constraints as growth-limiting factors also implies that the reduction of transaction costs in commodity markets and the development of effective market linkages are highly important for enhancing supply responsiveness.

Economic policy experiments based on bio-economic modelling indicate that farmers react most strongly to modifications in transaction costs and adjustments in prices of major cash crops and vital inputs (Sissoko, 1998). These policy measures contribute to the adoption of more efficient and sustainable production technologies that entail higher yields and less soil degradation, especially for wealthier farm households. A comparison of both types of instruments reveals that supply response is generally strongest for changes in transaction cost, but at the expense of high budgetary demands (Kruseman, 2000).

Different climatic conditions are likely to have different implications for crop and technology choice by poorer and wealthier farmers and therefore generate specific welfare effects. The distributional implications are highly dependent on local market conditions. Humid weather conditions tend to improve yields and lead to the increased use of variable inputs and thus contribute to higher incomes derived from tradeables (mainly produced by farm types A and B) whereas prices of non-tradeables (produced mostly by farm type C) may be reduced. Consequently, substitution between cash and food crops is likely to occur and market-oriented farmers will increase their reliance on externally purchased inputs. Similarly, in dry weather conditions, prices of non-tradeables will rise and more attention is given to food security objectives.

Economic policies for price reform and market development are also likely to generate different types of benefits for farmers in various conditions of climate variability. It may be expected that input price policies are most effective as a device for adjusting technology choice and input efficiency under wet and normal weather conditions, whereas output prices have a more direct effect on production and exchange relationships in normal and less favourable weather conditions. Changes in transaction costs tend to influence the product mix and the net demand or supply position of the household on local commodity markets. The latter effect is particularly important in dry years when farm households are more dependent on market purchases to satisfy consumption requirements.

Table 21.2. Policy Instruments and their Implications.

Policy Measure	Production Effects	Substitution Effects	Welfare Effects
Input price subsidy	Technology choice and input efficiency	Input intensificaton (fertiliser use)	Costs reduction
Output price support	Production volume and output size	Input substitution (labour/agrochemicals)	Net income increase
Reduction in Transaction costs	Product mix (cash/food crops)	Improved tradeability (more market exchange)	Lower marketing Margins

Policy incentives generate simultaneous changes in production systems (land use, crop and technology choice), market exchange and farm revenues. Table 21.2 provides an overview of the most important direct effects. Changes in relative input-output prices imply a movement along the production possibilities frontier (e.g. substitution), while adjustments in transaction costs lead to a shift in the production function. Final welfare effects depend on changes in farmers' net supply or demand position. Output price support is thus most beneficial for surplus farmers, while commercial farmers then use more external inputs benefit most from input subsidies.

7. RESPONSE TO POLICY INCENTIVES

Different categories of farmers respond in different ways to policy instruments in varying rainfall conditions. The intensity of supply response mainly depends on the possibilities available to adjust resource allocation, both internally within the farm household and externally through the adjustment of market exchange relationships. The direction of farm households' response to policy incentives can be explained by referring to their resource endowments and available assets (e.g. absorption capacity for climate shocks) and their relations with factor and commodity markets. Wealthier households with better access to rural financial and labour markets are more able to cope with climate risks (i.e. through borrowing or engagement in off-farm employment). Moreover, different types of supply response reactions to price changes are expected amongst net buying and net selling farm households.

Table 21.3. Supply Response to Policy Instruments.

Year	Performance	Farm Household A			Farm Household B			Farm Household C		
		Input price Subsidy	Output price Increase	Reduc-tion in T.C	Input price Subsidy	Output price increase	Reduc-tion in T.C	Input price Subsidy	Output price Increase	Reduc-tion in T.C.
Humid	Income	1.7	6.7	15.6	0.7	3.2	16.0	0.4	1.9	16.4
	OM Balance	1.4	0.5	-12.4	54.9	0.0	-56.7	-4.7	0.0	0.0
	Nutrient Balance	-5.7	0.0	-10.9	114.2	0.0	-114.2	0.0	0.0	0.0
	Food balance	0.0	-0.1	3.3	0.0	0.0	0.0	0.0	3.3	0.0
	Labour balance	0.0	0.0	-0.2	0.0	0.0	0.0	-0.5	0.0	0.4
Normal	Income	1.9	7.0	15.1	0.7	3.4	15.9	0.4	2.1	16.3
	OM Balance	10.3	-28.5	0.0	1.3	-4.8	0.0	-47.3	16.4	-5.6
	Nutrient Balance	-68.0	-131.9	0.0	20.7	-17.8	0.0	-40.5	48.0	-4.5
	Food balance	0.0	0.0	0.0	0.0	0.0	0.0	0.3	3.6	0.0
	Labour balance	-0.2	0.0	-0.2	-0.2	0.0	0.0	-0.2	-0.1	0.0
Dry	Income	0.8	6.1	18.9	0.6	3.4	16.8	0.5	1.8	16.4
	OM Balance	9.1	-12.8	-115.0	27.3	54.4	19.3	19.1	-12.2	-20.0
	Nutrient Balance	3.4	-17.6	-109.0	27.9	-43.3	-66.5	23.0	18.3	-20.8
	Food balance	0.0	0.0	0.0	0.0	0.0	-8.2	3.4	3.2	4.7
	Labour balance	0.2	-0.6	-0.7	-0.3	-0.4	0.3	-1.4	0.2	1.9

Note : OM = Organic Matter ; TC = Transaction costs

Table 21.3 provides an overview of the reaction of three types of farm households to lower input prices and transaction costs and to output price support in different rainfall conditions. These changes reflect adjustments in production (*i.e.* land use and/or technology choice), labour allocation and consumption due to discrete parametric changes in exchange conditions. We will now discuss internal and external adjustment mechanisms that explain the diversity in supply response. Attention is given to the implications for farm household income, food security and sustainability of the natural resource base (*i.e.* soil organic matter and nitrogen balances).

Land Use

The effectiveness of different economic incentives is clearly dependent on the prevailing weather regime. Farmers' expectations regarding humid or dry weather conditions have effects on crop choice and input applications. Price policies tend to generate stronger income effects in normal years, especially for more commercially oriented farmers (A and B) that specialise in profitable cotton production and expand livestock. Cotton cultivation is severely reduced under adverse weather conditions, since only far less input-efficient production options are available. Otherwise, reduced transaction costs generate most favourable income effects for poorer farm households. Even under dry weather conditions farmers can adjust their land use as regards cereal production (*i.e.* cowpea and sorghum), relying on more drought-resistant and labour-intensive cropping technologies that still provide acceptable net returns.

Income

Income effects differ widely with regard to household types, weather regimes and instrument choice. We notice that wealthier households (A and B) exhibit stronger income reactions to price policies under almost all climatic conditions compared to poorer households (C), since their higher involvement in market exchange enables them to rely more on external adjustment mechanisms. Moreover, households A and B possess more savings and have a lower discount rate that provides them with more internal flexibility. For the transaction costs scenario, farm household type C shows a substantially higher income response since narrower price bands provide increasing opportunities for participating in market exchange. However, the income effect is reduced somewhat in dry weather conditions when total production sharply declines, while wealthier households in net supply conditions are more able to reap the benefits from rising market prices for non-tradeables.

Looking at individual instruments, it is clear that farm household responsiveness to input price policies is notably high in humid and normal years, especially for those farmers that rely on purchased inputs (farm types A and to a lower extent type B). Output price policies generate higher responses in normal years. This is further enhanced by induced price adjustments that take place in humid or dry years due to excess or short supply. As expected, farm household response to output price support is considerably higher compared to input prices due to the more direct link between output prices and income. Both price policy instruments are considerably less effective compared to changes in transaction costs (*e.g.* a general price reduction), but the latter policy draws heavily on the public budget.

Nutrient balances

In terms of land use and natural resource balances, some contradictory effects may be noticed. In humid years, output price policies only marginally affect nutrient and organic matter balances, since most farmers are already inclined to select the most efficient production technologies. Whereas output price support is most effective in influencing farm household revenues, input price policies have more direct implications for nutrient balances. This is particularly the case for Type B farmers that shift to more sustainable technologies in both cotton and cereal production. Type C farmers, however, face labour and traction constraints that inhibit land use intensification. In some cases, input price subsidies lead to improving organic matter balances but deteriorating nutrient balances; this is basically explained by the improved availability of crop residues, which is only partially compensated for by changes in fallow practices.

Output price support becomes more effective in normal weather conditions but leads to negative nutrient balances in Type A and B households, whereas in type C households agro-ecological sustainability improves. Trade-offs between income and sustainability objectives mostly occur for net selling households where price increases lead to an expansion of cash crop production at the expense of fallow. Type C households that are net buyers of cereals prefer to intensify food production, relying on land use technologies that recycle crop residues. Moreover, output price policies demonstrate a strong positive effect on food balances, particularly for smaller farm households.

In dry weather conditions, output price changes generate opposite effects on agro-ecological balances in type B and C households. This is basically explained by the ability of type B households to preserve their livestock herd, while type C households have to sell cattle in order to maintain income stability. Consequently, type B farmers have to use more crop residues for animal feeding, while type C farmers are able to recycle crop residues in the fields, thus maintaining a better nutrient balance.

Labour use

Most price instruments have a limited impact on the labour balance (*i.e.* supply and demand of family labour) in humid and normal rainfall conditions. Lower input prices provoke factor substitution, especially by smaller farm types, whereas higher output prices are partly consumed as leisure in the larger farm types (indicated by the negative sign in the labour balance). Labour effects become somewhat more pronounced in dry years, when farm type C shows marked adjustments in labour use. Lower input prices lead to labour extensification, while higher output prices favour more intensive labour use in agricultural production. Reduced transaction costs enable these households to engage more in off-farm employment.

Food security

Price and market polices are particularly effective in improving food balances in type C households in dry weather conditions. Input price support might enable these poorer farmers to maintain reasonable fertiliser applications to stabilise cereal yields, while higher output prices for cash crops could partly compensate for yield losses. Reduced transaction costs prove to be particularly important in dry years to enable the purchase of food on the local market by net buying households.

However, specific instruments yield different implications for food security and sustainability of the resource base. Clear trade-offs are registered in humid years for transaction costs (in type A farms) and in dry years for some market interventions (in type B and C farms). These trade-offs occur when production and substitution effects work in opposite directions, *e.g.* when higher output volumes are reached at the expense of nutrient depletion. This is the case when market incentives favour the adoption of more extensive production technologies that reduce fallow practices or crop residue recycling procedures.

8. CONCLUSIONS

Highly variable weather conditions influence farm household resource allocation decisions as well as the effectiveness of agrarian policy instruments in influencing these decisions. Different types of farmers respond in different ways to economic incentives in situations of low or high rainfall. In this chapter we reviewed the available options for adapting land use systems and labour allocation for three typical households in the Southern region of Mali.

Changes in weather regimes and climate shocks can be partially offset by compensatory policy devices. The bio-economic model simulations demonstrate that price policies are particularly useful for enabling welfare-enhancing adjustments for wealthier farm households, although sometimes at the cost of deteriorating nutrient balances. Otherwise, poor farmers can benefit considerably from reductions in transaction cost. The latter effect is especially important in dry weather conditions when dependence on market exchange tends to be intensified. Input price subsidies remain important for improving the sustainability of land use. Moreover, market reform programmes continue to be a relevant instrument for reducing poverty and enhancing food security during drought periods.

Economic policies for drought mitigation should recognise differences in agrarian structure, taking into account diversity in supply response amongst farm types. Effective targeting may become feasible when differences in resource endowments, risk and market relations that give rise to particular coping strategies are fully considered. Contrary to most expectations, vulnerable households can benefit from market reform policies and thus structural adjustment programmes might provide an adequate framework for containing adverse effects from climate shocks (Benson & Clay, 1998).

Although price policies and market reforms improve farm household responsiveness even in dry weather conditions, additional instruments should be used to enhance food security and sustainable land use. Therefore, attention should be paid to improved financial systems (rural savings and credit) to enhance input demand, public investment for soil replenishment (*e.g.* phosphate rock), well-defined property rights regimes, and formal or informal insurance networks that provide options for income smoothing (Scoones & Toulmin, 1999; Sanders, 1996)

Chapter 22

CLIMATE CHANGE AND MANAGEMENT OF CATASTROPHIC RISK

Peter Hazell

Abstract: Climate change will increase the probability of extreme weather conditions leading to catastrophic income shortfalls. National governments need to review past interventions and develop innovative ways to assist rural communities in coping with, and recovering from, massive and large income reductions. This chapter illustrates some common problems in policy interventions with respect to crop insurance, credit provisioning, feed subsidies and emergency relief. They usually increase household dependence on government assistance and incite households to accept larger risks than necessary and reduce own efforts to manage income risks. In addition, they cause moral hazard problems and are ineffective in reaching the poorer households. This chapter proposes two promising alternative instruments for risk management. First, income insurance based on local rainfall conditions overcomes the problems of the previous policy interventions, is relatively easy to administer and can be implemented by the private sector. Second, early warning drought forecasts guide farmers as well as governments in anticipating disastrous weather conditions.

1. INTRODUCTION

Many of the chapters in this book are concerned with how climate change will impact on farming systems and livelihood strategies, particularly those of poor people living in low-rainfall areas. However, one aspect of climate change that has received relatively little attention is how rural households and communities will cope with catastrophic weather risks, like droughts and floods, whose frequency and severity seem destined to increase with global climate change. This paper discusses this problem and the policy options available for helping rural people manage these risks.

A.J. Dietz et al. (eds.),
The Impact of Climate Change on Drylands: With a Focus on West-Africa, 385–395.

Agriculture is by nature a risky business and farmers and rural societies around the world have evolved many strategies for managing and coping with risk. These include crop diversification, transhumant herding practices, food storage, savings, seasonal non-farm employment and kin support networks. While these strategies are reasonably effective in managing most of the variation or risk that arises (see, for example, Walker and Jodha, 1986), they are rarely sufficient to cope with catastrophic events which, while infrequent, cause severe losses when they occur. This is because nearly everyone living in an affected region incurs losses at the same time—the covariate nature of the risk—and strategies that require loss sharing within communities easily break down. Moreover, because most rural societies in developing countries are poor to begin with, they have limited reserves to fall back on when catastrophic events occur. As a result, these events correlate with peaks in poverty, hunger and some forms of environmental degradation. The problem has reputedly worsened with climate change, which may be contributing to more frequent and severe weather events. Population growth may also be exacerbating the problem. Not only are there increasing numbers of people to support on the same resource base in crisis years, but population pressure can also induce land use changes that increase exposure to losses in catastrophic years (e.g. cropping in drought prone areas better suited to pastoral systems).

Many governments have attempted to help farmers and rural people manage these weather related crop and livestock losses through debt forgiveness, livestock feed subsidies and crop insurance programs. But these interventions have often proven costly and ineffective, sometimes even counterproductive, for managing losses. As a result, many governments have had to turn to more direct forms of disaster assistance to relieve the problems of stricken areas, such as food aid and relief employment programs. The cost of disaster assistance is becoming a major financial burden for many governments and donors and is escalating as more people live in vulnerable areas and as global climate change increases the frequency and severity of many natural disasters. There is need to find alternative policies that can help rural people manage catastrophic losses more effectively while at the same time promoting economic development. This paper first reviews problems with existing approaches and then considers some new and more promising alternatives.

2. REVIEW OF PAST GOVERNMENT POLICIES FOR MANAGING CLIMATE RISK

Four of the most common public intervention programs to assist farmers manage climate risk in the developing countries have been crop insurance, debt forgiveness, livestock feed subsidies and emergency relief programs. We will review each in turn and then consider some common lessons.

Crop Insurance

Publicly provided crop yield insurance has been implemented in many countries to help farmers cope with weather related production risks (Hazell et al., 1986). Public crop insurance programs are typically multiple peril or all risk programs that compensate for all yield losses regardless of cause. Indemnities are typically based on the difference between actual yield and a pre-specified target yield, not on actual crop damage or input costs lost. In many countries, insurance is tied to loans from an agricultural development bank (ADB), with the bank paying the premium and collecting the indemnities. In some countries the insurance is compulsory for all farmers growing the insured crops or borrowing from the ADB.

The financial experience with publicly provided, multiple-peril crop insurance has been disastrous. In all cases, programs are heavily subsidized and governments not only pay part of the premium, but also most of the administration costs and they reinsure the agency by providing direct transfers in high-loss years. In order to be profitable, an insurer must structure its contracts so that the premiums collected exceed the payouts (indemnities plus administrative costs) on average over time, even if not each year. But in practice, public insurers invariably charge premiums that are less than half the average amount that they pay out and in many cases they collect a much smaller share than half (Hazell, 1992). Despite these high costs, there is little evidence to show that crop insurance has had any positive impacts on agricultural lending, agricultural production or farm income (Hazell et al., 1986).

Why has public crop insurance failed so badly? One of the most important reasons is that many of the risks covered by multiple risk insurance are subject to moral hazard problems, leading to large actuarial losses for the insurer. Moral hazards arise whenever the incidence or severity of a loss can be influenced by the insured's' behaviour, and many crop yield losses fall into this category. Farmers covered by multiple peril insurance have a reduced incentive, for example, to incur costs to prevent yield losses due to some pests and diseases once they know that the insurer will compensate for those losses.

Another important reason for failure is that public insurers are often mandated to extend their insurance to small farms and this can add enormously to administration costs.

A third reason for failure is that inappropriate incentive problems arise within insurance institutions when the government underwrites most of their programs. When insurers know that the government will automatically cover most losses, they have little incentive to pursue sound insurance practices when assessing losses. In fact, they may find it profitable to collude with farmers in filing exaggerated or falsified claims. Hazell (1992) reports that in Mexico, prior to closing the national agricultural insurance agency, it was not uncommon for inspectors to receive bribes of about 30 percent of the value of the indemnity payments made to farmers. When the insurer underwrites the loans of an ADB, these incentive problems can easily infect the bank too, leading to a serious loss of discipline in banking practices. Why, for example, should ADB staff try to collect loans from tardy borrowers if it is easier for them to obtain repayment from the insurer?

Another common reason for failure has been that governments undermine public insurers for political reasons. Hazell (1992) gives examples where insurers have had to pay out against exaggerated losses in election years.

Many crop insurance programs also tend to be too specialized, focusing on specific crops, regions and types of farmers, particularly when the insurance is tied to the loans of an ADB that has a mandate to serve particular target groups identified by the government. Without a well-diversified insurance portfolio, crop insurers are susceptible to co-variability problems and face the prospect of sizeable losses in some years. Since public insurers are rarely able to obtain commercial insurance or contingent loan arrangements, this specialization increases their dependence on the government.

It is possible to improve public crop insurance along the lines of good practice adopted by the private sector (Gudger, 1991). Unfortunately, this greatly limits the ability of insurers to serve small farms and non-commercial farms and to insure many important climate risks that these farmers face (Hazell, 1992).

Credit

In principle, if rural credit markets worked well, farmers would be able to borrow credit to buffer losses incurred in catastrophic years and repay in better years. Traditional credit markets do play this role to some extent, but they are hampered by the covariate nature of most catastrophic weather events. When these events occur, many people want to borrow at just the time when few people can afford to lend within the affected area. Traditional credit markets rely on personal relationships rather than contracts to enforce loan agreements and this constrains their spatial reach for spreading covariate risks in an effective manner (Binswanger, 1986).

At the same time, formal lending institutions such as ADBs and micro-finance organizations usually tie their credit to farm and non-farm business costs and investments and have little flexibility to lend to offset crop and livestock losses or to maintain consumption in catastrophic years. Without this flexibility, many governments have resorted to systematic rescheduling of publicly provided credit for farmers during catastrophic years, especially droughts. This approach has contributed to the chronically poor debt collection performance of many agricultural development banks (Von Pischke, 1986).

Livestock feed subsidies

Feed subsidies have been another popular government program in many drought prone pastoral societies, especially in the West Asia and North Africa (WANA) region.

Government feed programs allocate most of their resources to the costs and distribution of concentrates and other feeds, especially barley in the WANA region. These programs have been quite successful in protecting livestock numbers and production during droughts in several countries (Hazell et al., 2001). But they have also probably accelerated rangeland degradation in the long term by undermining the traditional process of adjusting flock size to inter-annual climatic variations. Herd sizes have increased sharply in recent years in most WANA countries, and grazing practices have changed so that many of the animals no longer leave the rangeland areas during the dry season but have their feed and water trucked in. This practice leads to overgrazing during the dry season, reduces the natural seeding of annual pasture species, disturbs the soil and contributes to wind erosion, particularly in areas near water and feed supply points (Hazell, et al., 2001).

Feed subsidies are an expensive burden on governments and although initially introduced during crisis years, they have a tendency to become permanent (Hazell et al., 2001; Pratt, Le Gall, and de Haan, 1997). They have also proved difficult to target, with the lion's share of the subsidized concentrates going to large herders and to commercial farms.

Emergency Relief

Given the shortcomings of the kinds of policies described above, many governments have turned to various forms of disaster assistance to relieve the problems of stricken areas. These include food aid, various types of employment programs and even cash or kind compensation for losses incurred (e.g. replacement of seed or livestock). Disaster assistance is a blunt instrument and it is difficult to target it at the truly vulnerable. It also takes time to launch and much hardship may occur before the disaster is officially recognized and relief efforts are sent into the field. There have been real advances in recent years in targeting and delivering assistance more effectively, often by devolving relief efforts to civil society and involving local communities in the design and implementation of targeted programs. These changes have helped avoid many of the bottlenecks that plague public distribution. However, relief has become more independent of government development efforts, exacerbating long term dependence on assistance and reducing the amount of resources available for development. Programs can also be designed to be "self-targeting", such as food-for-work (or cash-for-work) programs, which are only attractive to the truly needy.

The cost of disaster assistance is becoming a major financial burden for many governments and donors, often exceeding development assistance (Owens and Hoddinott, 1998). Moreover, the cost is escalating as more people live in vulnerable areas and as global climate change increases the frequency and severity of many natural disasters.

Lessons Learnt

Country experiences with different risk management interventions have revealed a couple of common problems that should be heeded in the future.

First, the subsidized nature of many risk management interventions encourages farming practices that increase both the extent of future losses and the dependence of local people on government assistance. For example, feed subsidies have led to much higher stocking rates in some drought prone areas, and subsidized crop insurance can encourage farmers to grow higher value but more drought prone crops in high risk areas.

Any good risk management aid should enable farmers to take greater risks in their quest for higher average returns. If farmers are risk averse, then they trade off some level of expected income for lower risk (e.g., through diversification strategies). The amount of expected income foregone to reduce risk can be viewed as a risk premium paid, or a production cost (Sandmo, 1971; Robison and Barry, 1987). If this cost can be reduced by the introduction of an improved risk management aid, then the farmer may be able to change strategy (e.g., specialize more in the most profitable activities) and obtain a higher average income for the same amount of risk. This change not only improves expected farm incomes, but can also lead to spill-over benefits to consumers at an aggregate level through lower prices as the supply function shifts downwards by the amount of the reduction in the risk premium per unit of output. This effect is very similar to the effect of a new cost-reducing technology and providing the new risk management aid is not subsidized, there is always a net gain in social welfare. But if the new risk management aid is subsidized, the effect is similar to a subsidy on any other farm input (e.g., fertilizer or credit). The reduction in unit costs is partly paid for by the subsidy and the dead weight loss of the subsidy is always greater than sum of the additional producer and consumer welfare that it generates (Siamwalla and Valdes, 1986).

What does this mean in practice? That subsidized risk management interventions can reduce risk costs to farmers to below their true social value, leading to excessive risk taking and increased exposure to future weather related losses. Not only is there a built-in dependence on future assistance from the government, the net social return to that assistance can be small or even negative as well. The bottom line is that wherever possible, public interventions should be limited to risk management interventions that farmers pay for themselves, although it might be necessary for governments to devise arrangements which allow deferred payment in instalments.

The second potential problem with poorly designed risk management policies is that they can lead to moral hazards. In the case of crop insurance, moral hazards lead to greater losses than necessary, increase the risk exposure of the insurer and make actuarial calculations of those risks difficult. Similar problems can arise if a government indiscriminately compensates for drought losses that could have been reduced or avoided by farmers or herders. Unless appropriately targeted, feed subsidy programs could, for example, lead to reduced incentive to exploit remaining grazing opportunities, particularly in more remote areas that require greater time and expense to reach. Debt forgiveness in drought years can also generate moral hazard problems. Once farmers' know that their debt will be forgiven, they have an increased incentive to borrow more than is prudent and a reduced incentive to minimize their costs during droughts. Such behaviour leads to greater losses than necessary and makes feed subsidies and credit programs more expensive than they need to be.

Moreover, once emergency assistance has been institutionalised and people know they can count on it, it has many of the longer-term effects of an insurance subsidy that inadvertently worsens future problems by encouraging people to increase their exposure to potential losses. For example, compensation for flood or hurricane damage to homes can lead to the building of more houses in flood and hurricane prone areas. Similarly, compensation for crop losses in drought or flood prone areas encourages farmers to grow more of the compensated crops even when they are more vulnerable to these weather risks than alternative crops or land uses.

What is needed are more effective mechanisms for enabling rural people to better manage their own catastrophic risks. While some disaster assistance should never be ruled out, especially for the poor, the burden on government might be substantially reduced through some new and innovative approaches.

3. NEW POSSIBILITIES FOR IMPROVED DROUGHT MANAGEMENT

A limitation of most drought management interventions is that they inadvertently subsidize inappropriate farming practices and encourage moral hazard problems. They also represent a recurring fiscal burden to governments, which can become institutionalised and are hard to sustain over the years. Two newly emerging approaches can avoid these problems by providing farmers and herders with the means of managing drought risks more effectively themselves with a minimum of government intervention.

Rainfall Insurance

New forms of catastrophe insurance are needed that are affordable, accessible to all kinds of people, compensate for household income losses not just specific crop losses, are practical to implement given the limited kinds of data available in developing countries and can be provided by the private sector without the need for government subsidies.

Area-based rainfall insurance offers a promising new alternative that in principle can meet all the requirements listed above (Skees et al., 1999). In this approach, rainfall insurance contracts are written against specific rainfall outcomes (e.g. drought or flood) at a local weather station. The rainfall events have to be defined at catastrophic levels and they have to be highly correlated with the value of regional agricultural production or income. For example, an insured event might be that rainfall during the most critical month of the growing season falls 70 percent below normal. In years when the insured event occurs, all the people who purchased the insurance would receive the same payment per unit of insurance. In all other years, no payments would be made.

Insurance is sold in standard units (e.g. $10 or $100), with a standard contract for each unit purchased called a Standard Unit Contract (SUC). Purchasers decide how many SUCs to buy. The insurance is sold on a full-cost basis and the price of the SUC is the premium. The insurance must be sold before season-specific information about the insured risk becomes available. This requires a purchasing deadline (such as a month before the normal arrival of the rainy season), after which new SUCs are not sold.

Area-based rainfall insurance has a number of attractive features.

- It avoids the moral hazard and adverse selection problems that plague crop insurance programs.
- It could be very inexpensive to administer.
- It uses only rainfall data, which is available in most countries for long periods of time.
- The insurance can be sold to anyone, including agricultural traders and processors, farm input suppliers, banks, shopkeepers, and agricultural workers. There is no need to be a farmer, or to keep livestock.
- It would be easy for the private sector to run.
- As long as the insurance is voluntary and unsubsidised, it will only be purchased when it is a less expensive or more effective alternative to existing risk management strategies.

- A secondary market for insurance certificates could emerge which would enable people to cash in the tradable value of a SUC at any time.

In designing an area-based rainfall insurance scheme, a number of difficulties need to be overcome, including:

- The insurer faces high risk because of the covariate nature of the insured risk. When a payment is due, all those who have purchased insurance against the same weather station must be paid at the same time. Moreover, if the insured risks at different rainfall stations are highly correlated, the insurer faces the possibility of having to make huge payments in one and the same year. To guard against this risk, the insurer can either diversify regionally by selecting weather stations and risks that are not highly (positively) correlated, or seek reinsurance in the international financial markets.
- Rainfall stations must be protected to prevent the possible tampering of rainfall measurements. Possible approaches include a) more secure, tamper proof stations and instruments, b) triangulation of readings from neighbouring weather stations and c) verification of low soil moisture by remote satellite sensing.
- The actuarial soundness of the insurance could be undermined by El Nino weather cycles that change the probability of the insured events. It may be necessary to adjust the cost of the insurance whenever an El Nino event is confirmed.
- The volume of insurance sold could be too small to be profitable. The insurance will only appeal to people whose economic losses are highly correlated with the insured rainfall event. If the basic risk (the uninsured part of a person's risk) is high, the insurance will not sell. Also, if the probability of the insured risk is high, the cost of the insurance could be prohibitive. To overcome these problems, the insurance should be limited to truly catastrophic rainfall events that significantly affect agricultural production in a region.

The private sector might be expected to take the initiative in developing rainfall insurance, but several set-up problems might require government intervention to jump start activity in developing countries. These include paying the research costs of identifying key catastrophic rainfall events that correlate strongly with agricultural production and income, educating rural people about the value of rainfall insurance, ensuring secure rainfall stations, establishing an appropriate legal and regulatory framework for rainfall insurance and underwriting the insurance in some way (perhaps through contingent loans) until a sufficient volume of business has been established that international re-insurers or banks are willing to come in and assume the underwriting role. These roles need not be costly, but could prove crucial in launching rainfall insurance. However, it is also important not to launch the insurance on a subsidized basis, so as not to distort incentives for private insurers or farmers and herders. Drought insurance of the kind proposed here has already been introduced in Morocco and Mexico.

Early Warning Drought Forecasts

In principle, the ability to provide early warning drought forecasts could be a powerful tool for avoiding many of the economic costs associated with the misallocation of resources that arise because farmers, herders and other decision makers have to commit resources each year before key rainfall outcomes are known. For example, decisions about planting crops (date of planting, seeding rate, initial fertilizer treatment, etc) often have to be made at the beginning of the wet season before knowledge about rainfall outcomes is available. The economic value of season specific forecasts really depends on the degree to which farmers can adjust their plans as the season's rainfall unfolds. Of course, the reliability of the forecasts and the ability of the farmers to adjust their initial decisions in response to this information are also critical. If decisions about planting and cultivation practices and the feeding, culling and seasonal movement of livestock can be sequenced, with key decisions being postponed until essential rainfall data is available, then forecast information will be less valuable. However, if most decisions have to be made up front each season, the scope for mistakes will be much larger and the potential economic gains from reliable forecast information will be greater. Stewart (1991) examines how the date of onset of the rainy season can provide a pretty reliable forecast of the ensuing seasonal rainfall pattern for Niamey, Niger and shows how this information could be used to adjust planting and input decisions for the season more optimally (his "response" farming approach). Barbier and Hazell (1998) use a stochastic programming model to show how many of the decisions in a typical agro-pastoral community in Niger can be optimally adjusted to rainfall outcomes.

Reliable drought forecasts could also enable governments and relief agencies to position themselves each year for more effective and cost efficient drought interventions. This possibility has already been realized and there are now several early warning drought systems already in place in Africa which have proved successful in giving advance notice of emerging drought situations. But these programs are really monitoring systems that track emerging rainfall patterns within a season rather than true weather forecasting systems that predict rainfall outcomes before they even begin.

Reliable multi-year rainfall forecasts are not yet possible, but seasonal (from 3 to 6 months out) forecasts have become more reliable, particularly where an important part of the year to year variation in seasonal rainfall can be attributed to the Pacific El Nino Southern Oscillation (ENSO) weather patterns. As the ability to model these phenomena at global and regional levels improves, it seems plausible to expect that more reliable seasonal forecasts will be available at local levels. This may prove to be one of the most exciting developments in drought management in the next few years. It seems likely that private weather forecasting services will expand and become more available to developing countries. But this is also an area where government could play a catalytic role and even subsidize many of the development costs without having to worry that this would distort resource management incentives at the farm level.

4. CONCLUSIONS

The need to improve methods for managing climate risks in developing countries has increased in recent decades as population growth and climate change have contributed to greater demands on the resource base and accentuated both the incidence and severity of catastrophic weather losses. Government interventions have typically been initiated in response to crisis situations with little thought to their long-term impacts on the way farmers and herders manage resources and the productivity of agricultural systems. Now there is accumulating evidence to show that once risk management interventions are institutionalised, they lead to changes in the way resources are managed, including the increased intensification of land uses. These changes can contribute to greater productivity and improved livelihoods. However, if these interventions are subsidized, they can also lead to the adoption of excessively risky farm management practices, with increased losses in the event of catastrophic weather events and a growing dependence on government assistance. Many risk management programs also contribute to moral hazard problems because they reduce incentives for prudent management by farmers. Risk management interventions need to be designed so that they assist farmers to manage risk more effectively and to improve their productivity and incomes, but without distorting incentives in inappropriate ways. The experience with feed subsidy and debt relief programs has had mixed results and while they have helped protect incomes and food security in drought years, they have had negative impacts on the way resources are managed. Better alternatives could take the form of area-based rainfall insurance, particularly if offered by the private sector, and the development of more accurate and accessible drought forecasting information.

Chapter 23

CLIMATE VARIABILITY AND EARLY WARNING SYSTEMS

Mama Konaté

Abstract: After a brief recapitulation of climate risks in Mali this chapter shows the promising usefulness of operational information systems to assess drought hazards as employed currently by the Malian government.

1. RAINFALL RISKS IN MALI

The rainfall situation in Mali is extremely variable. In the south (Sikasso Region) and west (parts of Kayes Region and around the capital city Bamako) average rainfall during the rainy season is between 1,000 and 1,300 mm. In the centre (Mopti, Ségou, Koulikoro and part of Kayes Regions) it is between 400 and 1,000 mm and in the north (Gao Region) it is below 400 mm and in most parts even below 200 mm. The agricultural growing season lasts for between 180 days in Sikasso and Koutiala, 150 days in Bamako, 110 days in Mopti and less than 30 days in Northern Gao. At least 80% of the annual rainfall is normally recorded in one rainy season per year, which starts in late April in the extreme south (10 degrees North), around 20 May in Koutiala (12 degrees North), around 25 May in Bamako (13 degrees North), around 29 June in Mopti, Bandiagara, and Douentza (15 degrees North) and around 19 July in Gao (16 degrees North). In the south the rains normally continue until mid October, in the north they stop much earlier, in late August. August is the wettest month everywhere. Map 23.1 shows the average annual rainfall in Mali, map 23.2 the length of the rainy season and map 23.3 the average start of the rainy season.

Inter-annual variability is high, as is the case everywhere in the drylands. Long-term data for Bamako shows differences of between less than 700 mm to more than 1500 mm, with a long-term mean of 1025 mm (see Figure 23.1). The total amount of rainfall in the growing season is

A.J. Dietz et al. (eds.),
The Impact of Climate Change on Drylands: With a Focus on West-Africa, 397–402.

important to farmers, as is the start of the rainy season and the distribution of rainfall within the rainy season. The balance between water available in the root zone and the loss of water due to potential evapo-transpiration conditions is of crucial importance for plant growth in tropical semi-arid zones. The water needs of a crop may vary significantly between two locations with similar amounts and distributions, if the atmospheric demands at the two locations are different.

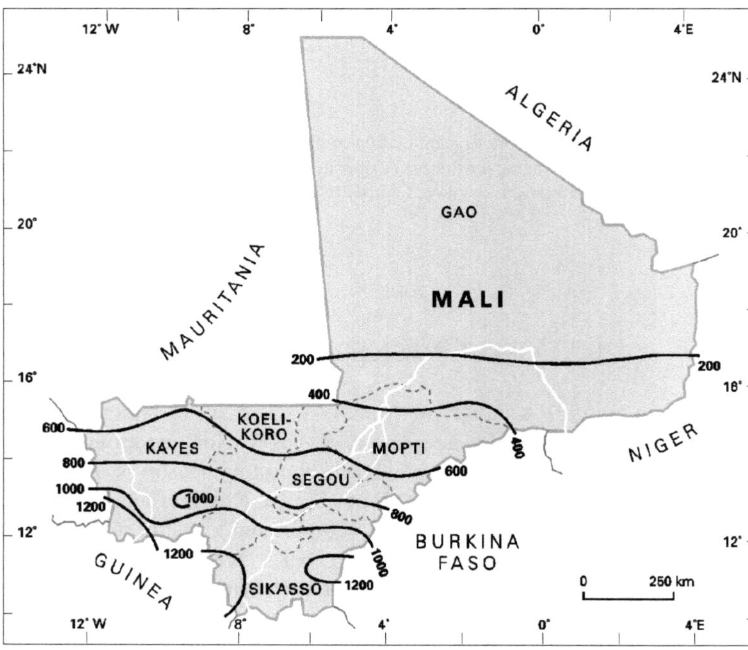

Map 23.1. Average annual rainfall in Mali.

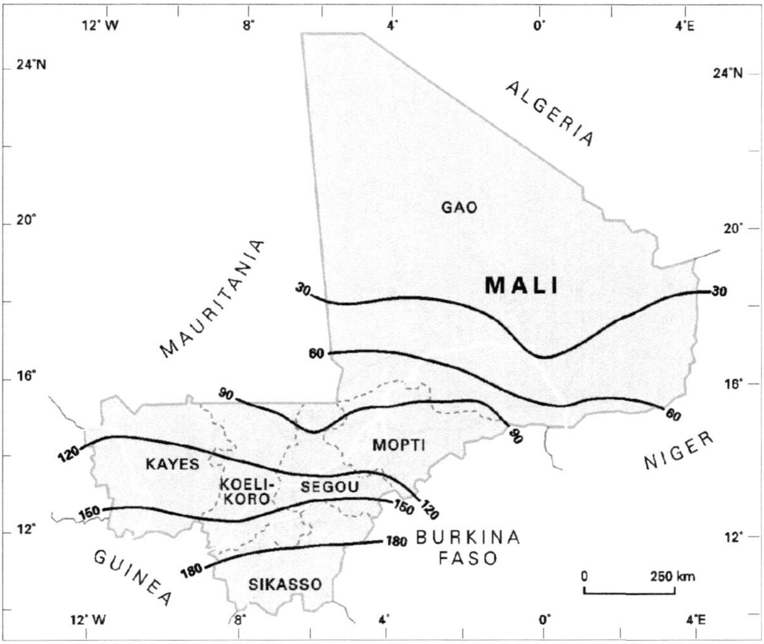

Map 23.2. Length of the rainy season.

2. OPERATIONAL INFORMATION SYSTEMS TO ASSESS DROUGHT HAZARDS.

In response to the severe drought that affected the Sahel during the 1970s, the National Meteorological Service in Mali implemented a project in 1982 aimed at providing farmers with meteorological information and agro-meteorological advice for agricultural practices. A multi-disciplinary, demonstrative and participatory approach was used to implement the project. The information and advice were related to seasonal forecasting, short-term forecasting, the reference agro-climatic calendar and the right periods for ploughing, planting, weeding, applying fertiliser, etc.

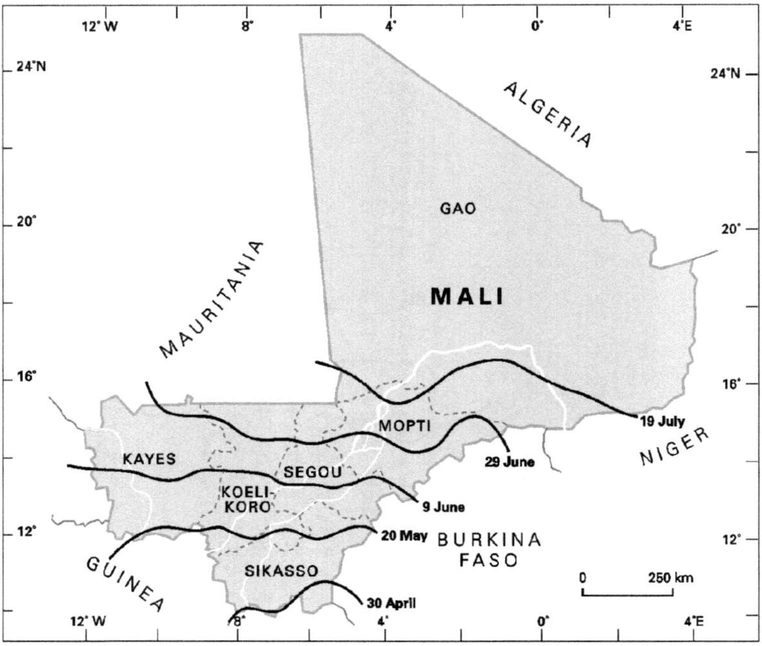

Map 23.3. Average start of the rainy season.

A multi-disciplinary working group comprising representatives from technical service providers in the rural sector, the hydrology sector and the Broadcast Company was co-ordinated by the National Meteorological Service. This group meets every ten days and analyses data coming from meteorological and hydrological stations, satellites (vegetation index maps), and agricultural and livestock workers (data on crops, pests and diseases, pastures, water holes, etc. At the end of the meeting a bulletin is issued. The bulletin gives the synoptic meteorological pattern, an analysis of the rainfall situation, the situation of rivers and their impacts on crops, pastures and water holes. The agro-hydro-meteorological conditions expected for the next ten-day period are also given in the bulletin. The bulletin is then broadcast on radio and TV and sent to decision makers in order to help them follow the evolution of the agricultural season and take appropriate and timely decisions and actions as necessary. The existing early warning system for food shortages also uses the information, together with other indicators, to assess the food situation in the country.

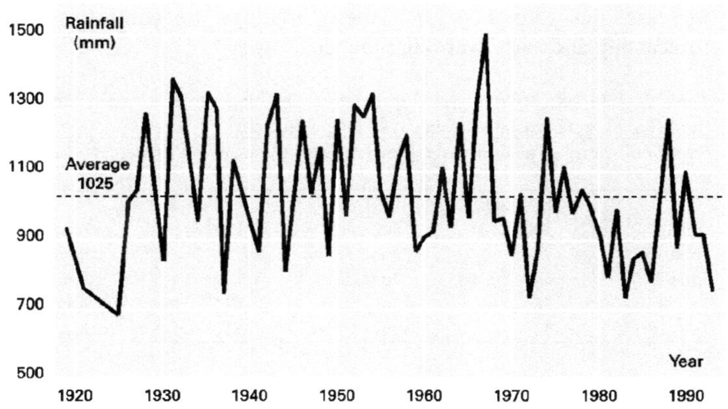

Figure 23.1 Inter-annual variability of the annual rainfall in Bamako.

An important tool for the assessment of imminent agro-meteorological conditions in a particular year is the statistical probability of a certain length of the agricultural growing season, given the start of the rainy season. Two examples are given here, one for Bamako, in the southern sub-humid zone and one for Bandiagara in the central, semi-arid zone. See tables 23.1 and 23.2.

Table 23.1. Probability of a growing season lasting longer than a certain number of days, for different rainy season starting days , for Bamako.

Start of the rainy season	Probability of a growing season longer than (in percentage)			
	>130 days	>150 days	>170 days	>190 days
May 5	99	86	42	7
May 15	96	68	20	2
May 25	86	42	7	0
June 4	68	20	2	0
June 14	42	7	0	0

Evaluations of the effectiveness of the Malian approach show that yields became 20 to 30% higher, fewer seeds were lost, the use of pesticides was reduced and the cost-benefit ratio was 1:10. As a result of improved farming performance, rural depopulation was reduced, social cohesion strengthened, the position of rural women alleviated and the environment more effectively preserved. For Malian agriculture, the 1990s was not a bad decade and many farmers look to the future with more confidence than during the 1970s and 1980s. However, climate change may threaten these positive developments. The National Meteorological Service and its partners are more aware now of the necessity to study the long-term changes in rainfall conditions and to increase the efforts to use these findings to continue improving forecasting and early warning practices.

Table 23.2. Probability of a growing season lasting longer than a certain number of days, for different rainy season starting days , for Bandiagara.

Start of the rainy season	Probability of a growing season longer than (in percentage)			
	>70 days		>70 days	
June 8	100	June 8	100	June 8
June 18	98	June 18	98	June 18
June 28	93	June 28	93	June 28
July 8	78	July 8	78	July 8
July 18	53	July 18	53	July 18

Chapter 24

CLIMATE CHANGE PREPAREDNESS IN WEST AFRICA

Ton Dietz, Ruerd Ruben, Jan Verhagen,
with Saa Dittoh, M. Konaté, David Millar, Edward Ofori-Sarpong, Hassane Saley and Ndiaye Cheikh Sylla

Abstract: This last chapter presents the results of expert prioritisation of policy recommendations that emerged from the ICCD programme. Experts gave the highest priority to developing an adequate early warning system with an efficient strategy to communicate with households and institutions. It is necessary for a better understanding of climate change and its effects and for the development of technologies adapted to location and sector specific conditions. In addition, high priority was given to maintaining social security mechanisms, understanding migration strategies and regulating land and water entitlements. Careful attention is needed for potential conflicts when resources become scarce. Local government and non-government organisations need support to monitor economic changes and to implement local policies. Agricultural research plays an important role in developing technologies that perform well under drought conditions. International agreements on climate change implications may be exploited for example by redefining subsidy practices. Finally there is plenty of scope for improving scientific research on climate change by extending research networks, by fine-tuning the ICCD models, and by expanding the geographic area of research.

1. POLICY PRIORITIES

At the final ICCD workshop in Wageningen (April 26-27, 2001) three groups of experts, chaired by African scholars, listed a number of

A.J. Dietz et al. (eds.),
The Impact of Climate Change on Drylands: With a Focus on West-Africa, 403–408.

policy priorities for African governments for interested donor countries and for interested scholars to enable them to become better prepared for climate change (rising temperatures, lower rainfall, and less reliable rainy seasons). Later, 22 of these experts gave priorities to 16 major policy statements. Some of these experts had participated in the ICCD research activities and others were invited guests, from Africa (Saa Dittoh, David Millar, Francis Obeng and Edward Ofori Sarpong from Ghana, M. Konaté from Mali, Hassane Saley from Niger and Ndiaye Cheikh Sylla from Senegal) and from elsewhere (Peter Hazell from IFPRI, Washington; Thea Hilhorst from IIED, London; Chris Reij from Free University, Amsterdam, The Netherlands). We will give the outcome of the prioritisation exercise:

Highest priorities

1. In all Sahelian countries adequate early warning systems should be (further) developed and operational assistance should be given to governments to enable them to develop effective indicators and to communicate early warning messages more efficiently to the relevant institutions and to the farmers and pastoralists in the region. This Early Warning Data should also guide famine relief operations. The experiences in Mali (since 1983) could be used as a guideline.

2. The development of knowledge of climate change and climate variability and of adaptations in the Sahel region should be improved (as well as knowledge of the world's drylands in general). This should facilitate the integration of scientific disciplines that deal with the issue (agro-biological sciences, geography, anthropology and economics in particular) partly on the basis of a more effective operationalisation of relevant concepts and partly by joint empirical (field-)work.

3. Adaptive technologies should be developed and tested, for agriculture, (agro-) pastoralism, sylviculture and horticulture; the adequacy of more 'northern' technologies should be tested in more 'southern' areas, supposing that the semi-arid zone moves southward.

4. Existing social security mechanisms should be maintained and new ones developed. More knowledge is needed about

the functioning of social security networks and mechanisms during and after catastrophic events (e.g. droughts; floods; locust invasions).

5. More attention should be paid to migration (and related remittance and remittance investment practices) and the importance of migration for both rural and urban economies should be redefined. Agricultural policy in the region should be more aware of the role of geographical mobility, not only in pastoral systems but in arable systems as well. Agricultural policy should also take (growing) urban demand as a point of departure and when assessing urban demand more attention should be paid to the (growing) importance of non-local sources of this demand.

6. The governments in the Sahel region should develop policy positions on land and water issues: ownership, access, control, investments and benefits, with specific attention for aspects of equity and for the livelihood of mobile persons/groups. It is important to map areas that are still relatively under-utilised in the sub-humid zone and to develop policy guidelines for sustainable land and water utilisation in these hitherto relatively 'empty' zones. It is also important to look specifically at the land and water 'entitlement' changes in the peri-urban areas in the region. In looking at the possibilities for policy interventions, careful attention needs to be paid to implications for inter-ethnic relations and potential violence.

Medium priorities

7. Attempts to decentralise policy formulation should be strengthened and operationalised/ implemented. The financial strength of the local government system needs to be built up as well as the capacity to monitor land use changes and to implement regulatory arrangements for local-level situations.

8. Public investments in two types of infrastructure should be given priority, namely education and water. Education is partly needed to make people less dependent on agriculture. Water investments are needed to make agriculture and animal husbandry less dependent on rainfall (irrigation, water

harvesting technology, water for animals), to make people less dependent on rainfall for their drinking water and to enable water-dependent forms of industrialisation and energy production. In developing groundwater dependent forms of irrigation (and other water 'production') groundwater levels should be better monitored and groundwater depletion prevented.

9. The policies on subsidies should be redefined to take advantage of international agreements, e.g. on 'carbon sinks'.

10. More research is needed on drought-resistant, early-maturing crops and varieties (agriculture should become less vulnerable). The exchange of information is important within the region and with other institutions in drylands elsewhere in the world, which are leaders in the development of dryland agro-technology. The role of ICRISAT could be strengthened and its geographical coverage could become wider.

11. Non-governmental organisations, community-based organisations and farmers' organisations should be strengthened and these institutions should be involved in the formulation and implementation of (government and donor) policies. The institutional capacity of these non-governmental agencies is crucial in coping with deteriorating situations. The strengthening of government institutions (central and local) should never undermine the resilience of the local-level non-governmental institutions.

Lower priorities (but still important)

12. Scientific models should be developed to facilitate a better understanding of adaptations over time and responses to periods of droughts (some experts at the workshop regarded this as a low priority because they think these models already exist, e.g. the pathway approach, and the major task ahead is to subject 'models' to additional testing).

13. Micro-credit systems (and micro-insurance systems) should be strengthened to enable people to cope better with drought shocks. This could be one of the new approaches of priority no. 4.

14. Further co-operation is needed between research centres in the region and between those centres and research centres abroad to enable the more efficient and more rapid dissemination of results (this is an extension of priority 10).

15. The (econometric) models, which were used in the ICCD research, need to be re-evaluated in terms of parameters used and modules included.

16. The attention for the impact of climate change on drylands should take a larger area into account (include the arid as well as the humid areas).

2. IMPLICATIONS FOR RESEARCH.

In order to reinforce policy-making in response to climate variability, a number of strategic research areas can be identified that are of critical importance for the development of adequate risk-coping or mitigation strategies.

At four different levels some important areas for further research can be identified:

a Plot-level research

- critical indicators for water availability and their impact on (potential) yields;
- water availability from rainfall and groundwater level;
- rainfall variability and yield risks;
- crop substitution patterns (sorghum-millet).

b Research at Farm Household level

- yield differences between farmers facing similar drought risks;
- food security strategies through diversification or specialisation;
- food security based on farm and non-farm income sources;
- livestock keeping for production and insurance purposes;
- energy requirements and forest rehabilitation;
- relationship between ethnicity and selected development pathways.

c Research at village and regional level

- spatial 'mobility' of crops under changing rainfall regimes;

- adjustment of farming practices by migrant populations;
- relationships between tenure change and land use patterns;
- rainfall risks and implications for land concentration;
- external (food) aid and the disruption of mutual assistance systems.

d Research on (inter)national level
- impact of price distortions on incentive regimes for food security;
- impact of climate variability on market prices;
- prospects for (inter)national insurance systems.

References

Adger, N. (1999). Social Vulnerability to Climate Change and Extremes in Coastal Vietnam. World Development 27, pp. 249-270.

Adolph, B., Y. et al. (1993). Coping with uncertainty: Challenges for agricultural development in the Guinea Savanna Zone of the Upper West Region, Ghana. Wageningen: International Centre for Development Oriented Research in Agriculture (ICRA).

ADRK (1995). Diagnostic de la zone d'intervention de l'ADRK pour l'amorce d'un programme de gestion des terroirs villageois. Kaya: ADRK.

Afrique Agriculture (1998), Bilan et perspective de la libéralisation de la filière coton dans la zone franc. Afrique Agriculture, No. 260 (June 1998), pp. 20-23.

Afrique Agriculture (1999a), Cameroun : récolte moyenne et privatisation en suspens. Afrique Agriculture, No. 271 (June 1999), pp. 24-25.

Afrique Agriculture (1999b), Maintien de la fertilité dans le Sud du Mali. Afrique Agriculture, No. 271 (June 1999), pp. 28.

Agossou, Y (1998), La campagne cotonnière 1997/98 au Togo. CORAF (1998), L'utilisation des intrants en cultures cotonnière et maraîchères (Dakar, Sénégal, 25-28 janvier 1998), pp. 63-73.

Agoua, C., S.B. Orou Yo, & M. Fadonougbo (1997), Contribution des producteurs-observateurs à la vulgarisation de la lutte étagée ciblée au Bénin. Campagnes 1995/96 et 1996/97. CORAF (1997), Réunion phytosanitaire de l'Afrique de l'Ouest et du Centre (Cotonou, Bénin, 27-31 janvier 1997), pp. 218-228. CORAF, Réseau Coton.

Albritton, D.L., L.G. Meira Filho, U. Cubasch, X. Dai, Y. Ding, D.J. Griggs, B. Hewitson & J.T. Houghton (2002). Technical summary, pp. 21-83. In: IPCC 2002a.

Al-Hassan, R., J.A. Famiyeh & A. de Jager (1997). Farm Household Strategies for Food Security in Northern Ghana: A Comparative Analysis of High and Low Population Farming Systems. In: Asenso-Okyere, W.K., George Benneh & Wouter Tims, eds. Sustainable food security in West Africa, pp.129-152. Dordrecht: Kluwer Academic Publishers.

Amin, S. (1974). Modern Migrations in Western Africa. Oxford: University Press.

Amin, S. (1995). Migrations in Contemporary Africa: A retrospective view. In: J. Baker & T.A. Aina, eds. The Migration Experience in Africa, pp. 29-40. The Scandinavian Institute of African Studies.

Ammar, S. & A. Valdes (1986). Should Crop Insurance be Subsidized? In: Hazell, P., C. Pomareda & A. Valdes, eds. Crop Insurance for Agricultural Development: Issues and Experience. Baltimore: Johns Hopkins University Press.

Antoine, P., D. Ouédraogo & V. Piché, eds. (1998). Trois générations de citadins au Sahel. Trente ans d'histoire sociale à Dakar et à Bamako. Paris: L'Harmatttan.

APM-AFRIQUE (1999a), Les filières coton d'Afrique subsaharienne en 98-99. L'Echo des cotonniers, No. Spécial (February 1999), pp. 3-14.

APM-AFRIQUE (1999b), La production cotonnière en Afrique francophone. L'Echo des cotonniers, No. 5 (May 1999), pp. 5.

Asenso-Okyere, W.K., S.Y. Atsu & I.S. Obeng (1993). Communal Property Resources in Ghana: Policies and Prospects. ISSER Discussion Paper 27. University of Legon, Ghana: Institute of Statistical, Social and Economic Research.

Atingdui (1988). Taxation and Structural Change in Ghana's Economy 1960-1980. (Ph.D. thesis, Department of Economic History, Lund University).

Baas, S., L. Yongong & J. Swift (2001). Guidelines for developing improved risk assessment, prevention and management approaches in pastoral areas of Northwestern China. [online] Sustainable Development Department of FAO. Available from: www.fao.org/sd/2001/IN0602a_en.htm.

Bailey, H.P. (1979). Semi-arid Climates: Their Definition and Distribution. In: Hall, A.E., G.H. Cannell & H.W. Lawton. Agriculture in Semi-Arid Environments. Springer, Ecological Studies No. 34, Berlin.

Baker, J. (2001). Migration as a positive response to opportunity and context: the case of Welo, Ethiopia. In: Bruijn, M. de, R. van Dijk, D. Foeken, eds. Mobile Africa, Changing Patterns of Movement in Africa and beyond, pp. 107-125. Leiden: Brill.

Baker, J., ed. (1997). Small towns in Africa: studies in rural-urban interaction. Uppsala: The Scandinavian Institute of African Studies.

Barbier, B. & P. Hazell (1998). Implications of Declining Access to Transhumant Areas and Sustainability of Agro-Pastoral Systems in the Semi-Arid Areas of Niger: A Bioeconomic Modeling Approach. In: McCarthy, N., B. Swallow, M. Kirk and P. Hazell, eds. Property Rights, Risk, and Livestock Development in Africa.

Barning, N.M. & J.B. Dambré (1994). Les styles d'exploitation. Une classification des exploitations dans la province du Sanmatenga, Burkina Faso. Kaya: PEDI.

Barral, H. (1977). Les populations Nomades de l'Oudalan et leur espace pastoral. Paris: ORSTOM.

Barret, P. (2003). Cooling a continent. Nature 421, pp. 221-223.

Bartle, P.F.W. (1980). Cyclical Migration and the Extended Community: A West African Example. Leiden: Afrika Studie Centrum.

Bascom, J. (1995). The New Nomads: An overview if Involuntary Migration in Africa. In: Baker, J. & T.A. Aina, eds. The Migration Experience in Africa, pp. 197-219. Uppsala: The Scandinavian Institute of African Studies.

Beauval, V. & G. Raymond (1992), L'avenir du coton au Bénin. Ministère de Développement Rural, Cotonou, Benin. February 1992, 154p..

Beck, U. (1992). Risk society. Towards a new modernity. London: Sage.

Beek, W. van & M. Peters (1999). Building Dams in Dogon. In: Venema, B. & H. van den Breemer, eds. Towards Negotiated Co-management of Natural Resources in Africa, pp. 87-113. Münster: Lit Verlag.

Beek, W. van & P. Banga (1992). The Dogon and their Trees. In: Croll, E. & D. Parkin, eds. Bush Base, Forest Farm, Culture Environment and Development, pp. 57-75. London: Routledge.

Benjaminsen, T.A. (1993). Fuelwood and desertification: Sahel orthodoxies discussed on the basis of field data from the Gourma Region of Mali. Geoforum 24, pp. 397-409.

Benjaminsen, T.A. (2001). The Malian Cotton Zone: Economic Success, but Environmental Failure? In: T.A. Benjaminsen & C. Lund, eds. Politics, Property and Production in the West African Sahel. Understanding Natural Resources Management, pp. 255-277. Uppsala: Nordiska Afrikainstitutet.

Benneh, G. (1973a). Small-scale farming systems in Ghana. Africa 43, pp.134-146.

Benneh, G. (1973b). Land tenure and farming systems in a Sissala village in Northern Ghana. Bulletin de l'Institut d'Afrique Noir 35, pp. 361-379.

Benneh, G. (1974). The perceived environment as a factor in the decision-making process of the the traditional farmer in Ghana. In: Laryea, A.M., ed. Environment and Development in West Africa. Accra: Ghana Academy of Arts and Science/UNESCO/Environmental Protection Council.

Benoit, M. (1982). Nature peul du Yatenga. Remarques sur le pastoralisme en pays mossi. Paris: ORSTOM.

Benson, C. & E. Clay (1998). The impact of drought on Sub-Saharan African Economies. Technical Paper 401. Washington: World Bank.

Berger, M. (1996), L'amélioration de la fumure organique en Afrique soudano-sahélienne. Agriculture et Développement, No. hors-série (1996), 59p..

Béroud, F. (1999), Coton et environnement : des idées reçues dont il faut se méfier. Afrique Agriculture, No. 271 (June 1999), pp. 59-61.

Bingen, R. J. (1994). Agricultural development policy and grassroots democracy in Mali: the emergence of Mali's farmer movement. African Rural and Urban Studies 1, pp. 57-72.

Binswanger, H. (1986). Risk Aversion, Collateral Requirements, and the Markets for Credit and Insurance in Rural Areas. In: Hazell, P., C. Pomareda and A. Valdes, eds. Crop Insurance for Agricultural Development: Issues and Experience. Baltimore: Johns Hopkins University Press.

Blaikie, P., T. Cannon, I. Davis & B. Wisner (1994). At risk. Natural hazards, people's vulnerability, and disasters. London & New York: Routledge.

Boateng, E.A. (1966). A Geography of Ghana. Cambridge: University Press.

Boer, Y. de (2001). On the Road to Tombouctou Climatological, Ecological And Human Influences on the Vegetation in Semi Arid Mali. (M.Sc. thesis, University Wageningen).

Bogner, A. (2000). The 1994 civil war in northern Ghana: the genesis and escalation of a 'tribal' conflict. In: Lentz, C. & P. Nugent, eds. Ethnicity in Ghana: The Limits of Invention, pp. 183-203. London: Macmillan Press Ltd.

Bohle, H.G., T.E. Downing & M.C. Watts (1994). Climate change and social vulnerability: Toward a sociology and geography of food insecurity. Global Environmental Change 4, pp.37-48.

Bonnard, P. (2000). Assessing Urban Food Security. Adjusting the FEWS Rural Vulnerability Assessment Framework to Urban Environments. FEWS/ARD.

Born, G.J. van den, M. Schaeffer & R. Leemans (1998). Identification of climate scenarios for the semi-arid and sub-humid regions, comparing existing climate scenarios for the dryland regions in West Africa, covering 1990-2050. Report ICCD 2. Utrecht: CERES; Wageningen: DLO/LUW/RIVM.

Born, G.J. van den, M. Schaeffer & R. Leemans (2000). Climate scenarios for semi-arid and sub-humid regions: A comparison of climate scenarios for the dryland regions in West Africa from 1990-2050. Bilthoven: RIVM.

Bosma, R., K. Bengaly, M. Traoré & A. Roeleveld (1996). L'élevage en voie d'intensification. Synthèse de la recherche sur les ruminants dans les exploitations agricoles mixtes au Mali-sud. Amsterdam: Institut Royal des Tropiques; Bamako: Institut d'Économie Rurale.

Boutillier, J.L. (1964). Les structures foncières en Haute-Volta. Études Voltaïques 5. Ouagadougou: IFAN-ORSTOM.

Boxel, J.H. van (2000). Climate change and precipitation: detecting changes. Meteorologica Colombiana 3, pp. 21-31.

Brandts, Aline, 2002, Dogon on the move. Cultivators in central Mali searching for ways to make a living. M.A. Thesis Utrecht University

Breman, H. & K. Sissoko (1998). Intensification Agricole au Sahel: Mythe ou Realite. Paris: Karthala.

Bretherton, F.P., K. Bryan & J.D. Woods (1990). Time-dependent greenhouse-gas-induced climate change, pp.173-193. In: IPCC 1990.

Breusers, M. (1999a). On the Move: Mobility, Land Use and Livelihood Practices on the Central Plateau in Burkina Faso. Hamburg: LIT.

Breusers, M. (1999b). The making of history in colonial Haute Volta: border conflicts between two Moose chieftaincies, 1900-1940. The Journal of African History 40.

Breusers, M. (1999c). The dynamics of land tenure arrangements on the Central Plateau in Burkina Faso. In: Proceedings of the seminar 'Acts of Man and Nature? Different Constructions of Natural and Social Resource Dynamics'. Wageningen: CERES.

Breusers, M. (2001a). Searching for livelihood security: land and mobility in Burkina Faso. The Journal of Development Studies, 37.

Breusers, M. (2001b). Pathways to negotiate climate variability. Land use and institutional change in the Kaya region, Burkina Faso. Leiden: African Studies Centre.

Breusers, M., S. Nederlof & T. van Rheenen (1998). Conflict or symbiosis? Disentangling farmer-herdsman relations: the Mossi and Fulbe of the Central Plateau, Burkina Faso. The Journal of Modern African Studies 36, pp. 357-80.

Brinkmann, A. (1990). Die sozio-ökonomische Bedeutung der Tierhaltung in kleinbäuerlichen Betrieben der Dagomba in Nordghana. Universität Göttingen: Diplomarbeit.

Broekhuyse, J.Th. (1974). Dévelopement du Nord des Plateaux Mossi. Amsterdam: KIT.

Broekhuyse, J.Th. (1983). Transformation du pays mossi. Amsterdam: KIT.

Broekhuyse, J. Th. (1985). Désertification et autosuffisance alimentaire. Une vue de la base. Kaya: Diocèse de Kaya; Oegstgeest: CEBEMO.

Broekhuyse, J.Th. & A.M. Allen (1988). Farming systems research on the northern Mossi Plateau. Human Organization 47, pp. 330-42.

Brons, J., F. Zaal, R. Ruben & L. Kersbergen (1999). Rural development and climate change in West Africa; A village level analysis in South Mali. Wageningen: WUR/DE; Amsterdam: UvA/AGIDS.

Brons, J., F. Zaal, R. Ruben & L. Kersbergen (2000). Climate Change, Agricultural Variability and Risk-coping Strategies. A farm and household level analysis in northern Burkina Faso. Wageningen: WUR/DE; Amsterdam: UvA/AGIDS.

Brons, J., S. Diarra, I. Dembele, S. Bagayoko & H. Djouara (1994). Diversite de Gestion de l'Exploitation: Etude sur les facteurs d'intensification agricole au Mali-Sud. Sikasso IER-DRSPR Document 94/33.

Brouwer, J. & J. Bouma (1997), Soil and crop growth variability in the Sahel. Highlights of research (1990-95) at ICRISAT Sahelian Center. Information Bulletin No. 49. ICRISAT/WAU, India/The Netherlands, 42p..

Brüntrup, M. (1997), Agricultural price policy and its impact on production, income, employment and the adoption of innovations. A farming systems based analysis of cotton policy in northern Benin. Development economics and policy, No. 9. Peter Lang GmbH, Frankfurt am Main, Germany, 515p..

Bruijn, M. de (1994). The Sahelian Crisis and the Poor, the Role of Islam in Social Security among Fulbe Pastoralists, Central Mali. Focaal 22/23, pp. 47-65.

Bruijn, M. de (1997). The hearthhold in pastoral Fulbe society, Central Mali. Social relations, milk and drought. Africa 67, pp. 625-651.

Bruijn, M. de (1998). Those who cannot travel: staying behind in Fulbe society in Central Mali. Paper presented at the International seminar on: "Crisis and Culture in Africa - with special Emphasis on Pastoral Nomads and farmers in the West African Sahel", 27-30 March 1998, Eklundshof, Uppsala, Sweden.

Bruijn, M. de & H. van Dijk (1994). Drought and Coping Strategies in Fulbe Society in the Hayre (Central Mali): A Historical Perspective. Cahiers d'Etudes africaines 34, pp. 85-108.

Bruijn, M. de & H. van Dijk (1995). Arid Ways, Cultural Understandings of Insecurity in Fulbe Society, Central Mali. Amsterdam: Thela Publishers.

Bruijn, M. de & H. van Dijk (1998). "Pathways" and "Habitus": Two Frameworks for the Analysis of Decision-Making in High-Risk Environments. ICCD Internal Paper (unpublished).

Bruijn, M. de & H. van Dijk (1999a). Fulbe Mobility: Migration and Travel into Mande. Mande Studies 1, pp. 41-62.

Bruijn, M. de & H. van Dijk (1999b). Insecurity and pastoral development in the Sahel. Development and Change 30, pp. 115-39.

Bruijn, M. de & H. van Dijk (2001a). Ecology and Power in the Periphery of Maasina: The Case of the Hayre in the Nineteenth Century. Journal of African History 42, pp. 217-238.

Bruijn, M. de & H. van Dijk (2001b). Moving People, Pathways in the Hayre-Seeno Area, Central Mali. ICCD/ASC working paper. Leiden: African Studies Centre.

Bruijn, M. de & H. van Dijk (in press). Risk Positions and Local Politics in a Sahelian Society. The Fulbe of the Hayre in Central Mali. In: Bill Mosely & B. Ikubolajeh Logan, eds. (Inter)National Political Economies and Local Ecologies: Rural African Livelihoods in a Political Ecology Context.

Bruijn, M. de, K. Sidibé & H. van Dijk (2001). Évaluation des appuis des OCF néerlandaises à la construction de la société civile au Mali. Rapport de synthèse. Ede, Comité de Pilotage « évaluation programme de co-financement néerlandaises », Leiden, Centre d'Études africaines, Bamako, Koni Expertise.

Bryceson, D.F. (1997b). De-agrarianisation: Blessing or Blight? In: Bryceson, D.F. & V. Jamal, eds. Farewell to Farms: De-agrarianisation and Employment in Africa, pp. 237-256. Hampshire: Ashgate Publishing Ltd.

Bryceson, D.F. (1997a). De-agrarianisation in Sub-saharan Africa: Acknowledging the Inevitable. In: Bryceson, D.F. & V. Jamal, eds. Farewell to Farms: De-agrarianisation and Employment in Africa, pp.3-20. Hampshire: Ashgate Publishing Ltd.

Campbell, D.J. & D.D. Trechter (1982). Strategies for coping with food consumption shortage in the Mandara Mountains Region of north Cameroon. Social Science and Medicine 16, pp. 2117-2127.

Cannon, T. (1990). Hazard Need Not a Disaster Make: Rural Vulnerability and the Causes of "Natural" Disaster. Lecture at the Institute of Social Studies in The Hague, 21 November 1990.

Cauquil, J. (1993), Maladies et ravageurs du cotonnier en Afrique au sud du Sahara. CIRAD, Montpellier, France, 92p..

Chambers, R. (1983). Rural Development: Putting the Last First. London, Lagos, New York: Longman.

Chambers, R. (1989). Editorial Introduction: Vulnerability, Coping and Policy. IDS Bulletin 20, pp. 1-7.

Chappellaz, J., J.M. Barnola, D. Raynaud, Y.S. Korotkevich & C Loius (1990). Ice-core record of atmospheric methane over the past 160,000 years. Nature 345, pp. 127-131.

Chauveau, J.P. (2000). La question foncière en Côte d'Ivoire et le coup-d'état. Ou: comment remettre à zéro le compteur de l'histoire. IIED Issue Paper 95. London: IIED.

Chen, M.A. (1991). Coping with seasonality and drought. New Delhi: Sage Publications India Pvt Ltd.

Chomitz, K.M., C. Griffiths & J. Puri (1998). Fuel prices, woodlands and woodfuel markets in the Sahel: an integrated economic-ecological model. Washington: Working Paper Development Research Group, World Bank/ US Environmental Protection Agency.

CIRAD (2002). Supplying fuelwood to towns and cities. [on line] Available on: http://www.cirad.fr.presentation/en/idr/bois.htm. [Accessed April 10th 2002].

Cissé, I. (1993). Les migrations agricoles au Mali. Cahiers du CIDEP 18.

Cissé, S & P.A. Gosseye (eds) (1991). Ressources naturelles et population, Série, Compétition pour des ressources limitées, le cas de la cinquième région du Mali: Rapport 1, Wageningen CABO, Mopti ESP.

RClarcke, T. (2001). Climate may have an infra-red iris. Nature science [online]. Available from: www.nature.com/nsu/010308/010308-9.html.

Claude, J., M. Grouzis & P. Milleville, eds. (1991). Un Espace Sahélien. La Mare d'Oursi. Burkina Faso. Paris: ORSTOM.

Cleveland, D.A. (1991). Migration in West Africa: A Savanna Village Perspective. Africa 61, pp. 222-246.

Club du Sahel (1995). The West African Long-Term Perspective Study. Newsletter No. 15, summer.

CMDT (1994). Annuaire Statistique – Resultats de l'Enquete Agricole Permanente. Bamako: CMDT Suivi Evaluation.

CMDT (1996), Le coton au Mali. Coton et Développement, No. 5 (January-March 1996), pp. 5-12 + 38-39.

Collier, P. & J.W. Gunning (1996). Policies towards commodity shocks in developing countries. Working Paper WP-96-84. Washington: IMF.

Collignon, B. & M. Vézina (2001). Independent Water and Sanitation Providers in African Cities. Full Report of a ten-country study. Washington: UNDP-World Bank Water and Sanitation Program.

CORAF (1998), L'utilisation des intrants en cultures cotonnière et maraîchères (Dakar, Sénégal, 25-28 janvier 1998), pp. 343-349

Corbett, J. (1988). Famine and household coping strategies. World Development 16, pp. 1099-1112.

Cretenet, M., D. Dureau, B. Traoré & B. Diallo (1994), Fertilité et fertilisation dans la région sud du Mali. Agriculture et Développement, No. 3 (August 1994), pp. 4-12.

Croll, E. & D. Parkin (1992). Anthropology, the Environment and Development. In: Croll, E. & D. Parkin, eds. Bush Base: Forest Farm, Culture, Environment and Development, pp. 3-11. London: Routledge.
CRS/Ghana (1992). Wa Diocesan Development Office Capacity Building Project. Concept Paper.
CTA (1996), Vers une gestion raisonnée des résidus de cotonniers. Spore, No. 66 (December 1996), pp. 8.
Curtin, P., S. Feierman, L. Thompson & J. Vansina (1978). African History. London: Longman.
Dakuo, D. (1998), Dégradation de la fertilité dans les zones cotonnières d'Afrique au sud du Sahara.
Davies, S. (1993). Are Coping Strategies a Cop Out? IDS Bulletin 24, pp. 60-72.
Davies, S. (1996). Adaptable livelihoods: coping with food insecurity in the Malian Sahel. Houndmills: Macmillan; New York: St. Martin's Press.
Davies, S. & N. Hossain (1997). Livelihood Adaptation, Public Action and Civil Society: A Review of the Literature. IDS Working Paper 57. Sussex: Institute of Development Studies.
Deaton, A. (1992). Understanding Consumption. Oxford: Clarendon Press.
DeConto, R.M. & D. Pollard (2003). Rapid Cenozoic glaciation of Antartica induced by declining atmospheric C02. Nature 421, pp. 245-249.
Degnbol, T. (1999). State bureaucracies under pressure. A study of the interaction between four extension agencies and cotton-producing farmers in the Sikasso region, Mali. (Ph.D. thesis, Roskilde University).
Dei, G.J.S. (1992). Hardships and Survival in Rural West Africa: A Case Study of a Ghanaian Community. Dakar: CODESRIA Monograph Series.
Dembélé, N.N. & J.M. Staatz (1999). The Impact of Market Reform on Agricultural Transformation in Mali. Nairobi. MSU Agricultural Economics Staff Paper pp. 99-29.
DEP/Ministère de l'Environnement et de l'Eau (1998). Politique et Stratégies en matière de l'Eau. Ouagadougou: DEP/MEE.
Département du Centre Nord (1978). Monographie du Département du Centre-Nord. Kaya.
Der, B.G. (1979). Agricultural Policy in Northern Ghana during the Colonial Era. Universitas 8, pp. 3-16.
Der, B.G. (1998). The Slave Trade in Northern Ghana. Accra: Woeli Publishing Services.
Dery, G. (1987). Inheritance and Marriage among the Dagaaba of Northern Ghana: Second Opinions to Existing Theories. Accra: Assembly Press.
Devereux, S. (1993). Goats Before Ploughs: Dilemmas of Household Response Sequencing During Food Shortages. IDS Bulletin 24, pp.52-59.
Deveze, J.D. (1994), Les zones cotonnières entre développement, ajustement et dévaluation. Réflexions sur le rôle du coton en Afrique francophone de l'Ouest et du Centre. Notes & Etudes, No. 53. CFD, Paris, France. April 1994, 29p.
Diabate, J.B. (1998), La campagne cotonnière 1997/98 au Mali. CORAF (1998), L'utilisation des intrants en cultures cotonnière et maraîchères (Dakar, Sénégal, 25-28 janvier 1998), pp. 31-42.
Diallo, A. (2000). L'effet du changement climatique sur la disponibilité en eau, l'agriculture et la sécurité alimentaire dans la cellule Géographique de Dori. Amsterdam: AGIDS/ UvA.
Diawara, M. (1997). "Dieu d'eau", eau du barrage. Les populations du plateau Dogon, face aux contraintes: Pluviométrie, terre et démographie. Africa 67, pp. 602-624.

Dietz, T. (1991). Crisis Survival Strategies: A summary of concepts and an example from the semi-pastoral Pokot in Kenya/Uganda. In: Stone, Jeffrey C., ed. Pastoral economies in Africa and long term responses to drought, pp. 86-108. Aberdeen: University African Studies Group.

Dietz, T., A. Arale Nunow, A. Wario Roba & F. Zaal (2001). Pastoral Commercialisation: On Caloric Terms of Trade and Related Issues. In: Mohamed Salih, M.A., T. Dietz & Abdel Ghaffar Mohamed Ahmed, eds. African Pastoralism: Conflict, Institutions and Government, pp. 194-234. London and Sterling, Virginia: Pluto Press.

Dietz, T. & A. van Haastrecht (1997). Livelihood stress in dryland zones. In: Naerssen van, T., M. Rutten & A. Zoomers, eds. The diversity of Development: Essays in Honour of Jan Kleinpenning, pp.43-56. Assen: Van Gorcum.

Dietz, T. & D. Millar et al. (1999). Coping with climate change in dryland Ghana: the case of Bolgatanga. Amsterdam/Tamale: internal NRP-ICCD report

Dietz, A.J. & D. Millar (1999a). Proceedings Workshop Climate change and farmers' coping strategies. Bolgatanga: (unpublished).

Dietz, A.J. & D. Millar (1999b). Impact of climatic variability on geographical and occupational mobility and social organisation in farming communities of north-eastern Ghana. (Ph.D. proposal for F. Obeng, University of Development Studies at Tamale).

Dietz, Ton, David Millar, and Francis Obeng, 2002, By the grace of God, the day will come when poverty will receive the final blow. The impact of NGOs supported by Dutch co-financing agencies on poverty reduction and regional development in the Sahel. Northern Ghana report. Working document to Steering Committee for the Evaluation of the Co-financing Programme. Amsterdam/Bolgatanga: University of Amsterdam/University for Development Studies Tamale (200 pp).

Dietz, T. & E. Veldhuizen (1998). Impact of climate change, water availability, agriculture and food security in semi-arid regions, with a special focus on West Africa. Report on the Identification phase. Amsterdam: UvA/AGIDS.

Dietz, T. & J. Verhagen (1998). The impact of Climate Change on Drylands: The Case of West Africa. Internal paper for the NOP-ICCD symposium. Garderen.

Dietz, T. & J. Verhagen (1999). The impact of climate change on drylands; the case of West Africa. Proceedings of the NOP symposium. Bilthoven: RIVM.

Dietz, T., J. Verhagen & R. Ruben (2001). Impact of Climate Change on Drylands with a focus on West Africa. Wageningen: ICCD. Report for Dutch Research Programme on Global Air Pollution and Climate Change, no 410 200 076 (133 pp).

Dijk, H. van (1994). Livestock transfers and Social Security in Fulbe Society in the Hayre, Central Mali. Focaal 22/23, pp. 97-112.

Dijk, H. van, D. Foeken & K. van Til (2001). Population Mobility in Africa: An Overview. In: Bruijn, M. de, R. van Dijk & D. Foeken, eds. Mobile Africa, Changing Patterns of Movement in Africa and beyond, pp. 9-27. Leiden: Brill.

Dinar, A. et al. (1998). Measuring the impact of Climate Change in Indian Agriculture. Technical Paper, 400. Washington D.C.: World Bank.

Djibo, M, M. Kioni, M. Moustapha & M. Ali (1999). Description de la cellule urbaine de Niamey (Niger) en functions des themes , problématique et objectifs de l'ICCD (Rapport provisoire).

DNSI (1991). Enquete Budget Consommation. Bamako: DNSI/PNUD-PADEM/CILLS-DIAPER.

Dreze, J. & A.K. Sen (1989). Hunger and Public Action. London: Clarendon.

DRPC (1990). Plan régional du secteur agro-pastoral, Région du Centre-Nord: Tome 2, Plan d'Action. Kaya: DRPC.

DRPC-Dori (1998). Schéma régional d'amenagement du territoire-STRAT. Sahel Tome 1. Direction Régionale du Plan et de la Coopération-Dori.

Dubourg, J. (1957). La vie des paysans mossi: le village de Taghalla. Les Cahiers d'Outre-Mer 10, pp. 285-324.

Ducellier, J. (1956). Rapport sur l'hydrogéologie de la région de Kaya (Haute-Volta). Gouvernement Général de l'A.O.F. Direction Fédérale des Mines et de la Géologie.

Dupriez, H. & P. de Leener (1990), Les chemins de l'eau. Ruissellement, irrigation et drainage. Manuel tropical. ENDA Tiers Monde, Dakar, Senegal, 380p..

Economist (1958). Government Revenue and Expenditure in Ghana. Universitas 3, pp. 109-112.

Ellis, F. (1992). Agrarian Policies in Developing Countries. Cambridge: Cambridge University Press.

Ellis, F. (1998). Household Strategies and Rural Livelihood Diversification. The Journal of Development Studies 35, pp. 1-38.

Ellis, F. (2000). Rural livelihoods and diversity in developing countries. Oxford: Oxford University Press.

Elobu, P., J.F. Orwanga, J. Ocan & G.W. Opolot (1995), Recovery of cotton (Gossypium hirsutum) from intercropping suppression by beans (Phaseolus vulgaris). In: G.A. Constable & N.W. Forrester (eds.) (1995), Challenging the future. Proceedings of the World Cotton Research Conference - 1 (Brisbane, Australia, February 14-17, 1994), pp. 206-209. CSIRO, Canberra, Australia.

ENDA-PRONAT (1998), Rapport Annuel du Projet Coton Organique au Sénégal. Campagne 1997/98. ENDA-Pronat, Dakar, Senegal. June 1998, 34p.

ENDA-PRONAT (1999), Rapport final. Campagne 1998/99. ENDA-Pronat, Dakar, Senegal. April 1999, 40p.

Escobar, A. (1995). Power and Visibility: Tales of Peasants, Women and the environment. In: Encountering Development: The Making and Unmaking of the Third World, pp.154-211. Princeton: University Press.

Eskelinen, R. K. (1979). Dogon Agricultural Systems. Sociological Aspect Relating to Development Interventions. In: Riita K. Eskelinen, John Aron Grayzel, John Van Dusen Lewis & Hamadoun Sidibé. Social Anthropological and Rural Economic Studies in Mali: The Dukulomba Forest, the Bandiagara Plateau. Binghamton: Research Foundation of the State University of New York.

Esteulle, B. (1996), Le coton au Ghana. Coton et Développement, No. 20 (October-December 1996), pp. 19-23.

Eyre-Smith, R. St. J. (1933). A Brief Review of the History and Social Organisation of the Peoples of the Northern Territories of the Gold Coast. Accra.

FAO (1980). Report on the second FAO/UNFPA Expert Consultation on Land Resources for Populations of the Future. Rome: FAO.

FAO (1997). FAO Production Yearbook. Rome: Food and Agricultural Organization.

FEWS (2000). Current Vulnerability Assessment [online]. FEWS Special Report, FEWS Sahel 1999/2000. Available from: www.fews.org.

FEWS (2002). FEWS Sahel, Burkina Faso Center, Mali Center, Niger Center [online]. Available from: www.fews.net/centers/mar 15 2002.

FEWS (2002). Mali rapport mensuel sur la sécurité alimentaire [online]. Washington D.C. : USAID. Available from: http://www.fews.net/ centers/files/Mali_200201fr.pdf.

Foeken, D. & S. Owuor (2001). Multi-sptaial livelihoods in Sub-Saharan Africa: Rural farming by urban households – The case of Nakuru town, Kenya. In: M. de Bruijn, R.

van Dijk & D. Foeken, eds. Mobile Africa. Changing patterns of movement in Africa and Beyond, pp. 125-139. Leiden: Brill.

Foeken, D.W.J. (1989). Aspects of seasonality in Africa. In: Foeken, D.W.J. & A.P. den Hartog, eds. Seasons, food supply and nutrition in Africa: Contributions to a workshop held in Wageningen on December 14, 1988, pp.6-29. ASC Research Report, Vol.43. Leiden: Afrika Studie Centrum.

Fok, M. (1999), Les exigences du marché : compétitivité, prix, qualité et image. Presentation at the seminar organised by the CIRAD Cotton Programme "Rôle et place de la recherche pour le développement des filières cotonnières en évolution en Afrique" (Montpellier, France, September 1-2, 1999). CIRAD, Montpellier, 9p.

Gaborel, C. (1999), Contributions de la recherche thématique en zones cotonnières. Portée et limites. Presentation at the seminar organised by the CIRAD Cotton Programme "Rôle et place de la recherche pour le développement des filières cotonnières en évolution en Afrique" (Montpellier, France, September 1-2, 1999). CIRAD, Montpellier, 8p.

Gallais, J. (1965). Le paysan dogon. Les Cahiers d'Outre-Mer VIII, pp. 123-143.

Gallais, J. (1975). Pasteurs et paysans du Gourma: la conditions sahélienne. Paris : Centre National de Recherche Scientifique.

Geest, K. van der (2002a). Impact Analysis of the Interventions of the Producer Enterprise Promotion Service Centre (PEPSC) in Northwest Ghana. Unpublished.

Geest, K. van der (2002b). We are Managing!? Vulnerability and responses to climate variability and change among rural households in Northwestern Ghana. (M.Sc. thesis, University of Amsterdam).

Gervais, R. (1990). Population et politiques agricoles coloniales dans le Mosi, 1919-1940. (Thèse de Doctorat, Université de Paris).

Geschiere, P. (2001). Liberalization and new struggles over belonging: towards a recomposition of the village in Africa? In: Andersson, J.A. & M. Breusers eds. Kinship structures and enterprising actors. Anthropological essays on development. Wageningen: Wageningen University.

Ghana Statistical Service (1993). Rural Communities in Ghana: Report of a national rural community survey carried out as part of the third round of the Ghana Living Standards Survey of 1991/92. Accra.

Ghana Statistical Service (2000a). 2000 Population and Housing Census: Provisional Results. Accra: Ghana Statistical Services.

Ghana Statistical Service (2000b). Poverty trends in Ghana in the 1990s. Accra: Ghana Statistical Services.

Gielen, G. (2001). De vleesvoorziening van Bamako, Mali. (Master thesis, University Utrecht).

GOM (1990). Notice de cercle, cercle de Koro, projet inventaire des ressources ligneuses et occupation agricole au Mali. République du Mali: MDRE, Direction Nationale des Eaux et Forêts.

GOM (1991). Développement régional du plateau dogon. Schéma directeur , Rapport de synthèse, annexes et cartes. République du Mali : Ministère de l'Agriculture, Direction National de l'Agriculture in collaboration with Essen, Agrar und Hydrotechnik Gmbh.

Gommes, R. & F. Petrassi (1994). Rainfall variability and drought in sub-Saharan Africa since 1960. FAO agrometeorological series 9.

Goody, J. (1967). The Social Organisation of the LoWiili. Oxford: University Press.

Goody, J. (1969). Comparative Studies in Kinship. London: Routledge and Kegan Paul.

Gottschalk, L. & I. Krasovskaia (1987). Water Resources in Gourma. A Desk Study. Uppsala: AB Hydroconsult.

Graaff, J. de (1996). The price of soil erosion. An economic evaluation of soil conservation and watershed development. Tropical Resource Management Papers 14.

Greeley, M. (1994). Measurement of Poverty and Poverty of Measurement. IDS Bulletin 25, pp. 50-57.

Griep, S. (2001). Masters over the field? Coping strategies of the Riimaybe in Debere. ICCD/ASC Working Paper. Leiden: Afrika Studies Centre.

Grigg, D. (1995). An Introduction to Agricultural Geography. London & New York: Routledge.

Gudger, M. (1991). Crop Insurance: Failure of the Public Sector and the Rise of the Private Sector. In: Holden, D., P. Hazell & A. Pritchard, eds. Risk and Agriculture: Proceedings of the Tenth Agricultural sector Symposium. Washington D.C.: World Bank.

Gueye, M. (1998), Productivité et valeur fertilisante de la matière organique d'origine animale : optimisation des pratiques de sa gestion dans le système de culture à base de coton au Sénégal. In : PAN-Africa (1998), Coton biologique en Afrique (Tambacounda, Sénégal, 24-28 novembre 1997), pp. 50-53.

Hall, S. (1999). Livestock Production Systems in Northern Ghana in relation to Environmental Conservation and Economic Development. In: Blench, R. ed. Natural Resource Management in Ghana and its Socio-economic Context, pp. 44-59. London: Overseas Development Institute.

Hampshire, K. & S. Randall (1994). Migration and household structure: flexibility as a survival strategy for Sahelian agro-pastoralists. Paper for the conference on population and environment, Bradford, DPPC, December 1994. London: UCL/Dept of Anthropology.

Hampshire, K. & S. Randall (2000). Pastoralists, agropastoralists and migrants: interactions between fertility and mobility. Population studies : a journal of demography 54, pp. 247.

Hardaker, J.B., R.B.M. Huirne & J.R. Anderson (1998). Coping with Risk in Agriculture. Wallingford: Cab International.

Hart, K. (1982). The political economy of West African agriculture. Cambridge: University Press.

Harts-Broekhuis, E.J.A. & A.A. de Jong (1993). Subsistence and survival in the Sahel. Responses of households and enterprises to deteriorating conditions and development policy in the Mopti region of Mali. Geographical Studies 168.

Hau, B. (1995). Progress with varietal improvement in CIRAD-CA. In: G.A. Constable & N.W. Forrester (eds.) (1995), Challenging the future. Proceedings of the World Cotton Research Conference - 1 (Brisbane, Australia, February 14-17, 1994), pp. 381-384. CSIRO, Canberra, Australia.

Hau, B. & M. Vaissayre (1999). Rapport de mission au Sénégal (17-28 mai 1999). CIRAD-CA, Montpellier, France, 47p.

Hawkins, S. (1997). To Pray or not to Pray: Politics, Medicine, and Conversion among the LoDagaa of Northern Ghana, 1929-1939. Canadian Journal of African Studies 31, pp. 50-85.

Hazell, P. (1992). The Appropriate Role of Agricultural Insurance in Developing Countries. Journal of International Development 4, pp. 567-581.

Hazell, P., C. Pomareda & A. Valdes, eds. (1986). Crop Insurance for Agricultural Development: Issues and Experience. Baltimore: Johns Hopkins University Press.

Hazell, P., P. Oram, & N. Chaherli (2001). Managing Droughts in the Low-Rainfall Areas of the Middle East and North Africa. Environment and Production Technology Division. Discussion Paper 78, 2001. Also forthcoming in: Löfgren, H. ed. Food and Agriculture in the Middle East. Research in Middle East Economics, 5.

Hearn, A.B. (1995). The principles of cotton water relations and their application in management. In: G.A. Constable & N.W. Forrester (eds.) (1995), Challenging the future. Proceedings of the World Cotton Research Conference - 1 (Brisbane, Australia, February 14-17, 1994), pp. 66-92. CSIRO, Canberra, Australia.

Hengsdijk, H., W..Quak, E.J. Bakker & J.J.M.H. Keetelaars (1996). A land use system and technology generator for the Koutiala region in South Mali. Wageningen: DLV Report Bo. 6.

Hesse, C. (1987). Livestock Market Data as an Early Warning Indication of Stress in the Pastoral Economy. Pastoral Development Network Paper 24f.

Hesse, C., A. Thiam, C. Fowler & J.J. Swift (1985). The Gurma Fulani. In: Swift, J.J. ed. Pastoral and Agro-pastoral Production Systems in Central Mali: Three Case Studies. Addis Ababa & Bamako: International Livestock Centre for Africa.

Hesseling, G. (1996). Legal and institutional incentives for local environmental management. In: Marcussen, H.S. ed. Improved Natural Resource Management – The Role of Formal Organisations and Informal Networks and Institutions. Occasional Paper no.17. International Development Studies, Roskilde University.

Hewitt, K. & I. Burton (1971). The hazardness of a place. Toronto: University Press.

Hijkoop, J. & P. van der Poel (1989). Mali-Sud. D'un aménagement anti-érosif des champs à la gestion de l'espace rural. Bulletin Institut Royal des Tropiques 317.

Hijkoop, J., P. van der Poel & B. Kaya (1991). Une lutte de longue haleine ... Aménagements anti-érosifs et gestion de terroir. IER/KIT, Bamako/Amsterdam, Mali/Pays-Bas, 154p.

Hill, A. G. (1991). Demographic responses to food shortages in the Sahel. Population 13, (special issue), pp. 168-192.

Hoefsloot, H., F. Van der Pol & L. Roeleveld (1993). Jachères améliorées. Options pour le développement des systèmes de production en Afrique de l'Ouest. Bulletin 333. Royal Tropical Institute, Amsterdam, The Netherlands, (86p).

Houghton, J. (2002). The physics of atmospheres. Cambridge & New York: Cambridge University Press etc.

Houghton, J.T., G.J. Jenkins & J.J. Ephraums (1990). Executive Summary and Introduction. In: Houghton, J.T., G.J. Jenkins & J.J. Ephraums, eds. Climate Change: The Intergovernmental Panel on Climate Change Scientific Assessment, pp.xi-xxxix. Cambridge.

Howorth C. & P. O'Keefe (1999). Farmers do it Better: Local Management of Change in Southern Burkina Faso. Land Degradation & Development 10, pp. 93-109.

Hulme, M. (1992). Rainfall changes in Africa: 1931-60 to 1961-90. International Journal of Climatology 12, pp. 685-99.

Hulme, M. (1994). Using Climate Information in Africa: Some Examples Related to Drought, Rainfall Forecasting and Global Warming. IDS Bulletin 25, pp. 59-68.

ICAC (1993). Survey of cotton production practices. ICAC, Washington D.C., USA. October 1993, 84p.

ICCD (1996). Impact of climate change on water availability, agriculture and food security in semi-arid regions, with special focus on West Africa (January 1997 - December 1999). CERES & DLV, Utrecht/Wageningen, The Netherlands, 16p.

INSD/RPG (1985). Résultats du Recensement Général de la Population 1985. Ouagadougou: Institut National de la Statistique et de la Démographie.

INSD/RPG (1991). Résultats de l'Enquete Démographique de la Population 1985. Ouagadougou: Institut National de la Statistique et de la Démographie.
IOB (2000). Revue de la coopération entre le Mali et les Pays-Bas, 1994-1998. Évaluations IOB, 285.
IOV (1992). Milieu en Ontwikkelingssamenwerking. Evaluatierapport Inspectie Ontwikkelingssamenwerking te velde. Den Haag: Ministerie van Buitenlandse Zaken.
IPCC (1990). Climate change, the IPCC scientific assessment. Houghton, J.T. G.J. Jenkins & J.J. Ephraums, eds. Cambridge University Press, Cambridge, New York, etc. 365pp.
IPCC (1996). Climate change 1995, the science of climate change. J.T. Houghton, L.G. Meira Filho, B,A, Callander, N, Harris, A, Kattenberg and K. Maskell, eds. Cambridge University Press, Cambridge, UK. 572pp.
IPCC (2000). Special Report on Emissions Scenarios. A Special Report of Working Group III of the intergovernmental Panel on Climate Change. Nakic´enovic´, N., J. Alcamo, G. Davis, B. de Vries, J. Fenhann, S. Gaffin, K. Gregory, A. Grübler, T. Yong Jung, T. Kram, E.L. La Rovere, L. Michaelis, S. Mori, T. Morita, W. Pepper, H. Pitcher, L. Price, K. Riahi, A. Roehrl, H.-H. Rogner, A. Sankovski, M. Schlesinger, P. Shukla, S. Smith, R. Swart, S. van Rooijen, N. Victor and Z. Dadi (eds.). Cambridge University Press, Cambridge, UK.
IPCC (2002 a). Climate change 2001, the scientific basis. J.T. Houghton, Y. Ding, D.J. Griggs, M. Noguer, P.J. van der Linden, X. Dai, K. Maskell and C.A. Johnson (eds.). Cambridge University Press, Cambridge, UK. 881pp.
IPCC (2002 b). Climate Change 2001: Synthesis Report, Contribution of Working Groups I, II, and III to the Third Assessment Report of the Intergovernmental Panel on Climate Change. Robert T. Watson (ed.). Cambridge University Press, Cambridge, UK. 397pp.
Izard, M. (1982). La politique extérieure d'un royaume africain: le Yatenga au XIXe siècle. Cahiers d'Études Africaines 22, pp. 363-85.
Janvry. A. de, M. Fafchamps & E. Sadoulet (1991). Peasant household behaviour with missing markets: some paradoxes explained. The Economic Journal 101, pp. 1400-1417.
Jonckers, D. (1987). La société minyanka du Mali. Paris: l'Harmattan.
Jonckers, D. (1995). Stratégies alimentaires et développement en région cotonnière du Mali-Sud. In: Devisch, R., F. de Boeck & D. Jonckers, eds. Alimentations, traditions et développements en Afrique intertropicale, pp. 219-247. Paris: L'Harmattan.
Kaiser, H.M., S.J. Rih, D.S. Wilks & R. Sampath (1993). Adaptation to global climate change at the farm level. In: Kaiser, H.M. & T.E. Drennen, eds. Agricultural dimensions of Global Climate Change. Delray Beach, Flo: St. Lucie Press.
Katary, A. (1998). La part du coton dans la déforestation en zone cotonnière au Bénin. CORAF (1998), L'utilisation des intrants en cultures cotonnière et maraîchères (Dakar, Sénégal, 25-28 janvier 1998), pp 333-342.
Keffing, S. (1998). Et demain l'agriculture? Options techniques et mesures politiques pour un développement agricole durable en Afrque subsharienne. Cas du Cercle de Koutiala en zone sud du Mali. Tropical Resource Management Papers 23.
Kemp, D. D. (1994). Chapter 3: Drought, Famine and Desertification. In: Global Environmental Issues: A Climatological Approach. London & New York: Routledge.
Kenworthy, J. (1991). The Recognition and Interpretation of Recurrent Drought in Africa. In: Stone, J.C., ed. Pastoral economies in Africa and long term responses to drought, pp. 7-21. Aberdeen: University African Studies Group.

Kerkhof, P. (1990). Agroforestry in Africa. A survey of Project Experiences. London: PANOS.

Kessler, J.J. & H. Breman (1995). L'évolution des systèmes de production agro-pastorale par rapport au développement rural durable dans les pays d'Afrique soudano-sahélienne. Document rédigé pour la FAO.

Klompenhouwer, C. & C. van Soest (1999). Stedelijke voedselvoorziening. Een onderzoek naar de groente- en fruitvoorziening van Bamako, Mali. (Master thesis, University Utrecht).

Koenig, D., D. Tiéman & M. Sow (1998). Innovation and Individuality in African Development. Changing Production Strategies in Rural Mali, Ann Arbor. The University of Michigan Press.

Koenig, D., D.Tiéman & M. Sow (1999). Aspects of Interetnic Relations in Contemporary Agricultural Migration and Settlement in Southern Mali. Mande Studies, 1, pp. 63-85.

Kohler, J.M. (1971). Activités agricoles et changements sociaux dans l'Ouest-Mossi. Paris: ORSTOM.

Koning, N., N. Heerink & S. Kauffman (1997). Integrated soil improvement and agricultural development in West Africa: why current approaches fail. Wageningen: International Soil reference and Information Centre.

Krings, T. (1991). Agrarwissen Bäuerlicher Gruppen in Mali / Westafrika. Berlin: Reimer.

Kristoferson, L. (1997). Seven energy and development myths – are they still alive? Renewable Energy for Development [online], 10. Available from: http://www.sei.se/red/red9707a.html. [Accessed April 3 th 2002].

assess the effectiveness of policy instruments. Agricultural Systems 58, pp. 465-481.

Kruseman, G. (2000). Bio-economic modelling for agricultural intensification. Mansholt Studies, 20.

Kruseman, G. & J. Bade (1998). Agrarian policies for sustainable land use: bio-economic modelling to

Kruseman, G., H. Hengsdijk, R. Ruben, P. Roebeling & J. Bade (1997). Farm household modelling system for the analysis of sustainable land use and food security: theoretical and mathematical description. Wageningen: DLV Report No. 7.

Kump, L.R., Reducing the uncertainty about carbon dioxide as a climate driver. Nature 419, pp. 188-190.

Kuyvenhoven, A., N. Heerink & R. Ruben (1999). Economic policy in support of soil fertility: Which interventions after structural adjustment? In: Smaling, E.M.A., O. Oenema & L.O. Fresco, eds. Nutrient disequilibria in agroecosystems: concepts and case studies. Wallingford: CAB International.

Lacape, J.M. (1996). Effets de la sécheresse sur le cotonnier et amélioration génétique de son adaptation au déficit hydrique. Revue bibliographique.CIRAD-CA, Montpellier, France. October 1996, 35p.

Ladouceur, P.A. (1979). Chiefs and Politicians : The Politics of Regionalism in Northern Ghana. London and New York: Longman Group Limited.

Lahmeyer, J. (2002). Availale on: www.ubu.ruu.nl/wesp/populstat/poulhome.html.

Lahuec, J.P. & J.Y. Marchal (1979). Mobilité et peuplement bissa et mossi. Travaux et Documents de l'ORSTO 103.

Lallemand, S. (1977). Une famille mossi. Recherches Voltaïques 17.

Lambert, S. & A. Sindzingre (1995). Droits de propriété et modes d'accès à la terre en Afrique: une revue critique. Cahiers d'Economie et Sociologie Rurales 36, pp. 95-128.

Laurent, P.J. & P. Mathieu (1994). Migration, environnement et projet de développement. Récit d'un conflit foncier entre Nuni et Mossi au Burkina Faso. Cahiers du CIDEP.

Leach, M., R. Mearns & I. Scoones (1999). Environmental Entitlements: Dynamics and Institutions in Community-Based Natural Resource Management. World Development 27, pp. 225-247.

Le Matinal (1999), Départements du Zou et des Collines : de graves menaces sur la production cotonnière de la campagne 1999/2000. Le Matinal (newspaper on Internet), October 30, 1999.

Lentz, C. & V. Erlmann (1989). A Working Class in Formation? Economic Crisis and Strategies of Survival among Dagara Mine Workers in Ghana. Cahiers d'Etudes Africaines 29-1, pp. 69-111.

Lentz, C. (1994a). 'They must be Dagaba first and any other thing second...': The Colonial and Post-Colonial Creation of Ethnic Identities in North-western Ghana. African Studies 53, pp. 57-91.

Lentz, C. (1994b). A Dagara Rebellion against Dagomba Rule? Contested Stories of Origin in North-Western Ghana. Journal of African History 35, pp. 457-492.

Lentz, C. (2000). Contested identities: the history of ethnicity in Northwestern Ghana. In: Lentz, C. & P. Nugent, eds. Ethnicity in Ghana: The Limits of Invention, pp. 137-161. London: Macmillan Press Ltd..

Levinson, D. & M. Ember, eds. (1996). Encyclopedyia of Cultural Anthropology. Volume 1. New York: Henry Holt & Company.

Lhoste P. (1977). Etude zootechnique, Inventaire du cheptel. Ouagadougou: IEMVT.

Lipton, M. (1986). Seasonality and Poverty. IDS Bulletin 17, pp. 4-8.

Longhurst, R. (1986). Household Food Strategies in Response to Seasonality and Famine. IDS Bulletin 17, pp.27-35.

Maas, P. (2001). L'influence des changemenst climatiques sur l'exploitation du mil à Wayre, village dogon du Mali centra. ICCD/ASC Working Paper. Leiden: Afrika Studies Centre.

Maatman, A. & C. Schweigman (1995). A Study of Farming Systems on the Central Plateau in Burkina Faso: Apllication of Linear Programming. Volume 1. Ouagadougou, Groningen: INERA, University of Ougadougou, University of Groningen.

Maatman, A., C. Schweigman, T. Thiombiano & J. van Andel (1992). Food security and sustainable agriculture on the Central Plateau in Burkina Faso: observations on a research agenda. Tijdschrift voor Sociaal-wetenschappelijk Onderzoek van de Landbouw 7, pp. 157-82.

Mahdavi, G. (1996). Nigeria : production et commercialisation. Coton et Développement, No. 19 (July-September 1996), pp. 7-11.

MARA/DSAP (1996). Résultats de l'Enquête Permanente Agricole, campagne 1995-1996. Ouagadougou: Ministère de l'Agriculture et des Ressources Animales/ Direction des Statistiques Agro-Pastorales.

Marc, L. (1909). Le pays mossi. Paris: Larose.

Marchal, J.Y. (1980). Croniques d'un cercle de l'A.O.F. Receuil d'archives du poste de Ouahigouya (Haute Volta), 1908-1941. Travaux et Documents de l'ORSTOM 125.

changing society. A case study in eastern Burkina Faso. Tropical Resource Management Papers 32.

Marchal, J.Y. (1983). Yatenga (Nord Haute-Volta): La dynamique d'un espace rural soudano-sahélien. Paris: ORSTOM.

Marchal, J.Y. (1987). En Afrique des savanes, le fractionnement des unités d'exploitation rurales ou le chacun pour soi. L'exemple des Moose du Burkina Faso. Cahiers des Sciences Humaines 23, pp. 445-54.
Martin, T., G. Ochou, F. Hala N'Klo, J.M. Vassal & M. Vaissayre (1997). Perte de sensibilité aux pyréthrinoïdes chez Helicoverpa armigera (Hübner). CORAF (1997), Réunion phytosanitaire de l'Afrique de l'Ouest et du Centre (Cotonou, Bénin, 27-31 janvier 1997), pp. 253-261. CORAF, Réseau Coton.
Martinelli, B. (1995). Trames d'appartenance et chaînes d'identité. Entre Dogon et Mosse dans le Yatenga et la plaine du Séno (Burkina Faso et Mali). Identités et appartenances dans les sociétés sahéliennes. Cahiers des Sciences humaines 31, pp. 365-405.
Marty, A. (1993). La gestion des terroirs et les éleveurs: un outil d'exclusion ou de négociation? Revue Tiers Monde XXXIV134, pp. 327-44.
Mathieu, P. (1993). Population, environnement et enjeux fonciers des politiques agricoles: le cas des pays du Sahel. In: Gérard, H., ed. Intégrer population et développement, pp. 433-48. Paris: l'Harmattan.
Matthews, G.A. (1989). Cotton insect pests & their management. Longman, United Kingdom, 199p.
Matthews, G.A. & J.P. Tunstall (eds.) (1994. Insect pests of cotton. CAB International, Wallingford, United Kingdom, 593p.
Mazur, R.E. (1984). Rural Out-Migration and Labor Allocation in Mali. In: Calvin Goldschneider ed. Rural Migration in Developing Countries: Comparative Studies of Korea, Sri Lanka and Mali, pp. 209-288. Westview press.
Mazzucato, V. & D. Niemeijer (2000). Rethinking soil and water conservation in a Lentz, C. (1993). Histories and political conflict: A case study of chieftancy in Nandom, Northwestern Ghana. Paideuma mitteilungen zur kulturkunde 39, pp. 177-215.
McAvaney, B.J et al. (2002). Model Evaluation. In: IPCC 2002a, pp. 471-523.
McCoy, R.F. (1988). Great things happen: A personal memoir of the first Christian missionary among the Dagaabas and Sissalas of northwest Ghana. Montreal: The Society of Missionaries of Africa.
MCD (1991). Le coton en Afrique de l'Ouest et du Centre. Ministère de la Coopération et du Développement, Paris, France, 353p.
Michel, B., I. Tereta, B. Bagayoko & N. Traoré (1997). Lutte étagée ciblée : l'expérience malienne. CORAF (1997), Réunion phytosanitaire de l'Afrique de l'Ouest et du Centre (Cotonou, Bénin, 27-31 janvier 1997), pp. 213-217. CORAF, Réseau Coton.
Millar, D. (1999a). Traditional Worldviews from a Cosmovision Perspective. In: Haverkorst, B. & W. Hiemstra, eds. Food for Thought: Ancient Visions and New Experiments of Rural People, pp.131-138. Leusden: ETC/COMPAS; Bangalore: Books for Change; London: Zed Books.
Millar, D. (1999b). When the Spirits Speak. In: Haverkorst, B. & W. Hiemstra, eds. Food for Thought: Ancient Visions and New Experiments of Rural People, pp. 139-154. Leusden: ETC/COMPAS; Bangalore: Books for Change; London: Zed Books.
Millar, D., J. Anglaaere & H.A.S. Pul (1997). Final Report of the Evaluation of Nandom Agricultural Project & Nandom Food Farmers Cooperative Union. Wa
Millar, D., R. Ayariga & B. Anamoh (1996). 'Grandfather's way of doing': Gender relations and the yaba-itgo system in Upper East Region, Ghana. In: Scoones, I. & C. Toulmin, eds. Sustaining the soil: indigenous soil and water conservation in Africa, pp.117-125. London: Earthscan.
Ministère de l'Agriculture et des Ressources Animales (1995a). Performances de l'agriculture 1984-1994. Ouagadougou.

Ministère de l'Agriculture et des Ressources Animales (1995b). Résultats de l'enquête permanente agricole. Campagne 1994-1995. Ouagadougou.
Ministère de l'Agriculture et des Ressources Animales (1996a). Enquête Nationale de Statistiques Agricoles E.N.S.A. 1993. Rapport d'Analyse. Ouagadougou: Direction des Statistiques Agro-Pastorales.
Ministère de l'Agriculture et des Ressources Animales (1996b). Résultats de l'enquête permanente agricole. Campagne 1995-1996. Ouagadougou.
Ministère de l'Agriculture et des Ressources Animales (1997). Résultats de l'enquête permanente agricole. Campagne 1996-1997. Ouagadougou.
Ministère de l'Agriculture et des Ressources Animales (1998). Résultats de l'enquête permanente agricole. Campagne 1997-1998. Ouagadougou.
Moerkamp, J. (1997). Ghana: Mensen, politiek, economie, cultuur (landenreeks). Amsterdam: KIT.
MoFA, UWR (1993). End of Year Performance Report 1992: The Potentials of Agricultural Production and Strategies for Future Development. Internal Paper.
Mortimore, M. & W.M. Adams (1999). Working the Sahel. Environment and society in Northern Nigeria. London and New York: Routledge.
Mortimore, M. (1989). Adapting to drought: Farmers, famines and desertification in West Africa. Cambridge: University Press.
Mortimore, M. (1998). Roots in the African Dust: Sustaining the Drylands. Cambridge: University Press.
Mosely, W. G. (2001). Sahelian 'White Gold' and Rural Poverty-Environment Interactions: The Political Ecology of Cotton Production, Environmental Change and Household Food Economy in Mali. (Ph.D. thesis, University of Georgia).
Muis, S. (2000). De melkvoorziening van Bamako, Mali. (MA thesis, University Utrecht).
Munro, J. (1994). Cotton and its production. In: G.A. Matthews & J.P. Tunstall (eds.) (1994), Insect pests of cotton, pp. 3-26. CAB International, Wallingford, United Kingdom
Myers, D. & S. Stolton (eds.) (1999). Organic cotton. From field to final product. IT Publications, London, United Kingdom, 267p
Niemeyer & Mazzucato (2000). Rethinking soil and water conservation in a changing society; a case study of Eastern Burkina Faso. (Ph.D. thesis University Wageningen. Also published as Tropical Resource Management Papers, 32.)
Niezen, R.W. (1990). The Community of Helpers of the Sunna, Islamic Reform among the Songhay of Gao (Mali). Africa, 60.
Nijenhuis, K. (1999). De Convention Locale Siwaa, Lokale afspraken over het gebruik van natuurlijke hulpbronnen in Zuid-Mali. (M.Sc. thesis, University of Amsterdam).
Nijenhuis, K. (2002). Decentralisation creating a new manoeuvring space and new power positions for local traditional authorities: An example from Southern Mali, Paper prepared for the APAD-conference, The Governance of Daily Life in Africa, Public and Collective Services and their Users, 22-25 May 2002, Leiden, the Netherlands.
Nijenhuis, Karin , fc, 'Migratory Drift of Dogon Farmers to Southern Mali', In: Mirjam de Bruijn & Han van Dijk, eds
Nijzink, L. (1999). More Grass Strips, Less Erosion: Intensification, the Market and Grass Strips on the Central Plateau, Burkina Faso. (M.A. thesis, University of Amsterdam).
Nikiéma, O. (1999). Riche ou pauvre, on réagit. Stratégies de survie de deux Nsiah-Gyabaah, K. (1994). Environmental degradation and desertification in Ghana: A Study of the North-Western Region. Hants: Ashgate Publishing Limited.

Ochou Ochou, G., T. Martin & F. Hala N'Klo (1998). Protection sur seuils en phase végétative du cotonnier : une approche de lutte intégrée raisonnée transférable en milieu paysan de Côte d'Ivoire. CORAF (1998), L'utilisation des intrants en cultures cotonnière et maraîchères (Dakar, Sénégal, 25-28 janvier 1998), pp. 201-216.

ODHD (1998). Rapport annuel sur le développement humain Durable au Mali. Bamako : Ministère de l'Economie, du Plan et de l'Intégration/PNUD/Banque Mondiale : Observatoire du Développement humain durable et de lutte contre la pauvreté au Mali.

catégories de paysans face aux contraintes de 1960 à nos jours à try et N'Goukan, Koutiala (Mali). (M.Sc. thesis, University of Amsterdam).

Okai, M. (1997). Agricultural Production, Food Security and Poverty in West Africa. In: Asenso-Okyere, W.K., G. Benneh & W. Tims, eds. Sustainable food security in West Africa, pp.14-34. Dordrecht: Kluwer Academic Publishers.

Onwueme, I.C. & T.D. Sinha (1991). Field Crop Production in Tropical Africa: Principles and Practice. Technical Centre for Agricultural and Rural Co-operation.

Ortiz, S. (1980). Forecasts, Decisions, and the Farmer's Response to Uncertain Environments. In: Barlett, P.F., ed. Agricultural Decision Making: Anthropological Contributions to Rural Development, pp. 177-202. New York: Academic Press.

Osie-Adjei, P. (2000). Rainfall variability and crop production strategies: a case study of Gambaga area in the Guinea savanna zone, Northern Ghana. University of Ghana at Legon, in collaboration with UDS Tamale and University of Amsterdam.

Ouedraogo, J.B. (1995). The girls of Nyovuuru: Dagara female migrants to Bobo-Dioulasso. In: Baker, J. & T.A. Aina, eds. The migration experience in Africa. Sweden: Nordiska Afrikainstitutet, pp. 303-320.

Owens , T. & J. Hoddinott (1998). Investing in Development or Investing in relief: Quantifying the Poverty Trade-offs Using Zimbabwe Household Panel Data. Mimeo: IFPRI.

Palutikof, J.P., C.M. Goodess, S.J. Watkins & T. Holt. Generating rainfall and temperature scenarios at multiple sites: examples from the Mediterranean. J. Climate 15, 3529-3548.

PAN-AFRICA (1998). Coton biologique en Afrique (Tambacounda, Sénégal, 24-28 novembre 1997). PAN-Africa, Dakar-Fann, Senegal, 111p.

Parab (1996). La recherche cotonnière au Bénin se dote d'un programme de création variétale. INRAB & CIRAD, Cotonou, Benin, 2p.

Parry, M.L. (1990). Climate change and world agriculture. London: Earthscan/IIASA/UNEP.

Payne, G. (1997). Urban Land Tenure and Property Rights in Developing Countries: A Review. IT Publications, ODA.

Peretti, M. (1977). Projet de Mise en place de l'ORD du Sahel. Ouagadougou: Ministere du Développement Rural.

Petit, V. (1998). Migrations et société dogon. Paris: L'Harmattan.

Pischke, J.D. von (1986). Can Crop Insurance Address Risks in Agricultural Lending? In: Hazell, P., C. Pomareda & A. Valdes, eds. Crop Insurance for Agricultural Development: Issues and Experience. Baltimore: Johns Hopkins University Press.

Plange, N.K. (1979). Underdevelopment in Northern Ghana: Natural Causes or Colonial Capitalism? Review of African Political Economy 15-16, pp. 4-14.

Poda, E.N. (1998). Is the Dagara an Acephalous Society? Hemispheres 13, pp. 53-62.

Pol, F. van der (1991). L'épuisement des terres, une source de revenus pour les paysans au Mali-Sud. In: Pieri, ed. Savanes d'Afrique, terres fertiles. Comment produire plus de façon durable en zone de savanes au sud du Sahara, pp.403-418. Paris: Ministère de la Coopération et du Développement.

Pol, F. van der (1992). Soil mining. An unseen contributor to farm income in southern Mali. Bulletin No. 325. Royal Tropical Institute, Amsterdam, The Netherlands, 48p.

Popkin, S. (1979). The Rational Peasant: The Political Economy of Rural Society in Vietnam. Berkely, Los Angeles and London: University of California Press.

Posnansky, M. (1984). Early Agricultural Societies in Ghana. In: Clark, J. D. & S. A. Brandt, eds. From hunters to farmers: the causes and consequences of food production in Africa, pp. 147-151. Berkely, Los Angeles and London: University of California Press.

Post, J. & M. Snel (2003). The impact of decentralised forest management on charcoal production practices in Eastern Senegal. Geoforum, 34, pp 85-98.

Prather, M., D. et al. (2002). Atmospheric Chemistry and greenhouse gases. In: IPCC 2002a, pp. 239-387.

Pratt, D.J., F. Le Gall, & C. de Haan (1997). Investing in Pastoralism: Sustainable Natural resource Use in Arid Africa and the Middle East. World Bank Technical Paper, 365.

Put & T. Dietz. (1998). Climate Change in Dryland West Africa and the Resultant Responses of the Local Population, their Institutions and Organizations. Proceedings of the seminar: Acts of Man and Nature: Different Constructions of Social and Natural Resource Dynamics. Bergen, the Netherlands. October 1998.

Put, M. & S. de Vos (1999). Klimaatonderzoek in West Afrika. Rooilijn.

Put, M. (1999). Climate change and climate variability in dryland West Africa. CERES/DLO/WAU/RIVM, December. Internal ICCD document.

Put, M., P. Jellema, E. Veldhuizen, F. Zaal & A. Verhagen (2001). Climate Change and Climate Variability in Dryland West Africa. Internal ICCD document.

Rattray, R.S. (1932). The Tribes of the Ashanti Hinterland. Vol. 1 & 2. Oxford: University Press.

Raymond G., G. Faure & C. Persoons (1991). Pratiques paysannes en zone cotonnière face à l'augmentation de la pression foncière (Nord-Togo et Mali-Sud). In: Pieri, ed. Savanes d'Afrique, terres fertiles. Comment produire plus de façon durable en zone de savanes au sud du Sahara, pp. 173-194. Paris: Ministère de la Coopération et du Développement.

Reardon, T., P. Matlon & C. Delgado (1988). Coping with household-level food insecurity in drought-affected areas of Burkina Faso. World Development 16, pp. 1065-74.

Ribot, J.C. (1995). The Causal Structure of Vulnerability: Its Application to Climate Impact Analysis. GeoJournal 35, pp. 119-122.

Ribot, J. (2001). Science, use rights and exclusion : a history of forestry in francophone West Africa. London: International Institute for Environment and Development. Issue paper [online], 104. Available from: http://www.iied.org/pdf/dry_ip104eng.pdf. [Accessed April 10th 2002].

Ribot, J.C., A. Njam and G. Watson (1996). Climate Variation, Vulnerability and Sustainable Development in the Semi-arid Tropics. In: Ribot, J.C., Antonio R. Magalhaes and Stahis S. Panagides, eds. Climate Variability, Climate Chage and Social Vulnerability in the Semi-Arid Tropics, pp.13-51. Cambridge: University Press.

Richards, P. (1985). Indigenous Agricultural Revolution: Ecology and food production in West Africa. London: Hutchinson & Co. (Publishers) Ltd.

Rijkes, P. (2000). The climatic conditions for agriculture in Bongo District, North-East Ghana. (M.A. thesis, University of Amsterdam, University Tamale).

RIM (1987). Refuge in the Sahel. Livestock Populations and Productions Systems in the Mali Fifth Region. A Rapid Assessment of Current Resources by Low Level Aerial Survey and Complementary Ground Studies. Jersey UK, Resource Inventory and Management Ltd.

Roberts, R. (1995). The Coercion of Free markets: Cotton, Peasants, and the Colonial State in the French Soudan, 1924-1932. In: Isaacman, A. & R. Roberts, eds. Cotton, Colonialism, and Social History in Sub-Saharan Africa, pp. 221-243. London & Portsmouth: Heineman & Currey.

Robison, L. & P. Barry (1987). The Competitive Firm's Response to Risk. New York: Macmillan.

Rosenzweig, C. & M.L. Parry (1994). Potential impact of climate change on world food supply. Nature 367, pp. 133-138.

Roupsard, M. (1987). Nord-Caméroun. Ouverture et développement. University of Yaoundé, Cameroon, 515p.

Rouville, C. de (1987). Organisation Sociale des Lobi. Paris: L'Harmattan.

Ruben, R. & A. van Ruijven (2000). Technical coefficients for bio-economic farm household models: a meta-modelling approach with applications for Southern Mali. Ecological Economics (fortcoming).

Ruben, R., H. Moll & A. Kuyvenhoven (1998). Integrating agricultural research and policy analysis: analytical framework and policy implications for bio-economic modelling. Agricultural Systems 58, pp. 331-350.

Ruddiman, W.F. (2000). Earth's climate, past and future. New York: W.H. Freeman and Company.

Runge-Metzger, A. & L. Diehl (1993). Farm Household Systems in Northern Ghana: A case study in farming systems oriented research for the development of improved crop production systems. Tamale: Nyankpala Agricultural Research Station. Special Report, No.9.

Rutgers van der Loeff, M. (2001). Between Hardship and Abundance: Caravaneers and Caravan Trade in Central Mali. (M.A. thesis, Dept of Geography and Planning, University of Amsterdam).

Ruthenberg, H. (1980). Farming Systems in the Tropics. Oxford: Clarendon Press.

Sachs, W. (1999). Chapter 4: Environment and Development: Story of a Dangerous Liaison. In: Planet Dialectics: Explorations in Environment and Developments, pp. 56-70. London: Zed Books.

Sahlins, M. (1974). Stone age economics. London: Tavistock.

Sanders, J.H., B.I. Shapiro & S. Ramaswamy (1996). The economics of agriculturasl technology in semi-arid Sub-Saharan Africa. Baltimore: The Johns Hopkins University Press.

Sandmo, A. (1971). On the Theory of the Competitive Firm under Price Uncertainty. American Economic Review 61, pp. 65-73.

Sangaré, B. & M. Seck (1994). Le coton malien à la conquête de nouvelles terres. Le Républicain, No. 106 (September 21, 1994).

Schaik, A. van & H. Reitsma (1992). Agriculture in dry-season environments. In: Reitsma, H., T. Dietz & L. de Haan, eds. Coping with semiaridity: How the rural poor survive in dry-season environments, pp. 19-36. Utrecht: Netherlands Geographical Society.

Schijf, A. (forthcoming). Pathways of Change in Semi-Urban Livelihoods in Nandom Town. (Provisional title). (M.A. thesis. University of Amsterdam, in collaboration with University of Development Studies, at Tamale, Ghana).

Schwartz, A. (1979). Colonisation agricole spontanée et émergence de nouveaux milieux sociaux dans le sud-ouest ivoirien: l'exemple du canton Bakwé de la sous-préfecture de Soubré. Cahiers ORSTOM, Série Sciences Humaines 16, pp. 83-101.

Scoones, I & C. Toulmin (1999). Policies for soil fertility management in Africa. London/Brighton: IIED/IDS.

Scoones, I. (1994). New Directions in Pastoral Development in Africa. In: Scoones, I., ed. Living with Uncertainty. New Directions in Pastoral Development in Africa , pp. 1-36. London: Intermediate Technology Publications Ltd.

Scoones, I. (1995). Investigating Difference: Applications of Wealth Ranking and Household Survey Approaches among Farming Households in Southern Zimbabwe. Development and Change 26, pp. 67-88.

Scoones, I. (1998). Sustainable Rural Livelihoods: A Framework for Analysis. IDS working paper 72. Sussex: Institute of Development Studies.

Scoones, I., ed. (1996). Hazards and Opportunities: Farming livelihoods in Dryland Africa. London: Zed Books/IIED.

Sékloka, E., M. Djaboutou, A. Hougni & J. Lançon (1999). Sélection participative du cotonnier au Bénin. INRAB/RCF, Cotonou, Benin.

Sément, G. (1988). Cotton. The Tropical Agriculturalist. CTA/MacMillan, (88p)

Sen, A. (1987). Hunger and Entitlements: Research for Action. Helsinki: World Institute for Development Economics Research.

Sène, E.H. (1993). Urban and peri-urban forests in sub-Saharan Africa : the Sahel. Unasylva [online], 173. Availabel from: <http://www.fao.org/docrep/u9300e/u9300e09.htm. [Accessed April 10th 2002].

Shepherd, A. (1981). Agrarian Change in Northern Ghana: Public Investment, Capitalist Farming and Famine. In: Heyer, J., P. Roberts & G. Williams, eds. Rural Development in Tropical Africa, pp. 168-192. London and Basingstoke: The MacMillan Press Ltd.

Sinou, A. (1993). Comptoirs et villes colonials du Sénégal. Saint-Louis, Gorée, Dakar, Paris, Karthala: ORSTOM.

Sissoko, K. (1998). Et demain l'Agriculture? Options techniques et mesures politiques pour un developpement agricole durable en Afrique subsaharienne. (Ph.D. thesis, Wageningen University).

Skees, J., P.Hazell, & M. Miranda (1999). New Approaches to Crop Yield Insurance in Developing Countries. Environment and Production Technology Division Discussion Paper, 55. Also published as: Nuovi approcci per le assicurazioni in agricoltura nei Paesi in via di sviluppo. Rivista di Politic Agraria XVIII, pp. 31-52.

Smith, G.A. (1984). Confederations of households: extended domestic enterprises in city and country. In: Long, N. & B. Roberts, eds. Miners, Peasants and Entrepreneurs. Regional Development in the Central Highlands of Peru, pp. 217-34. Cambridge: Cambridge University Press.

Snel, M. & J. Pomp (1999). Ghana & Benin: Rurale opdracht voor het colloquium Milieu en Ontwikkeling. Unpublished manuscript. Amsterdam.

Soepboer, W. (1999). The influence of Climate on Vegetation Dynamics. ALTERRA Interne Mededeling 574.

Sokona, Y. (2001). LPG introduction in Senegal [on line]. Dakar: ENDA. Available from: http://www.enda.sn./energie/Butane%20Senegal.htm. [Accessed April 8th 2002].

Songsore, J. (1985). Intra-regional commodity flows within a centre-periphery structure: the case of Wa town and North-western Ghana. Occasional Papers Series, No.19. Norwich: Centre for Development Studies.

Staatz, J.M., J. Dione & N.N. Dembele (1989). Cereal Markets Liberalization in Mali. World Development 17, pp. 703-718.

Stanneveld, A., B. Anamoh, J. Dakouo, A. van Voorst & W. Zoet (1994). Sustainable development from a farmer's perspective: Programme evaluation on sustainable land use in Ghana and Mali. The Hague: DGIS/CEBEMO.

Steenbrugge, J. van (2001). Movements in a new world: Fulbe mobility and survival in southern Mali. ICCD/ASC Working Paper. Leiden: Afrika Studies Centre.

Steenhuisen Piters, B de (1995). Diversity of fields and farmers: explaining yield variations in Northern Cameroon. (Ph.D. thesis, Wageningen University).

Stewart, J. I. (1991). Managing Climate Risk in Agriculture. In: Holden, D., P. Hazell & A. Pritchard, eds. Risk in Agriculture: Proceedings of the Tenth Agriculture Sector Symposium. Washington D.C: World Bank.

STRAT Sahel (1994). Schéma Régional d'Amenagement du Territoire -STRAT- Sahel. Description et analyse de la situation actuelle. Dori: Direction Régionale du Plan et de la Coopération.

STRAT Sahel (1998). Schéma Régional d'Amenagement du Territoire -STRAT- Sahel. Version provisoire. Dori: Direction Régionale du Plan et de la Coopération.

Stroosnijder, L. & T. van Rheenen, eds. (2000). Agro-silvo-patoral land use in Sahelian villages. Germany: Catena Verlag (forthcoming).

Struif Bontkes, T. (1999). Modelling the dynamics of agricultural development: a process approach. The case of Koutiala (Mali). (Ph.D. thesis, Wageningen University).

Struijf Bontkes, T. (1999). Modelling the dynamics of agricultural development: a process approach. The case of Koutiala (Mali). Tropical Resource Management Papers 25.

Sutton, I. (1989). Colonial Agricultural Policy: The Non-Development of the Northern Territories of the Gold Coast. The International Journal of African Historical Studies 22, pp. 637-669.

Swift, J. (1986). The Economics of Production and Exchange in West African Pastoral Societies. In: M.A. Adamu & D. Kirk-Greene, eds. Pastoralists of the West African Savanna, pp. 175-190. Manchester: Manchester University Press.

Swift, J. (1989). Why are Rural People Vulnerable to Famine? IDS Bulletin 20, pp. 8-15.

Swift, J. (1993). Understanding and preventing Famine and Famine Mortality. IDS Bulletin 24, pp.1-16.

Tacoli, C. (2001). Urbanisation and Migration in Sub-saharan Afica: Changing patterns and trends. In: Bruijn, M. de, R. van Dijk, D. eds. Mobile Africa, Changing Patterns of Movement in Africa and beyond, pp. 141-153. Leiden: Brill.

Taylor, J.E. & I. Adelman (1996). Village Economies: the desgin, estimation and use of villagewide economic models. Cambridge: Cambridge University Press.

Teal, F. (1986). The foreign exchange regime and growth: a comparison of Ghana and the Ivory Coast. African Affairs 85, pp. 267-82.

Tefft, J. (2000). Cotton in Mali: The "White Revolution" and Development. In: Bingen, R.J., D. Robinson & J. M. Staatz, eds. Democracy and Development in Mali, pp. 213-241. Michigan Sate University Press.

Tellegen, N. (1997). Brewing Women and Drinking Men: The Costs of Beer in Rural Malawi. In: Glas, M. van der, H. Goeren, W. Pompe & F. Zaal, eds. Under Pressure: Essays on Development Research, pp. 37-56. Utrecht: ISOR/CERES.

Tengan, Alexis B. (1999). Dagara Bagr: Ritualising myth of social foundation. Africa 69, pp. 595-633.

Tengan, Alexis B. (2000). Hoe-farming and social relations among the Dagara of Northwestern Ghana and Southwestern Burkina Faso. European university studies: Ser.19, Secion B, Volume 54. Frankfurt am Main: Peter Lang GmbH.

Tengan, E. B. (1994). The social structure of the Dagara: the house and the matriclan as axes of Dagara social organization. The Victor Series 3.

Thomas, R.G. (1973). Forced Labour in British West Africa: The Case of the Northern Territories of the Gold Coast 1906-1927. Journal of African History 16, pp. 79-103.

Tiffen, M. (1995). Population Density, Economic Growth and Societies in Transition: Boserup Reconsidered in a Kenyan Case-study. Development and Change 26, pp. 31-66.

Ton , P. (1993). Katoen : de witte motor. Analyse van plannen tot wijziging van het overheidsbeleid in de katoensektor in Benin. Msc-thesis. University of Amsterdam, The Netherlands. March 1993, 134p.

Ton, P. (1998). Organic cotton experiments in northern Benin. Presentation at the World Cotton Research Conference - 2 (Athens, Greece, September 6-12, 1998). AGIDS, Amsterdam, The Netherlands, 8p.

Ton, P. (1999). Organic cotton and sustainability in northern Benin. In : D. Foguelman & W. Lockeretz (eds.) (1999), Organic agriculture the credible solution for the XXIst Century (Mar del Plata, Argentina, November 15-19, 1998), pp. 74-80. IFOAM, Tholey-Theley, Germany.

Toulmin, C. (1986). Access to Food, Dry Season Strategies and Household Size amongst the Bambara of Central Mali. IDS Bulletin 17, pp. 58-66.

Traoré, P. (1996). Aptitudes paysannes à la gestion des terroirs à Kaibo-Nord (Province du Zoundwéogo). Mémoire de Maîtrise. Ouagadougou: Université de Ouagadougou, Antenne Sahélienne.

Tripp, R. & K. Marfo (1997). Maize Technology Development in Ghana During Economic Decline and Recovery. In: Byerlee, D & C.K. Eiches, eds. Africa's Emerging Maize Revolution. London: Boulder.

Tuurey, G. (1982). An introduction to the Mole Speaking Community.Wa: Catholic Press.

Udry, C. (1990). Credit markets in Northern Nigeria: credit as insurance in a rural economy. The World Bank Economic Review 4, pp. 251-269.

UNDP (2001a). Human Development Report; Making New Technologies Work for Human Development. New York: UNDP.

UNDP (2001b). Human Development Report 2001 [online]. Available from: www.un.og.sn/docs/cca_sen2001.

UNDP/FAO (1967). Land and Water Survey in the Upper and Northern Regions, Ghana. Final Report, Vol.3, Soil surveys (incl. map). Rome: UNDP/FAO.

United Nations (2000). United Nations Statistical Yearbook 2000. New York:UN.

Vaissayre, M. (1995). Ecological attributes of major cotton pests : implications for management. In: G.A. Constable & N.W. Forrester (eds.) (1995), Challenging the future. Proceedings of the World Cotton Research Conference - 1 (Brisbane, Australia, February 14-17, 1994), pp. 499-510. CSIRO, Canberra, Australia.

Vaissayre, M., T. Martin & J.M. Vassal (1998). Pyrethroid resistance in the bollworm Helicoverpa armigera in West Africa. Presentation at the World Cotton Research Conference - 2 (Athens, Greece, September 6-12, 1998). CIRAD-CA, Montpellier, France.

Veluw, K. van (1987). Traditional Poultry Keeping in Northern Ghana. ILEA Newsletter 3, pp. 12-13.
Verhagen, A. & H. van Keulen (1999). Analysis of rainfall variability and agricultural risk in sub-Saharan West Africa. Poster presentation at 20-23 September 1999, Reading, UK. Food and Forestry: Global Change and Global Challenges.
Verkoren, O. & A.C.M. van Westen (2000). Gedaanteverandering van de Afrikaanse stad. In: Verkoren, O. & E.J.A. Harts-Broekhuis, eds. Tropisch Afrika, pp.165-196. Assen: Van Gorcum.
Vodounnon, S. & P. Silvie (1997). Principaux ravageurs et maladies du cotonnier au Bénin. INRAB/RCF, Cotonou, Benin, (17p.)
Vos S. de (1983). Beschrijvende Statistiek: Een syllabus, geschreven ten behoeve van het eerstejaars-college statistiek voor sociale geografen. Amsterdam: Sociaal Geografisch Instituut (unpublished).
Waal A. de (1989). Is Famine Relief Irrelevant to Rural People? IDS Bulletin 20, pp. 63-67.
Walker, T.S., & N.S. Jodha (1986). How Small Farm Households Adapt to Risk. In: Hazell, P., C. Pomareda & A. Valdes, eds. Crop Insurance for Agricultural Development: Issues and Experience. Baltimore: Johns Hopkins University Press.
Walle, D. van de & K. Nead (1995). Public Spending and the Poor: Theory and Evidence. Baltimore/London: The Johns Hopkins University Press.
Warner, M., R. Al-Hassan & J. Kydd. (1999). A Review of Changes to Farming Systems in Northern Ghana (1957 to 1994). In: Blench, R. ed. Natural Resource Management in Ghana and its Socio-economic Context, pp.85-113. London: Overseas Development Institute.
Watts, M.J. & H.G. Bohle (1993). The Space of vulnerability: the causal structure of hunger and famine. Progress in Human Geography 17, pp. 43-67.
Webb, J.L.A. (1995). Desert Frontier. Ecological and Economic Change along the Western Sahel, 1600-1850. Madison, Wisconsin: The University of Wisconsin Press.
Webb, P. & T. Reardon (1992). Drought Impact and Household Response in East and West Africa. Quarterly Journal of International Agriculture 31, pp. 230-246.
Weiss, H. (1997). "Flight to Survive": Migrations during Times of Drought and Famine in Early Colonial West Africa. Working Paper no. 13. Helsinki: Institute of Development Studies.
Wenner, M. & Mooney, T.J. (1995). Livestock trade and marketing costs in the Burkina Faso – Ghana corridor. USAID and Abt. Associates Inc., Development Alternatives Inc., USA.
Wennink, B. (1998).Production du fumier dans le parc à bétail en zone soudanienne. Note synthétique d'information. INRAB/R-D, Parakou, Benin. December 1998, (10p.)
Wilks, I. (1989). Wa and the Wala: Islam and polity in northwestern Ghana. Cambridge: University Press.
World Bank (1988), Cotton development programs in Burkina Faso, Côte d'Ivoire and Togo.
World Bank, Operations Evaluations Department, Washington D.C., USA, (126p.)
World Bank (2000). World Development Report 2000-2001 Attacking Poverty. Washington D.C.: World Bank.
World Bank (2002). World Development Report 2002. Building Institutions for Markets. Washington D.C.: World Bank.

Zaa4l, F. & R. Oostendorp (2002). Explaining a miracle: intensification and the transition towards sustainable small-scale agriculture in drylands Machakos and Kitui Districts, Kenya. World Development 30, pp. 1271-1287.

Zaal, F. & T. Dietz (2002). The provisioning of African cities, with Ouagadougou as a case study. In: Baud, I.S.A. & J. Post, eds. Re-aligning actors in an urbanizing world: institutions and governance in an international perspective, pp. 263-284. Ashgate: Aldershot.

Zaal, F. (1998). Pastoralism in a Global Age: Livestock Marketing and Pastoral Commercial Activities in Kenya and Burkina Faso. (Ph.D. thesis, University of Amsterdam).

Zanen, S. (1996). Ontwikkeling: Idee en Handeling. De Confrontatie van Boeren- en Donorstrategieën in de Praktijk van Ontwikkelingssamenwerking (Nederland-Burkina Faso). (Ph.D. thesis University of Leiden).

Zondag, R. (2001). Douentza, a description of the dynamics of a rural centre in Central Mali. ICCD/ASC Working Paper. Leiden: Afrika Studies Centre.

Zutphen, T. van (1991). The struggle against desertification in Burkina Faso. In: Savenije, H. & A. Huijsman, eds. Making Haste Slowly: Strengthening Local Environmental Management in Agricultural Development, pp. 117-29. Amsterdam: KIT.

Other internet references:

Are We Having a Drought Yet? http://enso.unl.edu/ndmc/plan/define.htm

Asenso-Okyere, W.K., S.Y.Atsu An Overview of Vulnerability Analysis and Mapping (VAM). http://www.wfp.it/vam/vamhome.htm

Bodomo (1998). Dagaare Language and Culture. http://www.erols.com/johnston/lawra_language_culture.htm

Ghana national economy indicators. http://www.newafrica.com/economy

Guidelines for developing improved risk assessment, prevention and management approaches in pastoral areas of Northwestern China. www.fao.org/sd/2001/IN0602a_en.htm

Historical currency exchange rates. http://www.oanda.com/convert/fxhistory

Minimum Energy Requirements: http://www.fao.org/docrep/x8200e/x8200e03.htm#P2_33

Regional Trends and Variations of Temperature and Precipitation. http://www.grida.no/climate/ipcc/regional/305.htm

The Regional Impact of Climate Change (IPCC). Executive Summary. http://www.grida.no/climate/ipcc/regional/007.htm

Understanding and defining drought. http://www.enso.unl.edu/ndmc/enigma/def2.htm

United Nations Food and Agriculture Organisation (FAO). http://www.fao.org; FAOSTAT AGRIC.

Index

Colour Section

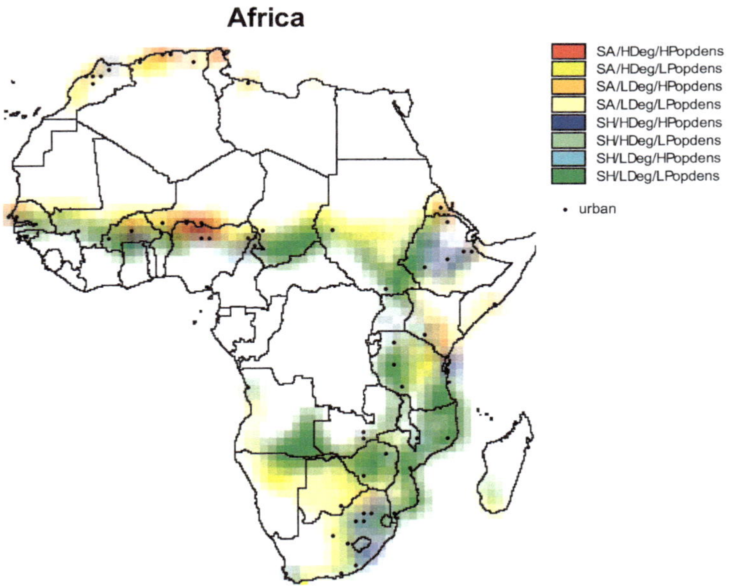

Map 2.1. Drylands of Africa p. 22

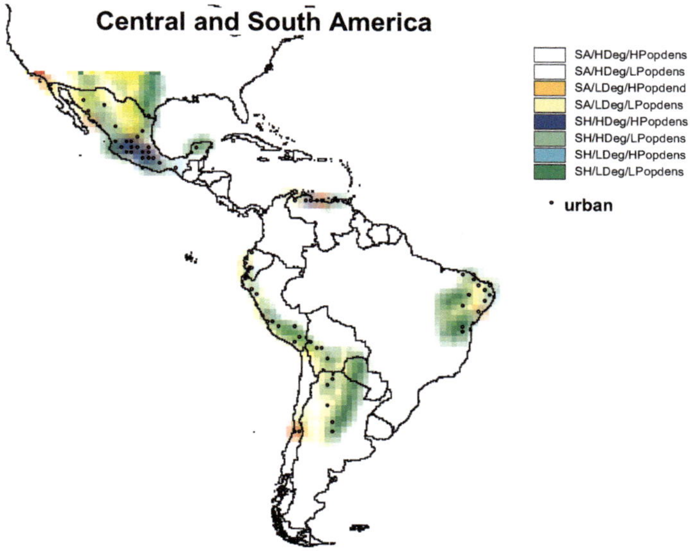

Map 2.2. Drylands of Central and South America p. 23

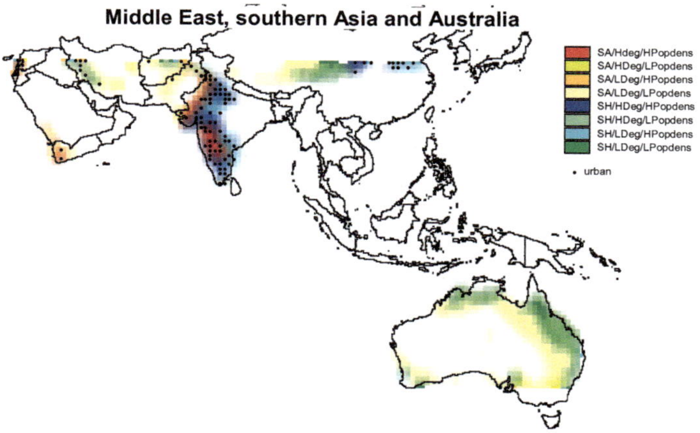

Map 2.3. Drylands of Asia and Australia p. 23

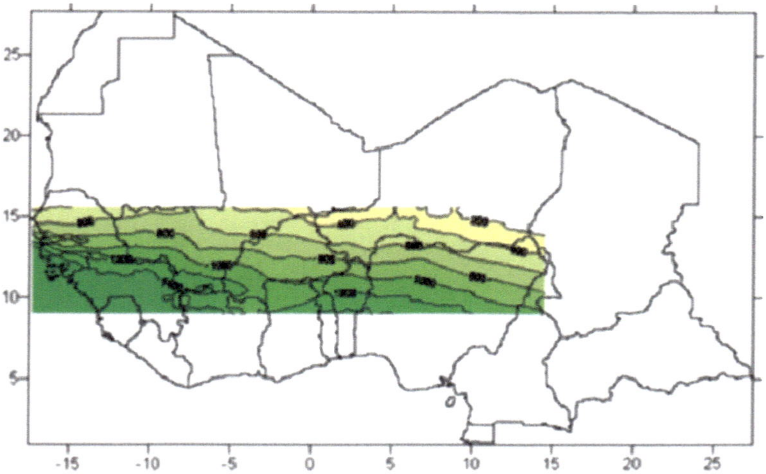

Map 3.1. Dryland West Africa: Rainfall situation 1960-'70 p. 29

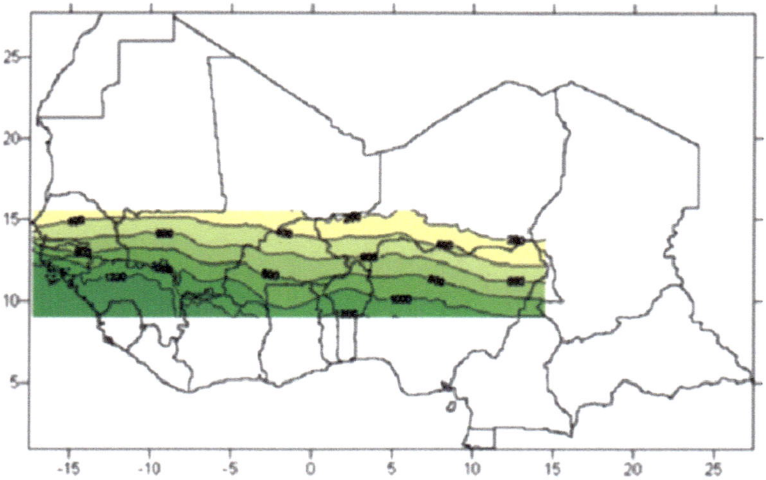

Map 3.2. Dryland West Africa: Rainfall situation 1975-'85 p. 29

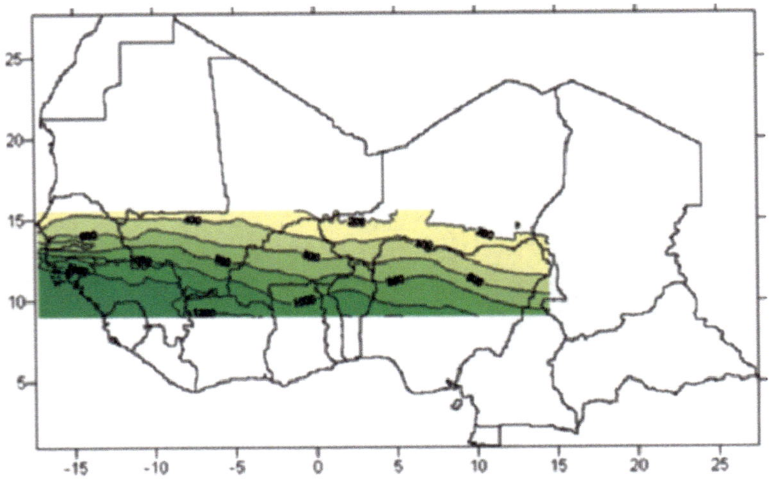

*Map 3.3. Dry*land West Africa: Rainfall situation 1983-'93 p. 30

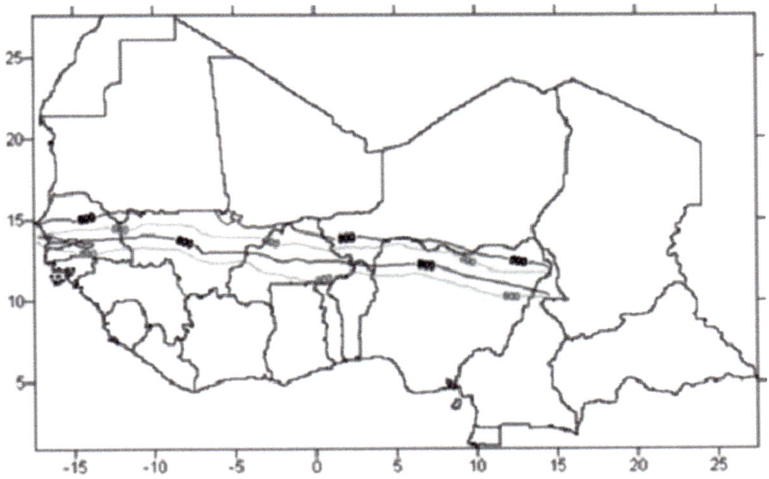

Map 3.4. West Africa: changes in 500 and 800 rainfall iso-lines, 1960-70 and 1975-'85 p. 30

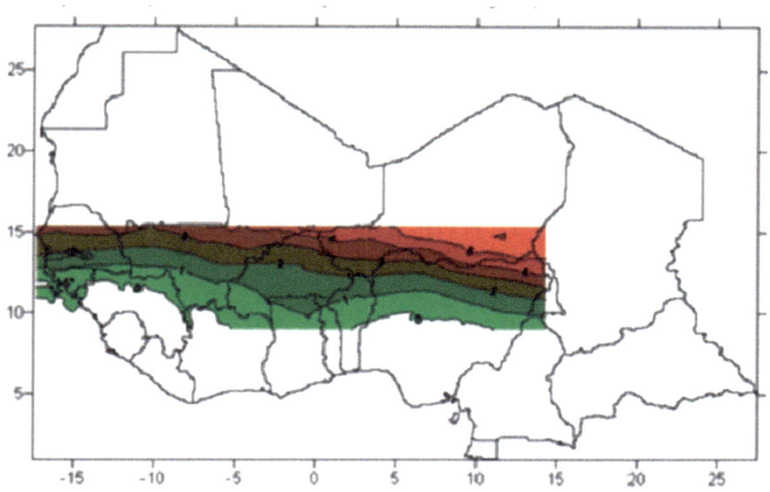

Map 3.5. Dryland West Africa: drought risks for millet/sorghum production 1960-'70 p. 31

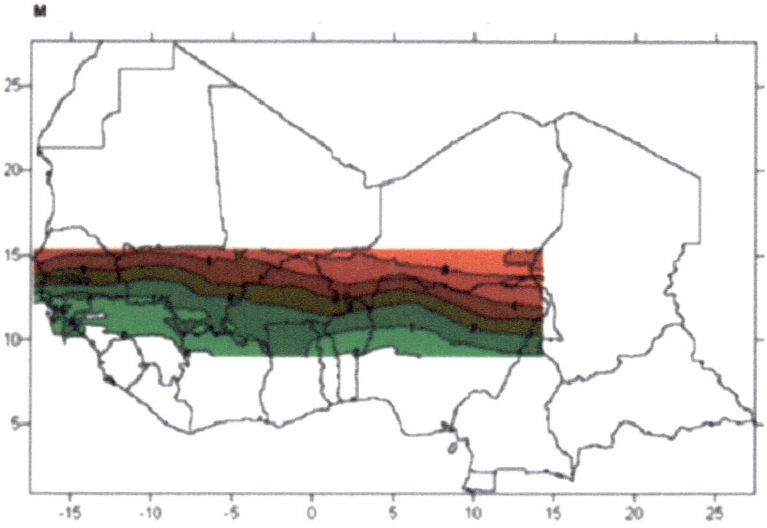

Map 3.6. Dryland West Africa: drought risks for millet/sorghum production 1975-'85 p. 31

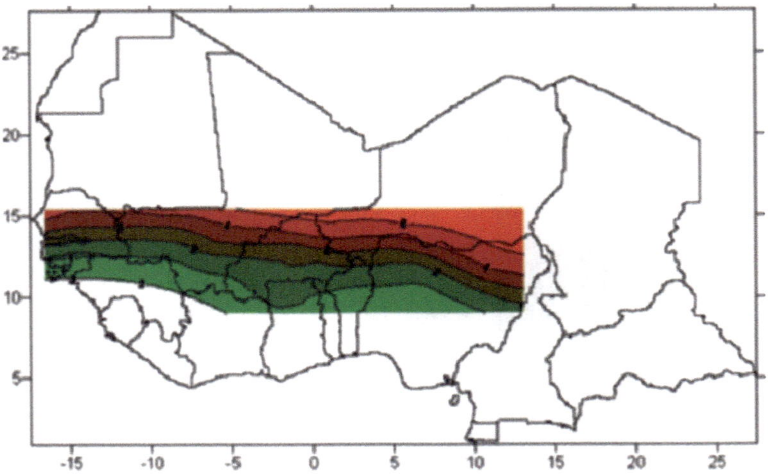

Map 3.7. Dryland West Africa: drought risks for millet/sorghum production 1983-'93 p. 32

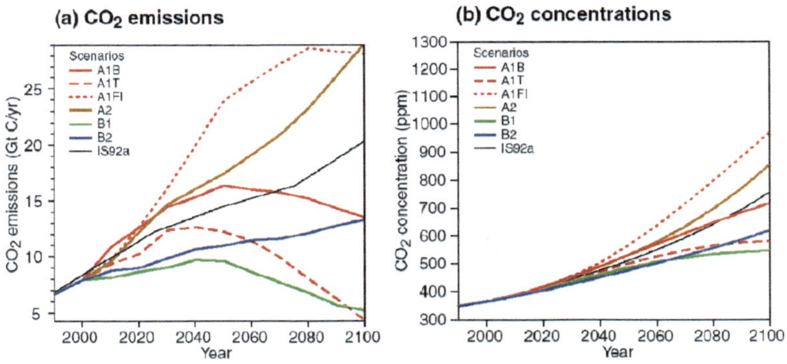

Figure 4.1.Carbon dioxide emissions according to six SRES scenarios and the IS92a scenario (a) and the resulting concentrations of carbon dioxide (b). (Source: IPCC 2002a) p. 36.

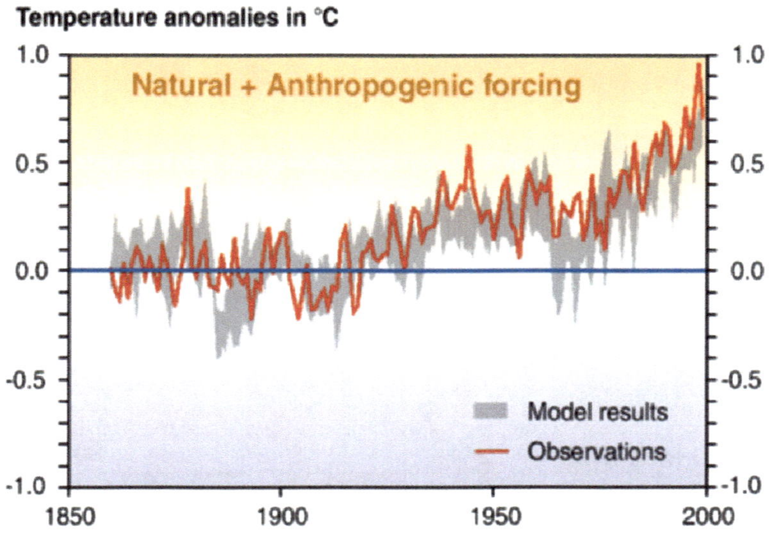

Figure 4.2. Observed and modelled variation in globally averaged temperature anomalies. The model used both natural forcing (e.g., variations in solar activity, volcanism) and anthropogenic forcing (increased concentrations of greenhouse gases). (source: IPCC 2002b) p. 37.

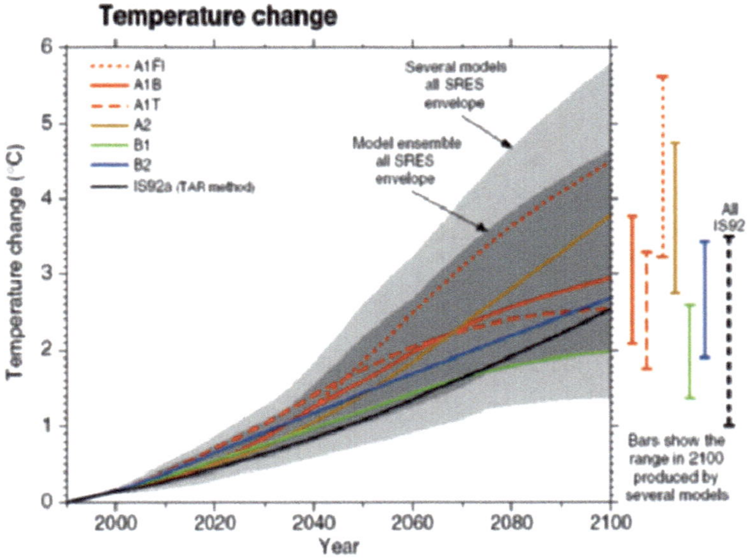

Figure 4.3. Predictions of global mean temperature using different climate models and emission scenarios (Source: IPCC 2002a) p. 39.

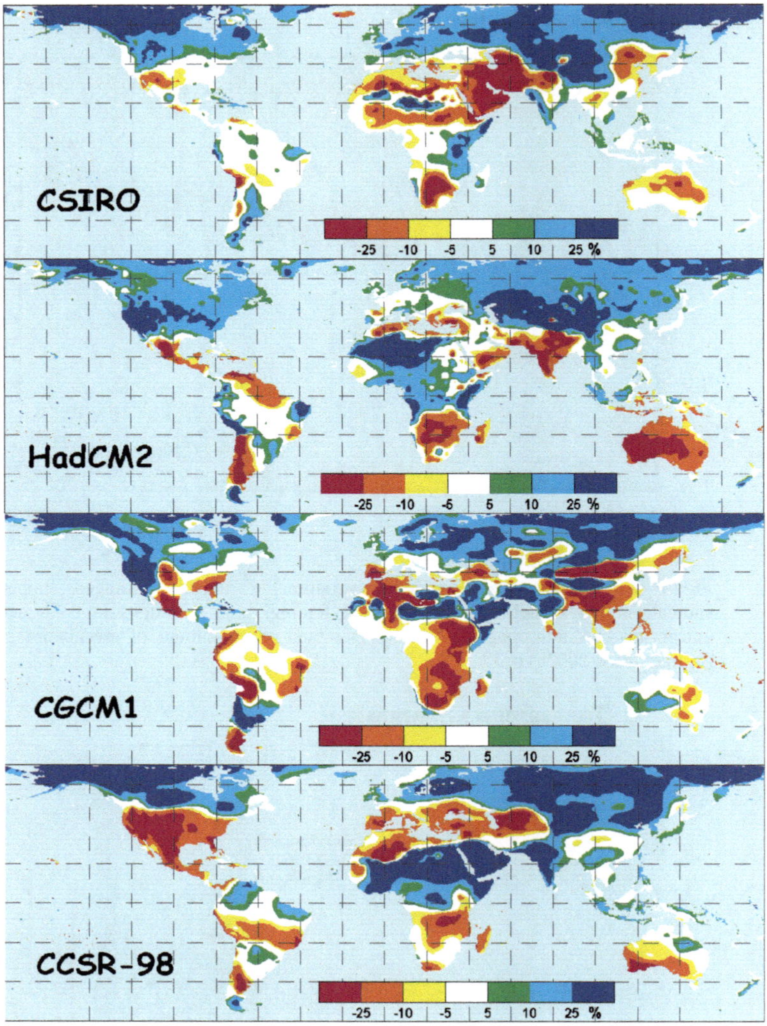

Figure 4.4. The relative increases in precipitation as predicted using four different climate models p. 41.

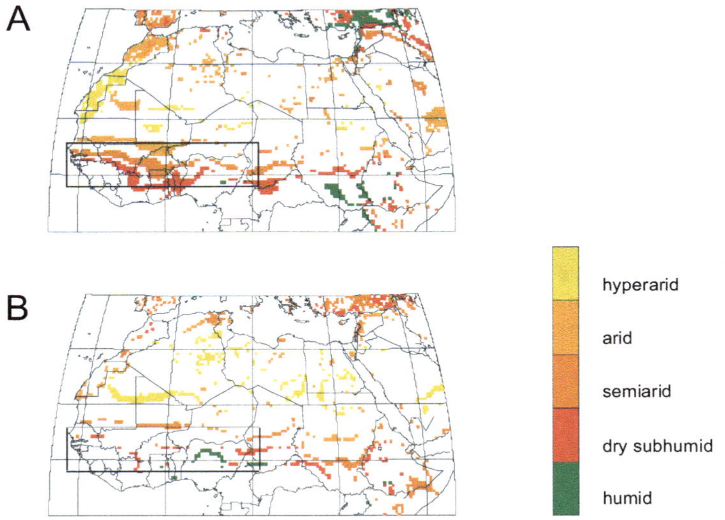

Map 5.1. Change in aridity classification, 1990-2050, for A: baseline A scenario, MPI-GCM; B:baseline A scenario, GFDL-GCM (ICCD study area is indicated by the box, coloured cells subject to change and show classification by 2050) p. 47.

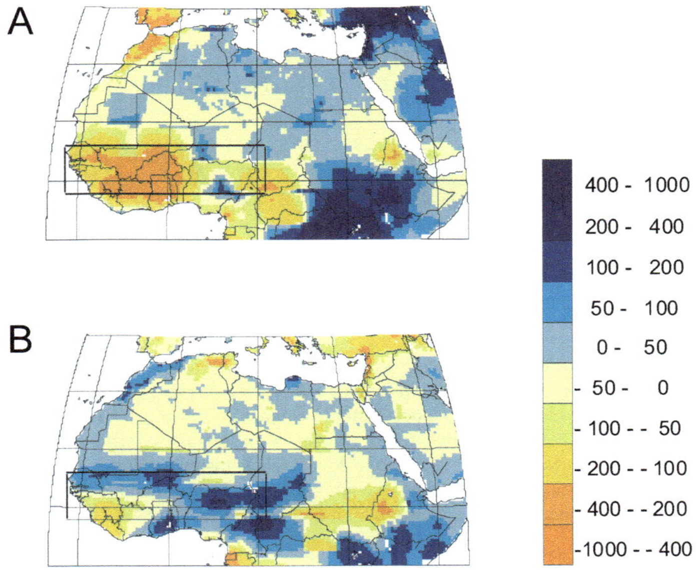

Map 5.2. Change in precipitation (mm), 1990-2050, for A: baseline A scenario, MPI-GCM and B: baseline A scenario, GFDL-GCM (ICCD study area is indicated by the box) p. 47.

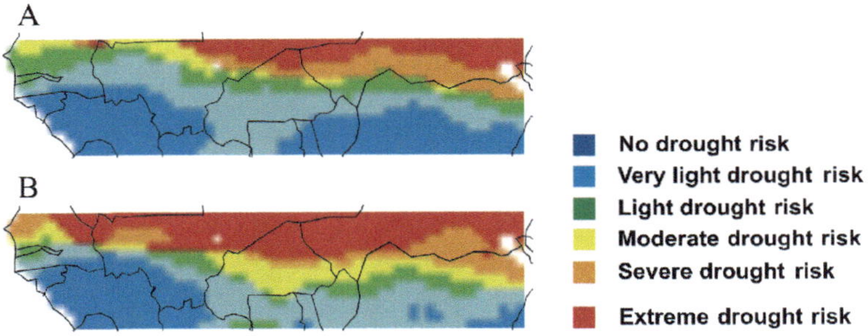

Map 6.1. Changes in drought risk for 1990-2050, A depicts the current situation (based on 30-year average) B displays the projected drought risk for baseline A scenario, MPI-GCM p. 55.

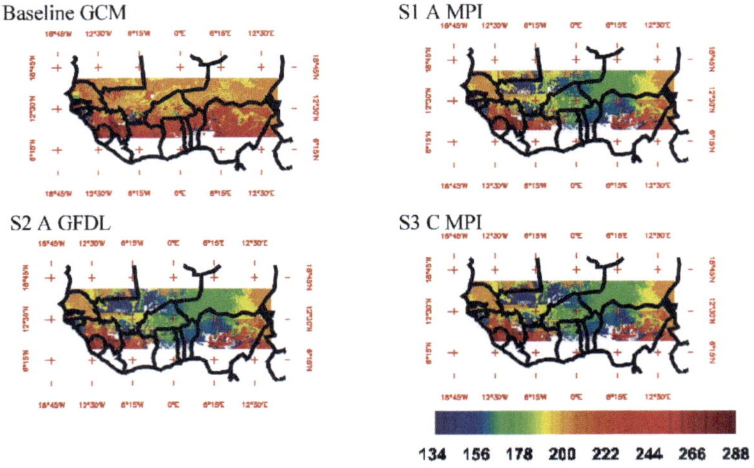

Map 6.2. Start of the growing season, day of year, related to highest water limited production level. Baseline GCM Is the 1960 – 1990 average. S1 A MPI Scenario 1 uses baseline A with the Max Planck Institute climate model. S2 A GFDL Scenario 2 uses baseline A with the Global Fluid Dynamics Laboratory climate model. S3 C MPI Scenario 1 uses baseline C with the Max Planck Institute climate model. Baseline A combines a medium population growth with a medium economic growth; Baseline C combines a medium population and high economic growth p. 59.

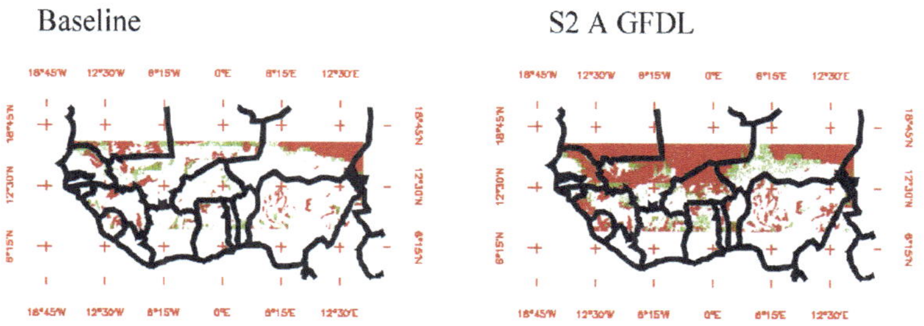

Map 6.3. Probability of yield exceeding 2250 kg ha-1 for the 1960-1990 baseline and S2 A scenario using GFDL GCM, day of planting is 190. (red = 1, white = 0). Baseline Is the 1960 – 1990 average. S2 A GFDL Scenario 2 uses baseline A with the Global Fluid Dynamics Laboratory climate model. Baseline A combines a medium population growth with a medium economic growth p. 59.

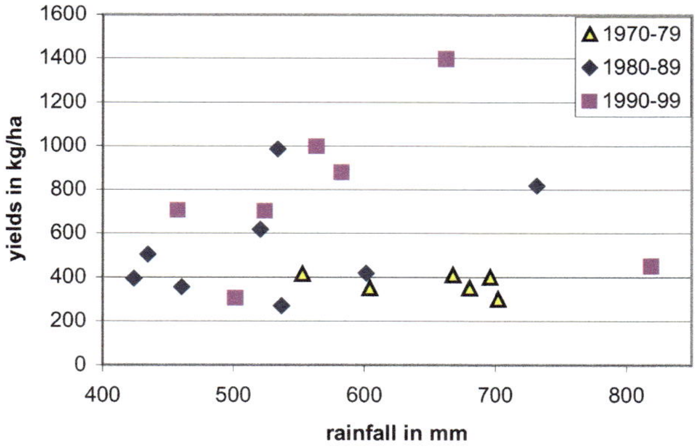

Figure 7.1. Relationship between rainfall in mm and groundnut yields in kg per ha, per decade, 1970-1999 p. 74.

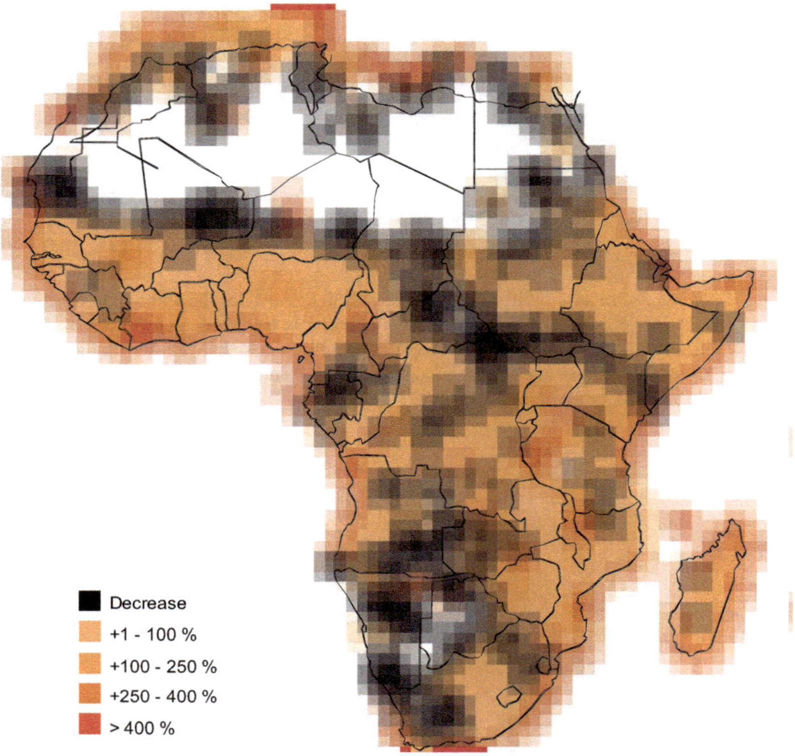

Map 8.1. Changes in population density (inh km^{-2}) between 1960 and 1994 (in %) p. 80.

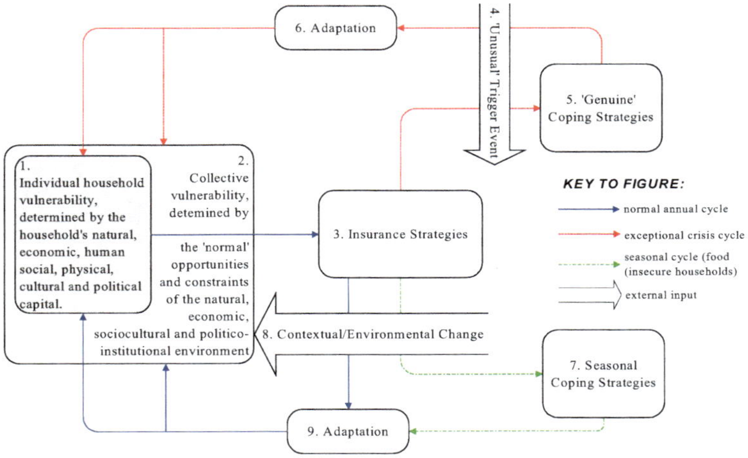

Figure 11.5. Conceptual framework: Farm household vulnerability and responses to normal opportunities and constraints, unusual events and changing conditions p. 144.

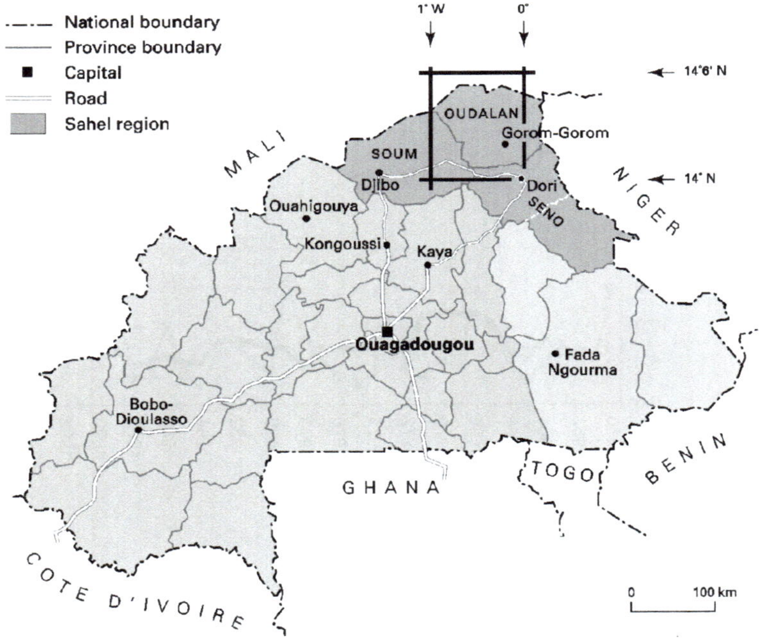

Map 15.1. The study area in Burkina Faso p. 245.

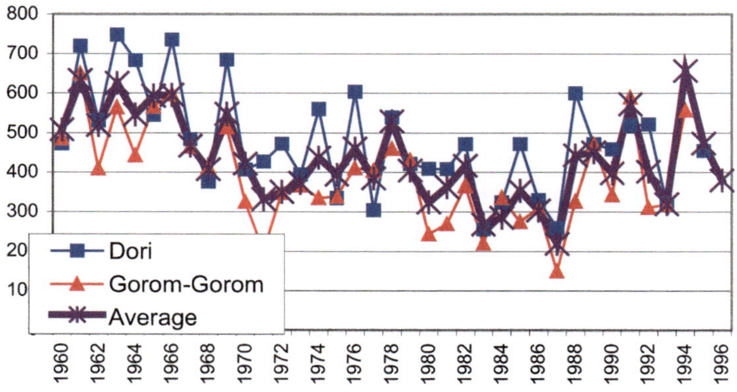

Figure 15.1. Rainfall in mm per year, Dori, Gorom-Gorom and average for four Sahel stations, period 1960-1996 p. 246.

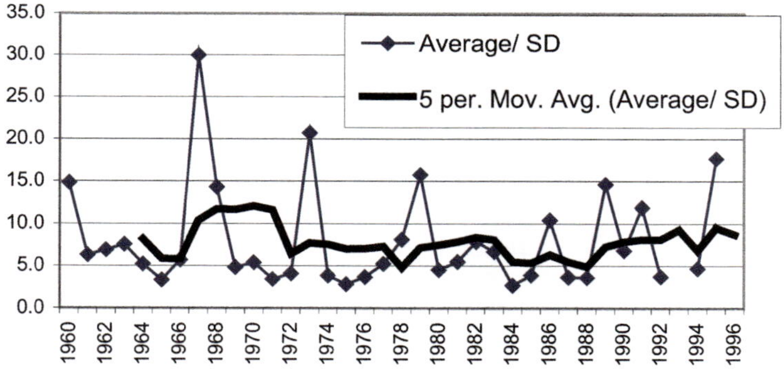

Figure 15.2. Average annual rainfall divided by SD and five year moving average for four Sahel stations during the period 1960-1996 p. 246.

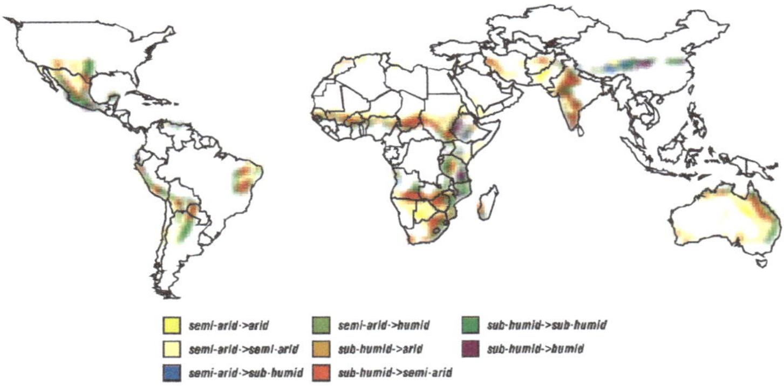

Map 19.1. The world's drylands: changes in aridity classification 1960-'90 compared with UNESCO's classification (ca 1930-'60). source: IIASA (1930-1960 Global Climate Database), CRU/LINK (1960-'90 Global Climatology) p. 337.

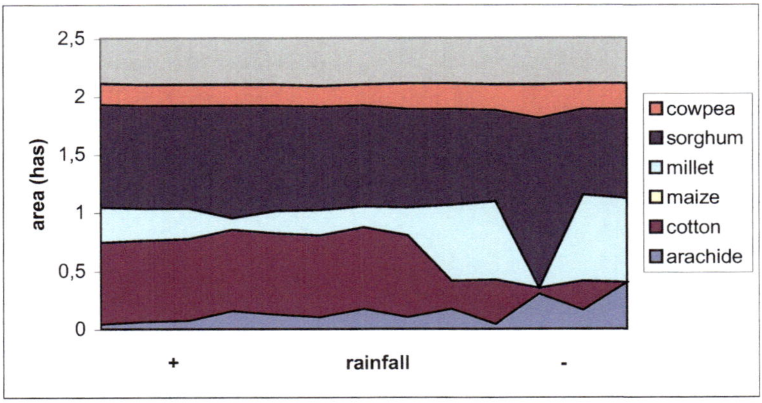

Figure 21.4. Cropping pattern of Household type C in different rainfall conditions p. 377.

ENVIRONMENT & POLICY

1. Dutch Committee for Long-Term Environmental Policy: *The Environment: Towards a Sustainable Future.* 1994 ISBN 0-7923-2655-5; Pb 0-7923-2656-3
2. O. Kuik, P. Peters and N. Schrijver (eds.): *Joint Implementation to Curb Climate Change.* Legal and Economic Aspects. 1994 ISBN 0-7923-2825-6
3. C.J. Jepma (ed.): *The Feasibility of Joint Implementation.* 1995 ISBN 0-7923-3426-4
4. F.J. Dietz, H.R.J. Vollebergh and J.L. de Vries (eds.): *Environment, Incentives and the Common Market.* 1995 ISBN 0-7923-3602-X
5. J.F.Th. Schoute, P.A. Finke, F.R. Veeneklaas and H.P. Wolfert (eds.): *Scenario Studies for the Rural Environment.* 1995 ISBN 0-7923-3748-4
6. R.E. Munn, J.W.M. la Rivière and N. van Lookeren Campagne: *Policy Making in an Era of Global Environmental Change.* 1996 ISBN 0-7923-3872-3
7. F. Oosterhuis, F. Rubik and G. Scholl: *Product Policy in Europe: New Environmental Perspectives.* 1996 ISBN 0-7923-4078-7
8. J. Gupta: *The Climate Change Convention and Developing Countries: From Conflict to Consensus?* 1997 ISBN 0-7923-4577-0
9. M. Rolén, H. Sjöberg and U. Svedin (eds.): *International Governance on Environmental Issues.* 1997 ISBN 0-7923-4701-3
10. M.A. Ridley: *Lowering the Cost of Emission Reduction: Joint Implementation in the Framework Convention on Climate Change.* 1998 ISBN 0-7923-4914-8
11. G.J.I. Schrama (ed.): *Drinking Water Supply and Agricultural Pollution.* Preventive Action by the Water Supply Sector in the European Union and the United States. 1998 ISBN 0-7923-5104-5
12. P. Glasbergen: *Co-operative Environmental Governance: Public-Private Agreements as a Policy Strategy.* 1998 ISBN 0-7923-5148-7; Pb 0-7923-5149-5
13. P. Vellinga, F. Berkhout and J. Gupta (eds.): *Managing a Material World.* Perspectives in Industrial Ecology. 1998 ISBN 0-7923-5153-3; Pb 0-7923-5206-8
14. F.H.J.M. Coenen, D. Huitema and L.J. O'Toole, Jr. (eds.): *Participation and the Quality of Environmental Decision Making.* 1998 ISBN 0-7923-5264-5
15. D.M. Pugh and J.V. Tarazona (eds.): *Regulation for Chemical Safety in Europe: Analysis, Comment and Criticism.* 1998 ISBN 0-7923-5269-6
16. W. Østreng (ed.): *National Security and International Environmental Cooperation in the Arctic – the Case of the Northern Sea Route.* 1999 ISBN 0-7923-5528-8
17. S.V. Meijerink: *Conflict and Cooperation on the Scheldt River Basin.* A Case Study of Decision Making on International Scheldt Issues between 1967 and 1997. 1999 ISBN 0-7923-5650-0
18. M.A. Mohamed Salih: *Environmental Politics and Liberation in Contemporary Africa.* 1999 ISBN 0-7923-5650-0
19. C.J. Jepma and W. van der Gaast (eds.): *On the Compatibility of Flexible Instruments.* 1999 ISBN 0-7923-5728-0
20. M. Andersson: *Change and Continuity in Poland's Environmental Policy.* 1999 ISBN 0-7923-6051-6

ENVIRONMENT & POLICY

21. W. Kägi: *Economics of Climate Change: The Contribution of Forestry Projects*. 2000 ISBN 0-7923-6103-2
22. E. van der Voet, J.B. Guinée and H.A.U. de Haes (eds.): *Heavy Metals: A Problem Solved?* Methods and Models to Evaluate Policy Strategies for Heavy Metals. 2000 ISBN 0-7923-6192-X
23. G. Hønneland: *Coercive and Discursive Compliance Mechanisms in the Management of Natural Resourses.* A Case Study from the Barents Sea Fisheries. 2000 ISBN 0-7923-6243-8
24. J. van Tatenhove, B. Arts and P. Leroy (eds.): *Political Modernisation and the Environments.* The Renewal of Environmental Policy Arrangements. 2000 ISBN 0-7923-6312-4
25. G.K. Rosendal: *The Convention on Biological Diversity and Developing Countries.* 2000 ISBN 0-7923-6375-2
26. G.H. Vonkeman (ed.): *Sustainable Development of European Cities and Regions.* 2000 ISBN 0-7923-6423-6
27. J. Gupta and M. Grubb (eds.): *Climate Change and European Leadership.* A Sustainable Role for Europe? 2000 ISBN 0-7923-6466-X
28. D. Vidas (ed.): *Implementing the Environmental Protection Regime for the Antarctic.* 2000 ISBN 0-7923-6609-3; Pb 0-7923-6610-7
29. K. Eder and M. Kousis (eds.): *Environmental Politics in Southern Europe: Actors, Institutions and Discourses in a Europeanizing Society.* 2000 ISBN 0-7923-6753-7
30. R. Schwarze: *Law and Economics of International Climate Change Policy.* 2001 ISBN 0-7923-6800-2
31. M.J. Scoullos, G.H. Vonkeman, I. Thornton, and Z. Makuch: *Mercury - Cadmium-Lead: Handbook for Sustainable Heavy Metals Policy and Regulation.* 2001 ISBN 1-4020-0224-6
32. G. Sundqvist: *The Bedrock of Opinion.* Science, Technology and Society in the Siting of High-Level Nuclear Waste. 2002 ISBN 1-4020-0477-X
33. P.P.J. Driessen and P. Glasbergen (eds.): *Greening Society.* The Paradigm Shift in Dutch Environmental Politics. 2002 ISBN 1-4020-0652-7
34. D. Huitema: *Hazardous Decisions.* Hazardous Waste Siting in the UK, The Netherlands and Canada. Institutions and Discourses. 2002 ISBN 1-4020-0969-0
35. D. A. Fuchs: *An Institutional Basis for Environmental Stewardship: The Structure and Quality of Property Rights.* 2003 ISBN 1-4020-1002-8
36. B. Chaytor and K.R. Gray (eds.): *International Environmental Law and Policy in Africa.* 2003 ISBN 1-4020-1287-X
37. F.M. Brouwer, I. Heinz and T. Zabel (eds.): *Governance of Water-Related Conflicts in Agriculture.* New Directions in Agri-Environmental and Water Policies in the EU. 2003 ISBN 1-4020-1553-4

ENVIRONMENT & POLICY

38. G.J.I. Schrama and S. Sedlacek (eds.): *Environmental and Technology Policy in Europe. Technological Innovation and Policy Integration*. 2003
ISBN 1-4020-1583-6
39. A.J. Dietz, R. Ruben and A. Verhagen (eds.): *The Impact of Climate Change on Drylands*. With a Focus on West Africa. 2004 ISBN 1-4020-1952-1

KLUWER ACADEMIC PUBLISHERS – DORDRECHT / BOSTON / LONDON

GPSR Compliance

The European Union's (EU) General Product Safety Regulation (GPSR) is a set of rules that requires consumer products to be safe and our obligations to ensure this.

If you have any concerns about our products, you can contact us on ProductSafety@springernature.com

In case Publisher is established outside the EU, the EU authorized representative is:

Springer Nature Customer Service Center GmbH
Europaplatz 3
69115 Heidelberg, Germany

Batch number: 10755329

Printed by Printforce, the Netherlands